AF556801

Neurodevelopment, Aging and Cognition

Neurodevelopment, Aging and Cognition

Ivica Kostović
Stevo Knežević
Henryk M. Wisniewski
George J. Spilich
Editors

Birkhäuser
Boston • Basel • Berlin

Ivica Kostović
Dept. of Anatomy
University of Zagreb
School of Medicine
41000 Zagreb, Croatia

Stevo Knežević
Dept. of Neurology
University of Zagreb
University Hospital Rebro
41000 Zagreb, Croatia

Henryk M. Wisniewski
NYS Institute for Basic Research
in Developmental Disabilities
Staten Island, NY 10314
USA

George J. Spilich
Dept. of Psychology
Washington College
Chestertown, MD 21620
USA

Library of Congress Cataloging-in-Publication Data

Neurodevelopment, aging, and cognition / edited by Ivica Kostović . . .
[et al.].
p. cm.
Includes bibliographical references and index.
ISBN 0-8176-3599-8. - - ISBN 3-7643-3599-8
1. Alzheimer's disease. 2. Developmental neurology. 3. Cognitive neuroscience. I. Kostović, Ivica, 1943-
RC523.N42 1992 92-396
616.8–dc 20 CIP

Printed on acid-free paper.

ISBN 0-8176-3599-8
ISBN 3-7643-3599-8

Typeset by ARK Publications, Inc., Newton Centre, MA
Printed and bound by Quinn Woodbine, Woodbine, NJ
Printed in the U.S.A.

9 8 7 6 5 4 3 2 1

Contents

Part II. Neuroscience

Part III. Cognitive Science

Foreword

It was Oscar Wilde who defined the tragedy of old age by saying that "...as soon as you are old enough to know better, you don't know anything at all." As improvements in the quality of health care bring about longer life, our attention has turned from the prolonging of life to the maintenance of involvement in life. In developed nations, a full 100% increase in the ranks of the elderly has appeared and with the benefits of this prolongation have come new and greater needs of the elderly cohort. Our interest is in those processes that may lead to dementia among the elderly, for in dementia we see a thief that robs victims of their memories and their place in life.

This text was conceived and developed from an international conference on neurodevelopment, aging, and cognition; the purpose of this conference was to bring together for a few days a group of experts in these fields from around the world to generate a dialog on common themes and unresolved problems. Our hope was that by keeping the meeting small and informal, we could break through barriers of terminology unique to the areas of developmental neurobiology, neuroscience, cognitive science, and clinical medicine, and have a meaningful discussion on processes that affect the biological integrity and cognitive performance of the aging nervous system. The text is organized into four sections based upon a general theme: neurodevelopment, neuroscience, cognitive science, and clinical applications. Our intent was to begin the dialog about dementia from many perspectives; through this text, we hope to carry the conversation to you.

Every large undertaking owes its success to many people, and this book owes a debt to many individuals on both sides of the Atlantic. First, the editors, authors, and participants would like to express their appreciation to the Inter-University Centre of Dubrovnik, Yugoslavia, whose purpose is to bring scholars together from every corner of the globe to share knowledge in the beautiful and historical environment of the Old City of Dubrovnik. During the preparation of this book, tragic events in Croatia resulted in the destruction of the building in which the IUC was located. The editors dedicate this book to the ideals of the IUC.

We would also like to thank our sponsors: the national airline of Yugoslavia, JAT airlines; also the three German pharmaceutical firms of Merck, Hoechst, and Schering, along with Pliva pharmaceuticals and Astra Trading in Yugoslavia, for their assistance in sponsoring the conference. SIZ za znanost, the scientific grant committee for Croatia, also was generous in its support. Mr. Rafael-Rafo Rodin of Razvitak Turist of Metkovic provided expert help with local arrangements and sponsorship of social events. It would not be possible to fully express our appreciation to the young scientists of the Department of Neuroanatomy at the University of Zagreb who served as the hands of the conference and prevented little logistical problems from developing into big headaches: Nenad Bogdanović, Darko Chudy, Ivana Delalle, and Miloš Judaš.

Soon after the conference ended on one side of the Atlantic, a pile of manuscripts began to accumulate on the other side and there this book was slowly assembled from files written by no less than a dozen different word processing programs. Our thanks to the computing center at Washington College for technical assistance with the conversion of these files one format into another. Karen Utkewicz converted unscannable text, tables, and charts into a publishable format, and without her secretarial skills all would have been lost. Two people deserve special mention. My research assistant, Dianna Holden, spent an entire summer surrounded by piles of manuscripts. Her good humor, industry, and meticulous work habits made the organization of the many chapters, disks, and original artwork possible as she and I read, scanned, spellchecked, revised, mailed revisions to authors, incorporated revisions into text, and otherwise slowly pushed this text along the road towards you, the reader. Last but certainly not least, Allan Ross and James Doran of Birkhäuser Boston were perfect editors. Their counsel carried this project along; somehow they knew exactly when a phone call would generate work-producing guilt and when silence would have the same effect. To all these people, the editors express their appreciation.

George Spilich
Washington College
Chestertown, MD, USA

Contributors

Jocelyne Bachevalier, Department of Neurobiology and Anatomy, The University of Texas Medical School, Houston, TX, USA

B. Berger, Pediatrics Building, Hôpital Salpetrière, INSERM, Paris, France

Rita Sloan Berndt, Department of Neurology, University of Maryland, School of Medicine, Baltimore, MD, USA

Enrico Biagioni, Institute of Child Neurology and Psychiatry, University of Pisa, "Stella Maris" Foundation, Pisa, Italy

Dorothy Bishop, MRC Applied Psychology Unit, Cambridge, England CB2 2EF

Nenad Bogdanović, Department of Geriatric Medicine, Karolińska Institute, Huddinge University Hospital, S-141 B6 Huddinge, Sweden

Eva Braak, Zentrum der Morphologie, Kinikum der J. W. Goethe Universität, Theodor Stern Kai 7 D-6000, Frankfurt, Germany

Heiko Braak, Zentrum der Morphologie, Kinikum der J. W. Goethe Universität, Theodor Stern Kai 7 D-6000, Frankfurt, Germany

Anthony Cheng, MRC Neurochemical Pathology Unit, Newcastle General Hospital, Newcastle-upon-Tyne, England NE4 68E

Darko Chudy, Department of Neuroanatomy, Medical Faculty, Zagreb, Yugoslavia

Giovanni Cioni, Institute of Child Neurology and Psychiatry, University of Pisa, "Stella Maris" Foundation, Pisa, Italy

Carlo Cipolloni, Neonatal Intensive Care Unit, University of Pisa, Pisa, Italy

Gillian Cohen, Human Cognition Research Laboratory, The Open University, Walton Hall, Milton Keynes, England MK7 6AA

A. Convit, Nathan S. Kline Psychiatric Research Institute, Department of Psychiatry, New York University Medical Center, New York, NY 10016, USA

R. F. Cowburn, Department of Geriatric Medicine, Karolińska Institute, Huddinge University Hospital, Huddinge, Sweden; Astra Research Centre, Sodertalje, Sweden

Fergus I. M. Craik, Psychology Department, University of Toronto, Toronto, Ontario, Canada

Stephen Curran, Adult Psychiatric Unit, St. James University Hospital, Leeds, England

M. Dambska, Medical Research Center, Polish Academy of Sciences, Warsaw, Poland

M. J. de Leon, Department of Psychiatry, New York University Medical Center, New York, NY 10016, USA

Ivana Delalle, School of Medicine, University of Zagreb, Croatian Institute for Brain Research, Zagreb, Republic of Croatia, Yugoslavia

Louis Dingcong, Aging and Dementia Research Center, New York University Medical Center, New York, NY 10016, USA

James A. Edwardson, MRC Neurochemical Pathology Unit, Newcastle General Hospital, Newcastle-upon-Tyne, England NE4 68E

Steven H. Ferris, Aging and Dementia Research Center, New York University Medical Center, New York, NY 10016, USA

C. J. Fowler, Department of Geriatric Medicine, Karolińska Institute, Huddinge University Hospital, Huddinge, Sweden; Astra Research Centre, Sodertalje, Sweden

Emile Franssen, Aging and Dementia Research Center, New York University Medical Center, New York, NY 10016, USA

A. E. George, Department of Radiology, New York University Medical Center, New York, NY 10016, USA

P. S. Goldman-Rakic, Section of Neuroanatomy, Yale University, School of Medicine New Haven, CT 06510, USA

J. Golomb, Department of Neurology, New York University Medical Center, New York, NY 10016, USA

E. Goudsmit, Netherlands Institute for Brain Research, 1105 AZ, Amsterdam, The Netherlands

Ian Hindmarch, Robens Institute, University of Surrey, Surrey, England

Miloš Judaš, School of Medicine, University of Zagreb, Croatian Institute for Brain Research, Zagreb, Republic of Croatia, Yugoslavia

Alan Kluger, Aging and Dementia Research Center, New York University Medical Center, New York, NY 10016, USA

Stevo Knežević, Department of Neurology, University Hospital Center, Zagreb, Yugoslavia

Ivica Kostović, School of Medicine, University of Zagreb, Croatian Institute for Brain Research, Zagreb, Republic of Croatia, Yugoslavia

H. P. H. Kremer, Department of Neurology, Academic Hospital Leiden, Leiden, The Netherlands

K. Krnjević, Anaesthesia Research Department, McGill University, Montréal, Québec, Canada

Elizabeth Marshall, Department of Psychiatry, University of Newcastle-upon-Tyne, Newcastle, England

T. McRae, Department of Medicine, New York University Medical Center, New York, NY 10016, USA

Mortimer Mishkin, Laboratory of Neuropsychology, National Institute of Mental Health, Bethesda, MD, USA

Richard S. Nowakowski, Department of Neuroscience and Cell Biology, Robert Wood Johnson Medical School, University of Medicine and Dentistry of New Jersey, Piscataway, NJ 08854, USA

C. O'Neill, Department of Geriatric Medicine, Karolińska Institute, Huddinge University Hospital, Huddinge, Sweden; Astra Research Centre, Sodertalje, Sweden

Angela Pattschull-Furlan, Department of Psychology, University of Freibourg, Fribourg, Switzerland

Elaine K. Perry, MRC Neurochemical Pathology Unit, Newcastle General Hospital, Newcastle-upon-Tyne, England NE4 68E

Robert H. Perry, Department of Neuropathology, Newcastle General Hospital, Newcastle-upon-Tyne, England NE4 68E

Zdravko Petanjek, School of Medicine, University of Zagreb, Croatian Institute for Brain Research, Republic of Croatia, Yugoslavia

R. Ravid, Netherlands Institute for Brain Research, Meibergdreef, Amsterdam, The Netherlands

Barry Reisberg, Aging and Dementia Research Center, New York University Medical Center, New York, NY 10016, USA

Steven G. Sclan, Aging and Dementia Research Center, New York University Medical Center, New York, NY 10016, USA

G. Smith, Departments of Psychiatry and Radiology, New York University Medical Center, New York, NY 10016, USA

George J. Spilich, Psychology Department, Dunning Science Center, Washington College, Chestertown, MD, 21620, USA

D. F. Swaab, Netherlands Institute for Brain Research, 1105 AZ Amsterdam ZO, The Netherlands

Adolfo Toledano, Instituto de Neurobiologia, Santiago Ramon y Cajal, 28002 Madrid, Spain

W. Tsui, Nathan S. Kline Psychiatric Research Institute, Department of Psychiatry, New York University Medical Center, New York, NY 10016, USA

C. Verney, Pediatrics Building, Hôpital Salpetrière, INSERM, Paris, France

B. Winblad, Department of Geriatric Medicine, Karolińska Institute, Huddinge University Hospital, Huddinge, Sweden; Astra Research Centre, Sodertalje, Sweden

Henryk M. Wisniewski, New York State Institute for Basic Research in Developmental Disabilities, Staten Island, NY 10314-6399, USA

K. E. Wisniewski, New York State Institute for Basic Research in Developmental Disabilities, Staten Island, NY 10314-6399, USA

Thomas M. Wisniewski, New York University Medical Center, Department of Neurology and Pathology, New York, NY 10016, USA

Sandra F. Witelson, Departments of Psychiatry, Psychology, and Biomedical Sciences, McMaster University, Hamilton, Ontario, Canada L8N 3Z5

W. Witting, Netherlands Institute for Brain Research, 1105 AZ, Amsterdam ZO, The Netherlands

A. P. Wolf, Department of Chemistry, Brookhaven National Laboratory, Upton, NY 11973, USA

Y. Z. Xu, Anaesthesia Research Department, McGill University, Montréal, Québec, Canada

PART I
Neurodevelopment

Developmental Reorganization of the Human Association Cortex during Perinatal and Postnatal Life

Ivica Kostović, Zdravko Petanjek, Ivana Delalle, and Miloš Judaš

The development of synaptic connections (Kostović et al., 1988; Molliver et al., 1973) and electrophysiological activity of the human cerebral cortex (Dreyfus-Brisac, 1979) begin during the early fetal life, many months before the onset of cognitive cortical functions. However, in this early fetal phase, development of cortical connections is not a dominant neurogenetic event. Other events, such as proliferation, migration of neurons and initial growth of dendrites and axons are main neurogenetic processes in the early fetal human brain. After 13 weeks of gestation begins a new phase characterized by transient arrangement of cortical afferents, synapses, and neurons. The most pronounced pattern of transient organization is present between 22 and 34 weeks of gestation (Kostović, 1990). The behavioral states of the human premature infants change dramatically during this period (Dreyfus-Brisac, 1979; Leijon, 1982; Parmelee, 1975; Prechtl, 1974; Trevarthen, 1979; Wolff and Ferber, 1979). However, the correlative studies of structural and behavioral developmental events should be ihterpreted cautiously since simultaneous ontogenetic emergence may be pure coincidence (Prechtl, 1984). The presence of transient patterns of cortical organization raises the question of their subsequent perinatal and postnatal reorganization. The reorganizational events in the human cerebral cortex extend at least up to the third year of postnatal life (Kostović, 1990). The reorganization of cortex involves the disappearance of fetal layers (Kostović, 1990; Kostović and Rakic, 1990) and cells (Kostović and Rakic, 1980; Shatz et al., 1988), rearrangements of cortico-cortical fibers (Chalupa and Killackey, 1989; Dehay et al., 1988; Goldman-Rakic, 1982, 1987; Innocenti, 1981, 1982; LaMantia and Rakic, 1990), reduction of synapses (Huttenlocher and de Courten, 1987) and changes in chemical properties of thalamocortical pathways (Kostović and Goldman-Rakic, 1983; Kostović and Rakic, 1984). The final phase of cortical maturation occurs during childhood and adolescence and is characterized by the very gradual chemical maturation of associative neurons of layer III (Kostović, 1990; Kostović et al., 1988). This late phase of the human cortical development is important for our understanding of cognitive development. This report is concerned with the transformation of transient patterns of cortical organization and maturational changes during the late phase of cortical development. Special attention will be devoted to the developmental changes of circuitry elements (afferents, postsynaptic cells, spines and synapses) in associative cortical areas.

Transient Laminar Arrangement of Neuronal Circuitry Elements in the Late Human Fetus

The development of synapses in the human cerebral cortex begins after the formation of the cortical plate at the end of the eighth week of gestation (Kostović, 1990; Molliver et al., 1973). The earliest synapses are situated both superficially and deep to the cortical plate displaying bilaminar arrangement (Kostović and Molliver, 1974; Molliver et al., 1973). In this early fetal period very few synapses are found. The number of synapses increases significantly during the formation of the new deep synaptic zone, the so-called subplate zone (Kostović and Rakic, 1990) between 13 and 15 weeks of gestation. After 15 weeks of gestation the vast majority of synapses is located in the deep subplate zone (Kostović and Rakic, 1990). The subplate zone contains transient accumulation of growing afferents ("waiting compartment"), peptidergic, and GABAergic postmigratory neurons. At least two subcortical afferent systems "wait" in the subplate zone around 22 weeks of gestation: thalamocortical (Kostović and Goldman-Rakic, 1983; Kostović and Rakic, 1984) and basal forebrain fibers (Kostović, 1986). The possibility of their cellular interactions was suggested on the basis of common histochemical marker (Kostović, 1990; Kostović and Goldman-Rakic, 1983) and spatio-temporal overlap in the pattern of growth. The predominance of deep synaptogenesis together with the accumulation of thalamic and basal forebrain afferents in the deep "waiting" compartment just before the penetration of the cortical plate (Kostović, 1990; Kostović and Rakic, 1990) correlates well with the first intermittent high-amplitude electroencephalographic bursts in the cerebral hemisphere found between 20 and

24 weeks of gestation (Dreyfus-Brisac, 1979). The predominance of deep circuitry elements must have influence on the prospective dipole of cortical electrical activity. This transient deep concentration of presynaptic axons, synapses, and postsynaptic elements changes dramatically between 24 and 28 weeks of gestation when thalamocortical fibers penetrate the cortical plate (Kostović and Goldman-Rakic, 1983; Kostović and Rakic, 1984, 1990) and intensive synaptogenesis occurs within the cortical plate (Molliver et al., 1973). The massive invasion of the cortical plate by the thalamocortical synapse seems to be the crucial basis for cortical analysis of sensory inputs (Anand and Hickey, 1987).

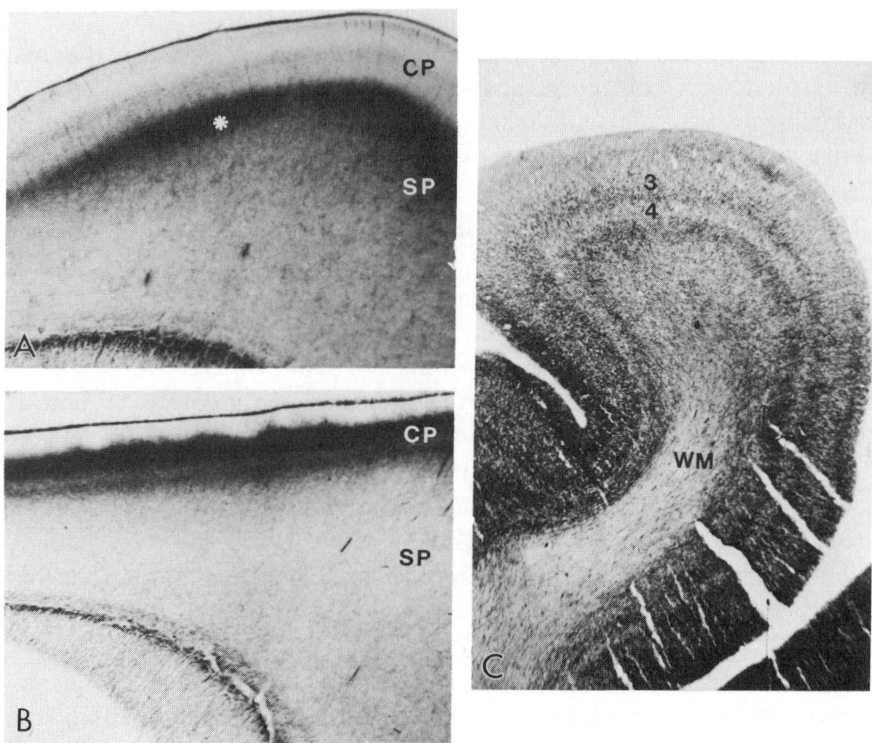

Figure 1. The reorganization of the laminar distribution of AChE reactivity in the human prefrontal cortex during prenatal and perinatal life. In a 22-week-old human fetus (A), AChE reactive fibers originating in the basal forebrain and thalamus accumulate transiently in the superficial part (asterisk) of the subplate zone (SP); at 28 weeks of gestation (B) strong AChE reactivity is evident in the cortical plate (CP) displaying a transient columnar pattern. In a 23-year-old adult (C) the AChE pattern is reversed: the granular layer IV (recipient of thalamocortical fibers) shows the weakest AChE reactivity.

In summary, the earliest organization of the cortex is characterized by transient laminar arrangement of major subcortical pathways in the waiting compartments below and above the cortical plate. The neuronal interaction seems to be predominantly trophic and histogenetic with predominant involvement of the deep subplate compartment. The early functional interactions are between postsynaptic elements in the subplate and marginal zone.

First (Perinatal) Reorganization of the Cerebral Cortex: Changes in Distribution of the Major Subcortico-Cortical and Long Cortico-Cortical Pathways

Relocation of thalamic and basal forebrain afferents indicates the onset of the first, perinatal reorganization of the human cerebral cortex. The reorganizational events are well documented at both structural and chemical level. The most prominent event in the laminar reorganization of the human cerebral cortex is the transformation and gradual dissolution of the subplate zone. In the primary visual cortex the subplate zone disappears during the last weeks of gestation (Kostović and Rakic, 1990). In the somatosensory cortex the subplate zone cannot be delineated as a cytoarchitectonic compartment in the newborn although numerous subplate neurons are present in the subjacent white matter (Kostović and Rakic, 1990). The dissolution of the subplate zone proceeds by both reorganizational and "regressive" events. The main reorganizational events are: relocation of waiting fibers in the cortical plate (Kostović and Rakic, 1990), incorporation of subplate neurons in the gyral white matter (Kostović and Rakic 1980, 1990; Shatz et al., 1988) and remodeling of subplate fibers during the formation of cortical convolutions (Kostović and Rakic 1990). The major "regressive" event is cell death (Cowan et al., 1984; Kostović and Rakic 1980; Shatz et al., 1988). The perinatal reorganization of chemical and transmitter-related markers is the most obvious in the thalamocortical system. During the late fetal period, some thalamic nuclei and recipient cortical layers show transient AChE reactivity (Kostović and Goldman-Rakic, 1983; Kostović and Rakic 1984). During the perinatal period, AChE reactivity of the layer IV as well as of the projecting thalamocortical nuclei gradually disappears, as is shown in Figures 2A and 2B.

The loss of AChE is an obvious sign of the change in chemical properties of thalamocortical axons. It is not known whether changes in chemical properties of thalamocortical axons have any significance for the

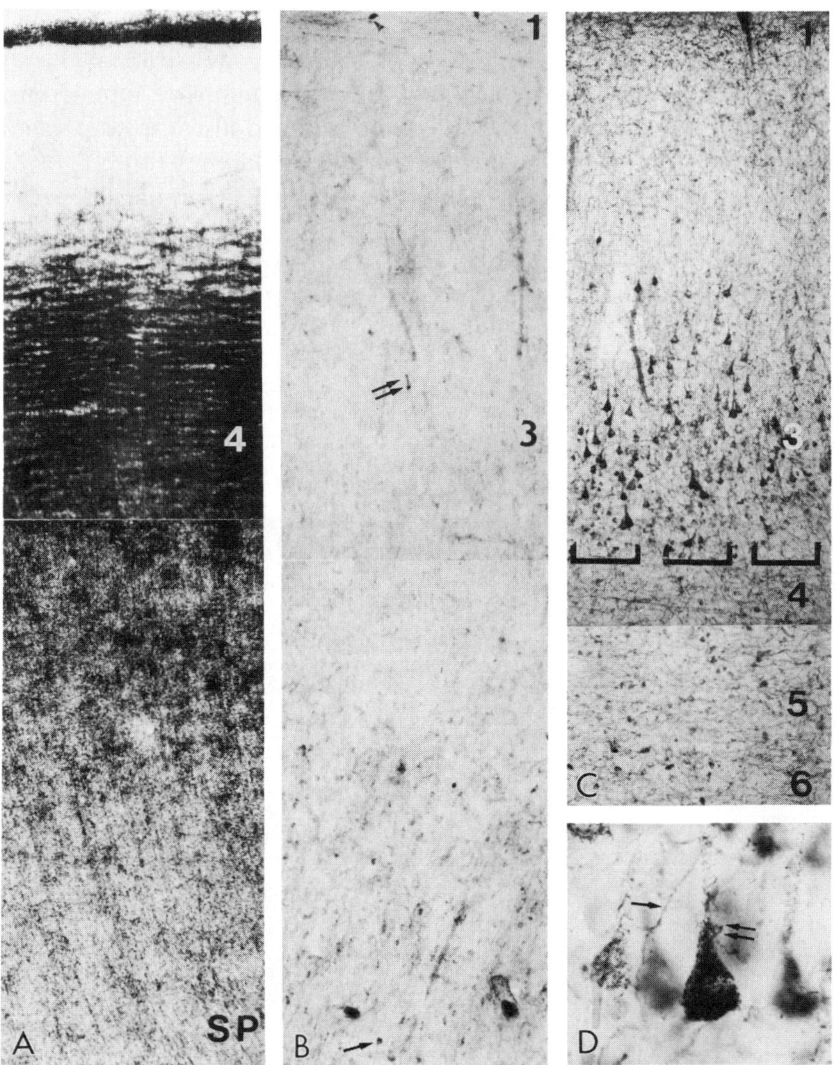

Figure 2. Developmental changes in laminar preference and cellular correlate of AChE staining in the developing human prefrontal cortex. Characteristic late fetal pattern (A) is represented by heavy staining in the neuropil of layer IV and the deep part of layer III (28 weeks of gestation). This neuropil-staining pattern disappears postnatally and AChE-reactive non-pyramidal neurons (arrows) are present at three months (B). Strongly reactive pyramidal perikarya (associative-commissural neurons) develop within layer III during the late postnatal life (C and D).

areal differentiation of the cerebral cortex. It is remarkable, however, that in sensory cortices histochemical changes of thalamocortical fibers occur concomitantly with the establishment of initial areal differences (Kostović and Rakic, 1984; Krmpotic-Nemanic et al., 1980). Recent experimental evidence in nonhuman primates indicates that diminished input from the appropriate thalamic nucleus during a critical period of development can influence the size of cytoarchitectonic areas (Rakic, 1988b).

The perinatal growth and relocation of cortico-cortical pathways in the human cortex is not known. Recent experimental studies in monkeys have shown that callosal cortico-cortical fibers reside in the transient subplate zone at the beginning of the last third of gestation (Goldman-Rakic, 1982). Ipsilateral cortico-cortical fibers also "wait" in the subplate zone. Subplate neurons also send their axons through the corpus callosum (Schwartz and Goldman-Rakic, 1983). However, the most remarkable phenomenon in development of long cortico-cortical connections is overproduction (exuberance). In primates, cortico-cortical axons are produced in excess during perinatal life (Chalupa and Killackey, 1989; La Mantia and Rakic, 1990). The final number of cortico-cortical axons is achieved by the process of competitive elimination during the postnatal life. This phenomenon is in accordance with the general principle of competitive elimination in the initial formation of the mammalian central nervous system (Rakic et al., 1986). The process of overproduction of long cortico-cortical axons and their subsequent elimination in the human cortex may be a relatively late phenomenon. On the basis of comparative data in the cat, Innocenti proposed that human callosal connections show exuberance up to the sixth month of postnatal life (Innocenti, 1981).

In the human frontal associative cortex, the peak of overproduction of commissural and associative axons seems to occur even later. Accordingly, one can expect a prolonged waiting period and prolonged existence of the subplate zone. Indeed, the subplate zone of the prefrontal cortex exists as long as six postnatal months (Kostović et al., 1989). Thus, prolonged persistence of the subplate zone in man may then be explained by the prolonged postnatal growth of the cortico-cortical pathways (Kostović and Rakic, 1990). Overproduction of the cortico-cortical axons during postnatal life is associated with two other postnatal developmental events: excessive synaptogenesis and overproduction of dendritic spines.

The Second Reorganization: The Reduction of Dendritic Spines and Synapses during Early Childhood

The dendritic spines are sites of synaptic input on pyramidal and spiny nonpyramidal neurons. The number of spines on pyramidal neurons reflects in some way the number of synapses. By counting the number of spines on the pyramids of layer III, one can obtain a reliable parameter of the innervating of callosal and associative neurons, because layer III pyramids give rise to the vast majority of long cortico-cortical pathways (Caminiti et al., 1985; Jones, 1981; Schwartz and Goldman-Rakic 1983). Michel and Garey (1984) and Garey et al. (1984) have found that in the human visual cortex initial spine proliferation is followed by their elimination. The maximum number of spines is reached at about five months of age and after that their number decreases, reaching the adult level at about two years. In the human prefrontal cortex we have found substantially prolonged overproduction of spines during early childhood and adolescence (Mrzljak et al., 1990). The first peak of spine production was seen as late as 2.5 years; this is shown in Fig. 3.

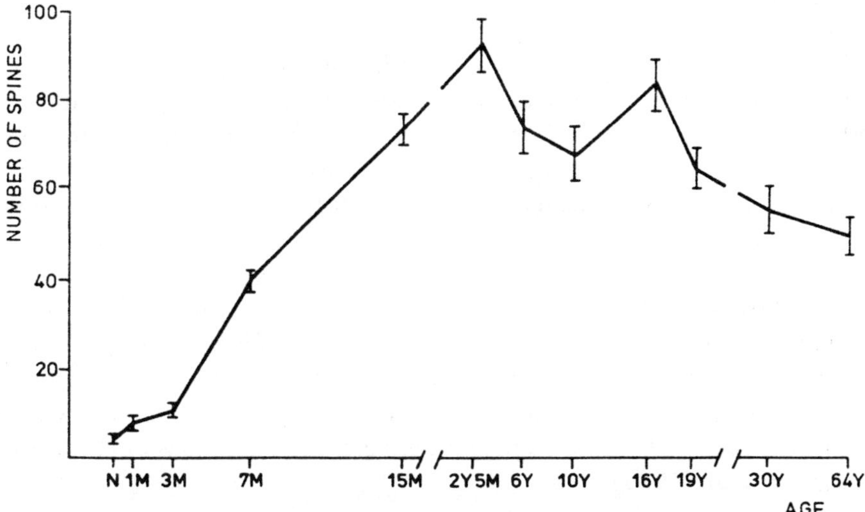

Figure 3. Developmental pattern of spine production on apical dendrites of layer III pyramidal (associative) neurons in the human prefrontal cortex. Note the overproduction of spines around 2.5 years of age.

The subsequent spine elimination indicates reorganization of the cortex at the fine structural level. Later onset of spine reduction in the prefrontal cortex as compared to the visual cortex may reflect a protracted process of synaptic development. However, one should keep in mind ambiguities of the Golgi method, which may account for differences among results of various research groups. The density of synapses in the human visual cortex is highest at about age eight months (Huttenlocher and de Courten, 1987) with gradual decline afterwards to an adult value, which is reached by age 11 years. These data correlate well with previously mentioned evidence on spine formation in the visual cortex where the rate of formation of spines slightly precedes synaptogenesis. If there exists a similar relationship in the prefrontal cortex, then the peak of frontal synaptogenesis might occur between two and three years of age, a developmental plateau lasts up to 16 years, and the adult value is reached after the reduction during adolescence. Although there are obvious difficulties in accurate quantification of both density and total number of synapses in permanently changing cortical tissue, one can safely accept the concept of three-dimensional changes in synaptic relationships during the postnatal period. The interpretation of these changes is limited by the absence of the complete data on postnatal development of cortical connections in the developing associative primate cortex. As stated earlier in this review, most investigators claim that exuberant long cortico-cortical projections are the main cause of postnatal overproduction of cortical synapses in the primate brain (Goldman-Rakic, 1987; Huttenlocher and de Courten, 1987; Rakic et al., 1986). However, other cortical connections may also contribute to the reorganization of cortical synapses as well. For example, short cortico-cortical pathways and intrinsic local circuits develop during early postnatal life and may contribute to the synaptic rearrangements during the period of second reorganization and possibly during early adolescence. The vast majority of local circuitry neurons are GABAergic and peptidergic (De Felipe et al., 1986). In spite of the fact that very little is known about perinatal and postnatal development of GABAergic and peptidergic neurons in the primate neocortex, one can predict that these intrinsic neurons will show transient overproduction of transmitter activity similar to that shown in experimental rodents (Cavanagh and Parnavelas, 1988; Naus et al., 1988; Van Eden et al., 1989).

Our studies of peptidergic development in the human frontal cortex show that peptidergic activity increased in transient fetal layers during late fetal development (Kostović et al., 1991) while multilaminar overproduction of peptidergic neurons occurs during postnatal life (Delalle and Kostović, 1991). The postnatal changes in the organization of peptidergic neurons and other elements of intrinsic cortical circuitry may be related to late stages of columnar and areal differentiation. Areal specification is a crucial event in the formation of the human cortex because it is the cortical map with numerous speech and associative areas that clearly sets us apart from other species. The complex process of areal specification passes through an early phase of laminar formation controlled by genetic factors (Rakic, 1988a) and a later phase in which epigenetic afferent inputs can be modified through interaction with afferent input

(Kaas, 1988; O'Leary, 1989; Rakic 1988b; Sur et al., 1990). The early phase of areal specification may be explained by the radial unit hypothesis. According to this hypothesis embryonic ventricular zones consist of proliferative units that provide a proto-map of prospective cytoarchitectonic areas (Rakic, 1988a). The proliferative units produce neurons for ontogenetic columns whose final number for each cortical area can be modified through interaction with thalamocortical sensory input (Rakic, 1982, 1988b). According to this hypothesis, basic laminar organization of a given cortical area is generated from the progenitors lining the cerebral ventricle. The late phase of areal specification can be accompanied by significant modification of the internal microcircuitry of the cortex (Sur et al., 1990). It is logical to assume that plastic changes of microcircuitry and short cortico-cortical connections contribute to the stabilization of the number of synapses and dendritic spines during the "second" reorganization of the cortex. The late occurrence of these changes in the human cortex may be expected on the basis of late structural and functional maturation of cortical association areas. For example, cytoarchitectonic organization of speech areas is not established until four years of age (Aldama, 1930; Judaš, 1987). Furthermore, damage of the speech area in the dominant hemisphere during the early childhood can be compensated by functions of the other hemisphere if the lesion occurs before the critical period (Vargha-Khadem et al., 1985). These data suggest that cortical speech areas develop a neurobiological substrate for the functional specialization during the postnatal life and that the period of specification correlates with final stage of structural maturation. However, the direct functional correlates of the postnatal reorganization of the associative cerebral cortex ("second" reorganization) are not known. In particular it is not clear what is the functional significance of the overgrowth phenomena during postnatal development in man. It is notable, however, that overgrowth phenomena develop during the period of initial elaboration of cognitive functions. According to Goldman-Rakic (1987) the critical mass of cortical synapses during the overgrowth period is important for the onset of cognitive function related to the delayed-response test (Diamond and Goldman-Rakic, 1989); the fully mature capacity may depend upon the elimination of excess synapses. In man, the onset of cognitive function may occur also during the overproduction of circuitry elements and thereafter may be protracted, occurring throughout childhood. The stabilization of the remaining synapses may be as important as elimination of synapses. It is very likely that afferent transmitters play a role in the stabilization of synapses and modulation of activity of cortical neurons.

Structural and Chemical Maturation after Three Years

This period follows the second reorganization of the cortex, after three years of age. Very little is known about structural and chemical changes during this period, which extends throughout childhood and adolescence. The reduction of the number of spines and synapses is evidently one important measurable parameter. Unfortunately we do not know what specific changes of cortical circuitry underlie these synaptic fluctuations. According to our data on development of the human associative prefrontal cortex, the period after four years is characterized by the protracted maturation of transmitter-related markers (Kostović et al., 1989). It is our hypothesis that maturation of transmitter-related innervating of associative pyramidal neurons of layer III is the major event of the late phase of cortical development in man (Kostović, 1990). The discrete AChE reactivity of layer III pyramidal neurons was first observed in a four-year-old specimen (Kostović, 1990; Kostović et al., 1988). After this period, AChE reactivity increases gradually (Figs. 2C and 2D) and reaches its peak intensity in young adults (Kostović et al., 1988; Mesulam and Geula 1988). This finding lead us to the hypothesis that AChE-rich prospective cholinergic elements play a role in the innervation of cortical associative neurons during cognitive development in man (Kostović et al., 1988). The AChE-rich innervation of layer III pyramidal neurons has been demonstrated only in the primate cortex so far. The fact that the development of cholinergic enzyme in layer III associative neurons parallels maturation of cholinergic receptors (Johnston et al., 1985) is in accordance with the idea that cholinergic innervation plays a role in the development of cognitive functions (Kostović et al., 1988). This idea is further supported by the general hypothesis that in adult humans the cholinergic system plays a modulatory role in cognitive functions (Bartus et al., 1985; Drachman and Leavitt, 1974; Mesulam and Geula, 1988). The AChE reactivity in layer III pyramidal associative neurons is probably associated with the postsynaptic, cholinoceptive nature of these neurons. If this is the case, the gradual increase of AChE reactivity of layer III pyramidal neurons is a result of increased cholinergic modulation by cholinoceptive neurons during cognitive development. It is very interesting that in aging and Alzheimer's disease cortical cholinergic markers decrease gradually, which results in the deterioration of cognitive functions (Davies and Maloney, 1976; Perry et al., 1978). Moreover, degenerative changes of cortical presynaptic fibers and postsynaptic neurons show the highest density in the third cortical layer (Rogers and Morrison, 1985).

At present, the comprehensive interpretation of cholinergic involvement in cognitive functions is not possible due to the fact that we do not know the dynamic interaction with other extrinsic and intrinsic transmitter systems in the cerebral cortex. The gradual developmental rise of AChE-rich innervation of associative pyramidal neurons during childhood and adolescence seems to be a good model for future investigation of chemical and structural processes underlying cognitive functions. It is notable that the maximal density of AChE-rich innervation of association neurons is established in adolescence during the last phase of cognitive development (Piaget 1954). The late production of synapses observed by us in the frontal cortex up to 16 years of age may also serve as an indicator of fine structural changes of the associative cortex during adolescence.

Summary

One of the major features of the late human fetal cortex is the transient arrangement of afferent axons, cortical neurons, and synapses. The transient arrangement of neurons is present in both primary and association cortical areas. In the association cortex, transient arrangement of neuronal circuitry elements lasts much longer than in primary cortical areas such as the primary visual cortex. It is very likely that transient arrangement of synapses and axons influence the transient expression of electrophysiological phenomena in the preterm infant. Transient accumulation of thalamocortical afferents is of particular significance in light of a possible external influence on cortical activity. The perinatal and early postnatal reorganization ("first reorganization") of transient fetal patterns involves relocation of thalamic and basal forebrain afferents, disappearance of the fetal layer (subplate zone), and changes at the transmitter-related level. Cortico-cortical and association fibers continue to grow during the postnatal period. The final number of cortico-cortical axons is achieved by the process of competition elimination. The most interesting aspect of postnatal reorganization ("second reorganization") occurs during the onset and initial phases of cognitive development: overgrowth of dendritic spines and overproduction of synapses. The maximal number of these circuitry elements is achieved at around 2.5 years, which is significantly later than in the visual cortex. The late production of spines on association layer III pyramidal neurons of the frontal cortex is probably a sign of late development of cortical microcircuitry and short cortico-cortical connections. The final phase of maturation is characterized by the protracted maturation of transmitter-related innervation. The postsynaptic cholinergic marker in pyramidal neurons was first observed after four years of age. This finding leads to the hypothesis that AChE-rich cholinergic elements play a role in the innervation of cortical associative neurons during cognitive development in man.

Acknowledgments. This work has been supported by U.S.–Yugoslav Joint Board Grant No. 855 and Scientific Funds of the Republic of Croatia. We wish to acknowledge the excellent technical assistance of Zdenka Cmuk, Danica Budinscak, and Bozica Popovic.

References

Aldama J (1930): Cytoarchitektonik der Grosshirnrinde eines 5 jahrigen und eines 1 jahrigen Kindes. *Z ges Neurol Psychiatr* 130:532–630

Anand KJS, Hickey PR (1987): Pain and its effects in the human neonate and fetus. *New Engl J Med* 317:1321–1329

Bartus RT, Dean RL, Pontecorvo MJ, Flicker C (1985): The cholinergic hypothesis: a historical overview, current perspective, and future directions. *Ann NY Acad Sci* 444:332–358

Caminiti R, Zeger S, Johnson PB, Urbano A, Georgopoulos P (1985) Corticocortical efferent systems in the monkey: A quantitative spatial analysis of the tangential distribution of cells of origin. *J Comp Neurol* 241:405–419

Cavanagh ME, Parnavelas JG (1988): Development of somatostatin immunoreactive neurons in the rat occipital cortex: immunocytochemical-autoradiographic study. *J Comp Neurol* 68:1–12

Chalupa LM, Killackey HP (1989): Process elimination underlies ontogenetic change in the distribution of callosal projection neurons in the postcentral gyrus of the fetal rhesus monkey. *Proc Natl Acad Sci USA* 86:1076–1079

Cowan WM, Fawcett JW, O'Leary DDM, Stanfield BB (1984): Regressive events in neurogenesis. *Science* 225:1258–1265

Davies F, Maloney AFJ (1976): Selective loss of central cholinergic neurons in Alzheimer's disease. *Lancet* 11:1403

DeFelipe J, Conley M, Jones EG (1986): Long-range focal collateralization of axons arising from corticocortical cells in monkey sensory-motor cortex. *J Neurosci* 6:3749–3766

Dehay C, Kennedy H, Bullier J, Berland M (1988): Absence of interhemispheric connections of area 17 during development in monkey. *Nature* 331:348–350

Delalle I, Kostović I (1991): Laminar distribution of NPY-immunoreactive neurons in human prefrontal cortex during development. *Eur J Neurosci Suppl* (in press)

Diamond A, Goldman-Rakic PS (1989): Comparison of human infants and rhesus monkeys on Piaget's AB task: evidence for dependence on dorsolateral prefrontal cortex. *Exp Brain Res* 74:24–40

Drachman DA, Leavitt J (1974): Human memory and the cholinergic system: A relationship to aging? *Arch Neurol* 30:113–121

Dreyfus-Brisac C (1979): Ontogenesis of brain bioelectrical activity and sleep organization in neonates and infants. In: *Human Growth, Vol. 3, Neurobiology and Nutrition*, Falkner F, Tanner JM, eds. London: Bailiere Tindall, pp. 157–182

Garey LJ, Michel AE, Leuba G (1984): Changes in spine density of human visual cortical neurons during development. *Behav Brain Res* 12:192–193

Goldman-Rakic PS (1982): Neuronal development and plasticity of association cortex in primates. *Neurosci Res Prog Bull* 20:520–532

Goldman-Rakic PS (1987): Development of cortical circuitry and cognitive function. *Child Development* 58:601–622

Huttenlocher PR, de Courten Ch (1987): The development of synapses in striate cortex of man. *Human Neurobiol* 6:1–9

Innocenti GM (1981): Growth and reshaping of axons in the establishment of visual callosal connections. *Science* 212:824–827

Innocenti GM (1982): Development of interhemispheric cortical connections. *Neurosci Res Prog Bull* 20:532–540

Johnston MV, Silverstein FS, Reindel FO, Penney JB Jr, Young AB (1985) Muscarinic cholinergic receptors in human infant forebrain: [3H]quinuclidinyl benzilate binding in homogenates and quantitative autoradiography in sections. *Dev Brain Res* 19:195–203

Jones EG (1981): Anatomy of cerebral cortex: Columnar input-output organization. In: *The Organization of the Cerebral Cortex*, Schmitt FO, Worden FG, Adelman G, Dennis SG, eds. Cambridge: MIT Press, pp. 199–235

Judaś M (1987): Perinatal development of cytoarchitectonics of prospective motor speech area in human frontal lobe. M.S. thesis, University of Zagreb, Zagreb, 1987

Kaas JH (1988): Development of cortical sensory maps. In: *Neurobiology of Neocortex*, Rakic P and Singer W, eds. John Wiley & Sons Ltd., S. Bernard, Dahlem Konferenzen, pp. 101–113

Kostović I (1986): Prenatal development of nucleus basalis complex and related fibre systems in man: a histochemical study. *Neuroscience* 17:1047–1077

Kostović I (1990): Structural and histochemical reorganization of the human prefrontal cortex during perinatal and postnatal life. In: *The Prefrontal Cortex: Its Structure, Function, and Pathology*, Uylings HBM, Van Eden CG, De Bruin JPC, Corner A, and Feenstra MGP, eds., *Progress in Brain Research* 85:223–240

Kostović I, Goldman-Rakic PS (1983): Transient cholinesterase staining in the mediodorsal nucleus of the thalamus and its connections in the developing human and monkey brain. *J Comp Neurol* 219:431–447

Kostović I, Molliver ME (1974): A new interpretation of the laminar development of cerebral cortex: synaptogenesis in different layers of neopallium in the human fetus. *Anat Rec* 178:395

Kostović I, Rakic P (1980): Cytology and time of origin of interstitial neurons in white matter in infant and adult human and monkey telencephalon. *J Neurocytol* 9:219–242

Kostović I, Rakic P (1984): Development of prestriate visual projections in the monkey and human fetal cerebrum revealed by transient cholinesterase staining. *J Neurosci* 4:25–42

Kostović I, Rakic P (1990): Developmental history of the transient subplate zone in the visual and somatosensory cortex of the macaque monkey and human brain. *J Comp Neurol* 297:441–470

Kostović I, Skavic J, Strinovic D (1988): Acetylcholinesterase in the human frontal associative cortex during the period of cognitive development: early laminar shifts and late innervation of pyramidal neurons. *Neurosci Lett* 90: 107–112

Kostović I, Lukinovic N, Judaš M, Bogdanović N, Mrzljak L, Zecevic N, Kubat M (1989): Structural basis of the developmental plasticity in the human cerebral cortex: The role of the transient subplate zone. *Metabolic Brain Disease* 4:17–23

Kostović I, Stefulj-Fucic A, Mrzljak L, Jukic S, Delalle I (1991) Prenatal and perinatal development of the somatostatin-containing neurons in the human prefrontal cortex. *Neurosci Lett* (in press).

Krmpotic-Nemanic J, Kostović I, Kelovic, Z, Nemanic G. (1980) Development of acetylcholinesterase (AChE) staining in human fetal auditory cortex. *Acta Otolaryngol* (Stockholm) 89:388–392

LaMantia AS, Rakic P (1990): Axon overproduction and elimination in the corpus callosum of the developing rhesus monkey. *J Neurosci* 10:2156–2175

Leijon I (1982): Assessment of behavior on the Brazelton scale in healthy preterm infants from 32 conceptional weeks until full-term age. *Early Human Develop* 7:109–118

Mesulam MM, Geula C (1988): Acetylcholinesterase-rich pyramidal neurons in the human neocortex and hippocampus: Absence at birth, development during the life span, and dissolution in Alzheimer's disease. *Ann Neurol* 24:765–773

Michel AE, Garey LJ (1984): The development of dendritic spines in the human visual cortex. *Human Neurobiol* 3:223–227

Molliver ME, Kostović I, Van der Loos H (1973): The development of synapses in the human fetus. *Brain Res* 50:403–407

Mrzljak L, Uylings HBM, Van Eden CG, Judaš M (1990): Neuronal development in human prefrontal cortex in prenatal and postnatal stages. *Progr Brain Res* 85:185–222

Naus CCG, Miller FD, Morrison JH, Bloom FE (1988): Immunocytochemical and in situ hybridization analysis of the development of the rat somatostatin-containing neocortical neuronal system. *J Comp Neurol* 269:448–463

O'Leary DDM (1989): Do cortical areas emerge from a protocortex? *Trends Neurosci* 12:400–406

Parmelee AH (1975): Neurophysiological and behavioral organization of premature infants in the first months of life. *Biol Psychiat* 10:501–512

Perry EK, Tomlinson BE, Blessed G, Bergman K, Gibson PH, Perry RH (1978): Correlation of cholinergic abnormalities with senile plaques and mental test scores in senile dementia. *Br Med J* 2:1457–1459

Piaget J (1954): *The Construction of Reality in the Child.* New York: Basic Books

Prechtl HFR (1974): The behavioral states of the newborn infant (a review). *Brain Res* 76:185–212

Prechtl HFR (1984): Continuity and change in early neural development. In: *Continuity of Neural Functions from Prenatal to Postnatal Life.* Spastics International Medical Publications, Prechtl HFR, ed. Oxford: Blackwell Scientific Publications Ltd.; Philadelphia: J.B. Lippincott Co., pp. 1–15

Rakic P (1982): Early developmental events: Cell lineages, acquisition of neuronal positions, and areal and laminar development. *Neurosci Res Prog Bull* 20:439–451

Rakic P (1988a) Specification of cerebral cortical areas. *Science* 241:170–176

Rakic P (1988b) Intrinsic and extrinsic determinants of neocortical parcellation: A radial unit model. In: *Neurobiology of Neocortex*, Rakic P, Singer W, eds. New York: Wiley, pp. 5–27

Rakic P, Bourgeois JP, Zecevic N, Eckenhoff MF, Goldman-Rakic PS (1986): Concurrent overproduction of synapses in diverse regions of the primate cerebral cortex. *Science* 232:232–235

Rogers J, Morrison JH (1985): Quantitative morphology and regional and laminar distributions of senile plaques in Alzheimer's disease. *J Neurosci* 5:2801–2805

Schwartz ML, Goldman-Rakic PS (1983): Prenatal development of callosal and intrahemispheric cortico-cortical input to prefrontal cortex in the rhesus monkey. *Soc Neurosci Abstr* 9:1058

Shatz CJ, Chun JJM, Luskin MB (1988): The role of the subplate in the development of the mammalian telencephalon. In: *Cerebral Cortex*, Jones EG, Peters A, eds. New York: Plenum, pp. 35–58

Sur M, Pallas SL, Roe AW (1990): Cross-modal plasticity in cortical development: differentiation and specification of sensory neocortex. *Trends Neurosci* 13:227–233

Trevarthen C (1979): Neuroembryology and the development of perception. In: *Human Growth, Vol. 3, Neurobiology and Nutrition*, Falkner F, Tanner JM, eds. London: Bailiere Tindall, pp. 3–96

Van Eden CG, Mrzljak L, Voorn P, Uylings HBM (1989): Prenatal development of GABAergic neurons in the neocortex of the rat. *J Comp Neurol* 289:213–228

Vargha-Khadem F, O'Gorman AM, Watters GV (1985): Aphasia and handedness in relation to hemispheric side, age at injury and severity of cerebral lesion during childhood. *Brain* 108:677–696

Wolff PH, Ferber R (1979): The development of behavior in human infants, premature and newborn. *Ann Rev Neurosci* 2:291–307

Prenatal Monoaminergic Innervation of the Cerebral Cortex: Differences between Rodents and Primates

B. Berger, C. Verney, and P. S. Goldman-Rakic

After the discovery of the cerebral monoaminergic systems by the Swedish group in the early sixties, the early prenatal ontogenesis of the monoaminergic neurons in the brainstem and the precocious development of their cortical projections were soon recognized (Lauder and Bloom, 1974; Nobin and Björklund, 1973; Olson and Seiger, 1972; Seiger and Olson, 1973) and presumed to be of great biological importance. The early arrival of aminergic afferents into the anlage of the cerebral cortex, concomitant to important intrinsic events in the neocortex (active stages of migration and differentiation) has led to the hypothesis that monoamines might exert an inductive or regulatory influence on the development of the cerebral cortex (see review in Berger and Verney, 1984). This hypothesis triggers new interest in view of the proposal that an involvement of the dopaminergic system in the pathophysiology of schizophrenia might be associated to an atrophy/aplasia process of developmental origin in the medial telencephalon (Roberts and Crow, 1987). Rodents have generally been used in the experimental models elaborated for understanding the anatomy and physiology of the cortical monoaminergic systems. However, in contrast with rodents, the cerebral cortex of primates is characterized by a protracted prenatal development, which allows for complex and prolongated interactive processes. Moreover, in adult primates, major changes are observed in the distribution of the cortical catecholaminergic innervation as compared to rodents. Besides a laminar and regional-specific rearrangement of the noradrenergic afferents (Lewis and Morrison, 1989; Morrison et al., 1982), there is a considerable expansion of the dopaminergic innervation (Berger et al., 1988; Lewis et al., 1987; Gaspar et al., 1989), which extends to the whole cerebral cortex instead of being concentrated, as in rodents, in the prefrontal cortex and the limbic areas, anterior cingulate and entorhinal cortices and the ventral subiculum-CA1 field of the hippocampus (Berger et al., 1985a; Berger et al., 1976; Björklund and Lindvall, 1984; Descarries et al., 1987; Van Eden et al., 1987; Verney et al., 1985). This modified dopaminergic innervation is also characterized by a regional and laminar-specific redistribution, the highest density of dopaminergic afferents being observed in the three motor areas and the anterior cingulate cortex (Berger et al., 1988; Gaspar et al., 1989; Lewis et al., 1987, 1988). We report here preliminary results concerning the prenatal development of the cortical monoaminergic innervation in human and rhesus monkey fetuses. Comparison with our previous data obtained in rat embryos (Berger et al., 1985b; Verney et al., 1982) and others (Kalsbeek et al., 1988b) reveals major differences in the prenatal distribution of the cortical dopaminergic system, confirming the species-specificity already observed in adulthood. These results justify the use of higher mammals as models for understanding the establishment and the role of the aminergic circuitry in the human brain.

Prenatal Development of the Monoaminergic Innervation in the Cerebral Cortex of Primates

Antibodies raised against serotonin and against dopamine-β-hydroxylase (DBH), the enzyme that converts dopamine into noradrenaline, were used to reveal the serotonergic and the noradrenergic fibers respectively. The dopaminergic innervation was visualized with antibodies raised against tyrosine hydroxylase. Tyrosine hydroxylase (TH), the first enzyme in the biosynthetic pathway of catecholamines, was tentatively used as a marker of the dopaminergic system since several studies from our group and others have shown that it seems to preferentially label the cortical dopaminergic axons (versus the norepinephrine-containing fibers) in human (Gaspar et al., 1989) and nonhuman primates (Noack and Lewis, 1989) and in rat embryos (Verney et al., 1982). In the cerebral cortex of human fetuses, a double label immunocytochemical study of the two markers shows that less than 5% of the TH-like immunoreactive (TH-LIR) fibers and terminals are also DBH-LIR.

Rhesus Monkey Fetuses

General cytoarchitectural features. The monoaminergic cell groups in the brainstem (NA, DA, and 5-HT) are generated over a short period of time, from day 27 to 43, within the first quarter of the 165-day gestational period (Levitt and Rakic, 1982). The genesis of the cortical neurons starts approximately at the same time (E38) but is more prolonged, lasting through the first half of gestation, from E40 to E80 for the anterior cingulate and motor cortex (Dekker and Rakic, 1980), E40 to E90 for the prefrontal cortex (Goldman-Rakic et al., 1983), E38 to E80 for the hippocampal formation (Rakic and Nowakowski, 1981). There are two exceptions: the neurons destined to the visual cortex are generated until day 102 (Rakic, 1974) and the hippocampal dentate gyrus continues to acquire granule cells during the first three postnatal months (Rakic and Nowakowski, 1981). At E70, the earliest age considered in this study, the fetal cerebral wall is comprised of six layers: the marginal zone (prospective layer I), the cell dense cortical plate, the large subplate, the intermediate zone (prospective white matter), and the densely cellular subventricular and ventricular zones. Two layers are of particular importance: the cortical plate, which will differentiate into the mature cortical layers, and the subplate (Kostović and Rakic, 1990), a transient zone that assumes in later fetal life an important role in the establishment of cortical circuitry.

Development of the aminergic projections. Three fetuses (70, 83, and 126 days old) and one newborn animal were examined. Mounted or free-floating coronal cryostat sections from formalin-fixed brains were processed for immunocytochemistry.

At *E70*, the aminergic fibers that arrived in the basal forebrain reached the cerebral cortex through the large cingulate fasciculus and the external capsule and penetrated simultaneously the intermediate zone, adjacent subplate, and the marginal zone (the prospective molecular layer). The latter afferent fibers entered the molecular layer directly and seemed to spread independently of the underlying fiber systems located in the intermediate zone. At this early age, monoaminergic fibers in the marginal zone occupied only the most superficial part under the pia.

The ingrowth of the monoaminergic fibers in the developing cerebral cortex exhibited a rostrocaudal, medio-lateral, and ventrodorsal gradient that partly corresponded to that of neuronal differentiation. A mediolateral gradient of differentiation was observed in the frontal cortex at E70, represented by the presence of densely packed fusiform neurons in the cortical plate of the lateral prefrontal cortex, which did not exhibit any laminar differentiation, whereas an incipient lamination could be observed medially in the prospective pregenual cingulate cortex with more differentiated neurons in layer VI. Correspondingly, the subplate contained more monoaminergic fibers in the medial than in the lateral frontal cortex. Rare fibers penetrated the cortical plate.

The motor cortex appeared less differentiated than the anterior cingulate cortex with more densely packed neurons in the cortical plate. In contrast to the cingulate cortex, in which aminergic fibers were quite prominent, in the motor cortex, few aminergic fibers were present in the subplate and even fewer in the cortical plate. Bundles of tangentially running aminergic axons were observed in the intermediate zone above the migrating neurons of the subventricular zone. They ran in both mediolateral and lateromedial directions as indicated by their growth cones. The same held true for those that penetrated the corpus callosum.

The primary visual cortex was the least differentiated cortical area, with a densely packed cortical plate and ventricular zone. Almost no aminergic afferents were observed, in contrast to the frontal cortex; one or two fibers could be detected in the subplate but none in the marginal zone or the cortical plate.

The ventrodorsal gradient was best visualized at the level of the insula and the entorhinal cortex. In the insula, both the subplate and the marginal zone contained a high density of aminergic fibers even higher than that of the medial cortical zones. In the entorhinal cortex where clusters of pyramidal neurons in the prospective layer III were just emerging at E70, the pattern of aminergic fibers was similar to that observed in the adult. A dense serotonergic innervation predominated over the TH-LIR terminals, contrasting with a poor noradrenergic input.

At *E83*, the mediolateral gradient of differentiation was still evident (Figs. 1a, 2a) but neuronal differentiation had progressed insofar as the cortical plate was less densely packed in all areas except the primary visual cortex. The immaturity of the visual cortex was not surprising as the neurons of the visual cortex continue to be generated until day 102 in the rhesus monkey (Rakic, 1974). Layers V–VI were emerging in the anterior cingulate cortex (Fig. 1a) and layer VI in the temporal cortex.

The monoaminergic innervation increased in density in all areas examined except the visual cortex. In the marginal zone, the long tangentially running aminergic axons were more numerous and now invaded the deeper part of this layer. Numerous TH- and serotonin-containing fibers were present in the subplate and the emerging deeper cortical layers in nonvisual areas (Figs. 1b, 1c, 2b, 2c). Dense monoaminergic fascicles ascended in the intermediate zone within the cingulum bundle (Fig. 1b, 1c). An asynchronous development of the three aminergic systems was observed in the motor cortex. The noradrenergic innervation was more developed than the dopaminergic and serotonergic innervation both in the

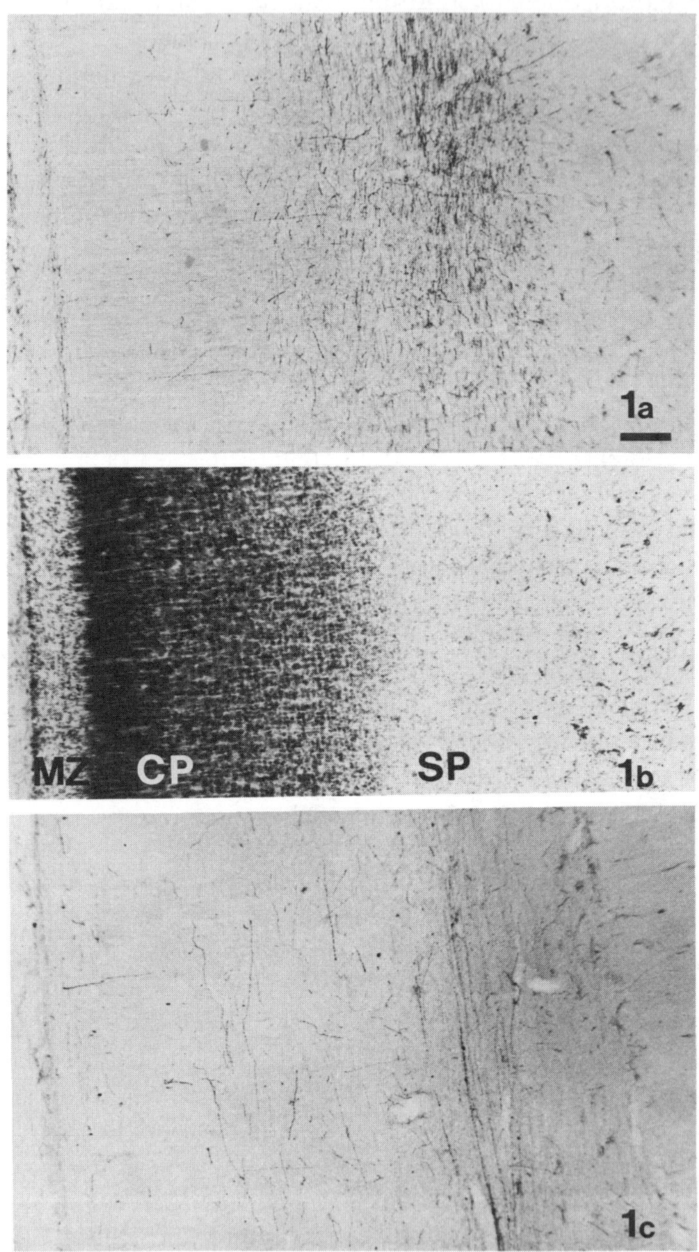

Figure 1. Pregenual cingulate cortex. (b) Nissl staining. Layers V–VI are emerging from the deep cortical plate. MZ: marginal zone; CP: cortical plate; SP: subplate. (a, c) distribution of TH- (1a) and serotonin-LIR (1c) fibers in the same area as in (b). From monkey cerebral cortex on embryonic day 83 (E83), at magnification bar = 100μ.

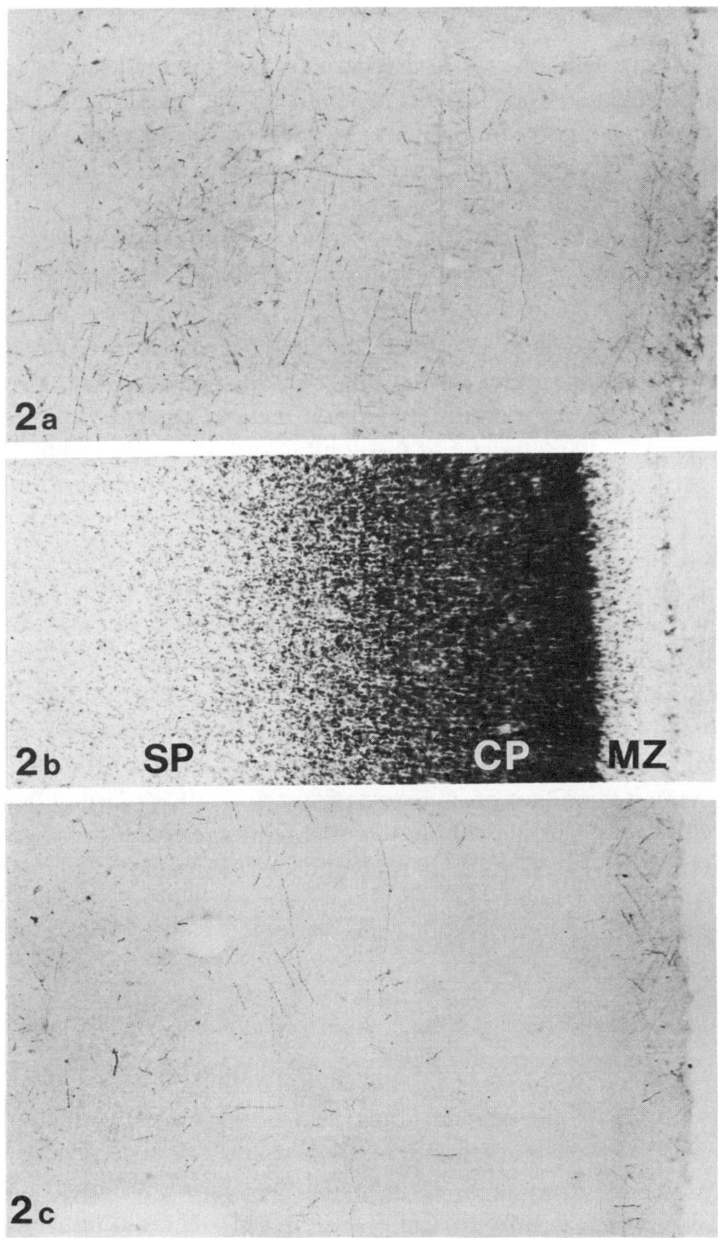

Figure 2. Lateral prefrontal cortex. (b) Nissl staining. The cortical plate is less differentiated than in Fig. 1b; it contains more densely packed neurons and does not exhibit any laminar differentiation. TH- (2a) and serotonin-LIR (2c) axons are correspondingly less numerous than in the pregenual cingulate cortex. From monkey cerebral cortex on embryonic day 83 (E83), at magnification bar = 100μ.

subplate and the lower cortical plate. This situation is reversed in the newborn.

At *E126*, considerable changes characterized the cerebral cortex (six weeks before birth). The main sulci could be identified: principal, cingulate, central, lateral, intraparietal, and temporal sulci as well as the calcarine fissure. Most of the cortical areas displayed a clarification of the laminar organization in the deeper cortical layers IV–V–VI. As the differentiation increased, so did the monoaminergic innervation, which occupied all emerging strata with the exception of the densely packed layer of the upper cortical plate.

A bilaminar type of dopaminergic innervation was clearly apparent in the granular cortices similar to that observed in the adult. In the primary visual cortex, the serotonergic innervation had not reached full development in layer IV, characteristic of the adult cortex (Kosofsky et al., 1984); it predominated in layers I and V–VI whereas the dopaminergic innervation was basically restricted to layer I as in the adult. The monoaminergic innervation also increased in the agranular cortices. The ventral cingulate cortex was densely innervated by dopaminergic fibers even in layer III. In the primary motor cortex, a clustering of TH-labeled terminals appeared in layer III similar to that observed in the adult (Berger et al., 1988). In contrast, in the supplementary motor cortex, dopaminergic fibers largely predominated in the deeper cortical layers V–VI and in the subplate, contrasting with the denser input of serotonergic afferents in the superficial layers, whereas the situation is reversed in the adult. A large number of neurons still populated the subplate at this fetal age. Many of them were labeled with DARPP-32, a dopamine and cAMP-regulated phosphoprotein selectively enriched in dopaminoceptive cells bearing dopaminergic receptors of the D-1 type (Berger et al., 1990).

Human Cerebral Cortex

The embryonic development of the central monoaminergic neurons in man is less understood than in nonhuman primates. Until recently, the precise timing of the generation of monoaminergic cell groups in the brainstem was not known; an early generation during the first quarter of gestation was suggested by their presence in ten-week-old fetuses (Olson et al., 1973; Pearson et al.,1980). In fact, we could observe the main catecholaminergic cell groups in the brainstem of six-week-old human embryos (Verney et al., 1991b). Also similar to nonhuman primates, the full complement of cortical neurons is acquired during the first half of gestation (Rakic, 1988).

The rostral telencephalon of four 20–24-week-old fetuses was available for immunocytochemical study. Prefrontal, anterior cingulate, and motor cortex were among the areas available for analysis. At this age the cortical anlage is composed of the marginal zone (prospective layer I), the cortical plate with the emerging cortical lamination, an extended subplate—three to four times thicker than the cortical plate—and the intermediate zone, the future white matter. Pyramidal neurons were clearly identifiable in the different areas in prospective layers VI and V (Marin-Padilla, 1970; Mrzljak et al., 1988); however, the cortical plate was less differentiated in the prefrontal cortex than in the anterior cingulate and motor areas.

A conspicuous feature of the cortical catecholaminergic innervation was its high overall density. The highest density of terminals was observed at the level corresponding to the head of the caudate nucleus: TH-LIR (dopaminergic) fibers were numerous in all areas, from the medial telencephalon (future cingulate cortex) to the ventro-lateral cortex, and even more in the dorsal part of the presumed motor cortices. They were present in all layers, showing various orientations and density. The lower cortical plate and the upper subplate were the most innervated laminae. The deep subplate exhibited a lower number of positive fibers but contained TH-LIR cell-bodies. The marginal zone was moderately innervated in its medial third. DBH-LIR (noradrenergic) afferents were also numerous, albeit less than TH-LIR fibers. They were also located preferentially in the lower cortical plate and upper subplate but a few vertical axons crossed the whole cortical plate. Only some short axons were observed in the marginal zone. More rostral sections, at the beginning of the lateral ventricle, were characterized by a lower density of both dopaminergic (TH-LIR) and noradrenergic (DBH-LIR) fibers. A bilaminar pattern of distribution in the marginal zone and the upper subplate, respectively, was obvious, but immunoreactive fibers were always more numerous in the inner than in the outer aminergic plexus. A TH- and a DBH-LIR bundle of tangentially running axons was clearly delineated in the intermediate zone (white matter) of all areas. They were especially apparent medially and ventro-laterally: numerous long immunoreactive fibers could be followed in the external capsule. To sum up, although the different layers of the cortex were not clearly delineated at this age, catecholaminergic fibers were present in the whole depth of the future cingulate and motor cortices, whereas a bilaminar pattern predominated in more rostral areas. The region-specific distribution of the dopaminergic innervation was similar to that observed in the adult human cortex (Gaspar et al., 1989). However, in spite of the high density of positive axons throughout the cortical anlage of the frontal lobe, the extensive dopaminergic terminal field observed in layer I in adulthood had no counterpart in the marginal zone of the developing cortex in the 24-week-old human fetuses.

Prenatal Development of the Monoaminergic Innervation in the Rodent Cerebral Cortex

In the rodent, all cortical neurons are generated between days 14 and 22 (Hicks and D'Amato, 1968). Since the term of pregnancy ranges from 21 to 22 days, it means that the last cortical neurons are generated close to birth. The formation of the cortical plate, which initiates the development of the main cortical layers starts on days 15–16 of embryonic life, dividing the cerebral wall into four layers: marginal zone, cortical plate, intermediate zone, and periventricular zone (Raedler et al., 1980). The lateral areas of the cerebral cortex develop 1–2 days before the medial areas with regard to cell density, cell migration, and gross neuropil development. At birth, only layers I and VI are well differentiated, but the cortical lamination will be achieved during the first postnatal week (Rice et al., 1985; Van Eden and Uylings, 1985). The monoaminergic nuclei in the brainstem are generated later than in primate, at the beginning of the second half of gestation from E12 to E16 (Altman and Bayer, 1981; Lauder and Bloom, 1974; Specht et al., 1981a). The three types of monoaminergic fibers reach the frontal cortex at almost the same time (Schlumpf et al., 1980; Specht et al., 1981b): the dopaminergic fibers on embryonic day 16 (Verney et al., 1982; Kalsbeek et al., 1988b), the noradrenergic (Levitt and Moore, 1979; Verney et al., 1984) and serotonergic fibers (Lidov and Molliver, 1982a) at E17, and then all systems extend caudalwards. However, the timing and organization of their cortical ingrowth differ markedly.

The ingrowth of the *noradrenergic* innervation proceeds caudalwards from the frontal pole in three sagittal bands: first a medial and a lateral one, the dorsal cortex becoming innervated with a two-day delay (Verney et al., 1984). Therefore, the medial and lateral bundles achieved their rostrocaudal extension (to the retrosplenial, temporal, and entorhinal cortices) at E20 when the dorsal axons just reached the parietal cortex; the dorsal occipital cortex was reached at birth. Positive NA fibers were first observed in the marginal and intermediate zones, just above and below the cortical plate. They gradually invaded the other cortical layers, also following a rostrocaudal gradient. Several morphological features suggested that NA axons arise from both superficial and deep plexuses to invade simultaneously the cortical plate. The noradrenergic innervation develops rapidly through the first postnatal week (Lidov et al., 1978) and displays at that time an axon pattern similar to that observed in the adult.

The ingrowth of the *serotonergic* innervation proceeds along both a rostrocaudal and a mediolateral gradient, more slowly than the noradrenergic innervation, so that very few axons are present in the occipital cortex (areas 17–18) at P6 (Lidov and Molliver, 1982b). Similar to the noradrenergic system, serotonergic axons enter the cortical anlage as two tangential and parallel sheets, one above and one below the cortical plate. This bilaminar distribution remains a prominent feature during the first postnatal week except in the barrel field of the primary somatosensory area, where 5-HT immunoreactive axons are clustered in small patches forming radial modules in layers IV and VI (D'Amato et al., 1987). This aspect contrasts with the sparse innervation of the parietal and visual association areas, but is a transient process; after postnatal day 12, the density of the overall serotonergic innervation increases while that in dense patches decreases (D'Amato et al., 1987). It is not until P14–P21 that 5-HT axons become distributed across the cortical layers with a pattern similar to that in the adult neocortex. The developmental gradient of axon growth contrasts with the neurogenetic gradients in the neocortex. The earliest generated cells, those of the lateral cortex, are apparently the last to be reached by the 5-HT neurons (Lidov and Molliver, 1982a).

The *dopaminergic* innervation of the rat cerebral cortex is much more restricted than the NA and 5-HT innervations, as most of the primary somatosensory and association areas are devoid of DA afferents except in layer VI (Björklund and Lindvall, 1984; Descarries et al., 1987). The developmental schema is also more complex, since two main contingents that develop pre- and postnatally, respectively, contribute to the cortical DA innervation. The first contingent reaches the frontal pole by E16, located solely in the intermediate zone (Kalsbeek et al., 1988b; Verney et al., 1982). From E17 onward, the dopaminergic innervation extends from the frontal pole toward the medial prefrontal, the deep supragenual and the rhinal cortices, and reaches the entorhinal and visual cortices at birth. This dopaminergic contingent, which remains confined in the intermediate zone, will distribute predominantly to the deeper cortical layers V–VI. It displays a striking rostrocaudal gradient of decreasing density from the prefrontal cortex, where it is almost fully developed at birth (Schmidt et al., 1982), to the visual cortex (Papadopoulos et al., 1989). The second class of DA afferents reaches its cortical targets only after birth, during the first and second postnatal weeks, and does not assume the adult distribution pattern until the fourth week (Berger et al., 1985b). It is distributed to the superficial cortical layers I–III, being mainly concentrated in the anterior cingulate cortex, which receives the highest contingent of dopaminergic terminals of the whole cerebral cortex; but it also sparsely innervates the superficial layers of the premotor, retrosplenial, and visual cortices (Berger et al., 1985a, Phillipson et al., 1986).

Discussion

Two main features emerge from these comparative studies. First, the precocious monoaminergic innervation of the primate cerebral cortex, suggested by the early generation of the monoaminergic neurons in the brainstem (Levitt and Rakic 1982; Olson et al., 1973; Pearson et al., 1980), is demonstrated directly for the first time. Although the precise timing needs to be determined, the first monoaminergic cortical afferents arrive at the cortex before the end of the first half of fetal life and their further development proceeds rapidly until birth, in contrast with the observations made in the rodent. Second, among the monoaminergic systems, the development of the cortical dopaminergic innervation seems species specific, both with respect to timing and to regional and laminar distribution.

The precocious development of the monoaminergic projections parallels the early genesis of the cortical target neurons. The presence of aminergic afferents in the cortical anlage, well before all neurons have been generated, supports previous hypotheses on their functional importance. Their immediate bilaminar distribution in the marginal zone and the subplate and intermediate zone sets the conditions for multiple interactions, all the more so since it parallels the synaptogenesis. Synaptogenesis is indeed an early event, proceeding concurrently in all areas (Rakic et al., 1986). In the motor cortex, synapses are first observed in the subplate and the marginal zone by embryonic day 53 in the monkey (Zecevic et al., 1989), and in the marginal layer by seven weeks in the human embryo (Larroche et al., 1981). Synaptic triads including an excitatory and an inhibitory synapse on the same dendritic spine are well known in the adult cerebral cortex and it has recently been demonstrated that dopaminergic as well as peptidergic terminals contribute to these triads (Goldman-Rakic et al., 1989; Verney et al., 1991a). For instance, in the marginal zone they might establish connections with the Cajal–Retzius cells and/or the apical dendrites of prospective pyramidal cells (Marin-Padilla, 1984). Interestingly, although the pyramidal neurons are still very immature in the E70-old monkey, they already issue subcortical projections from the prefrontal cortex to the caudate nucleus and the thalamus (Goldman-Rakic 1981, 1987).

The possible interactions between the monoaminergic terminals and the different cellular elements of the subplate deserve special consideration. The subplate layer individualized in cats by Marin-Padilla (1978) and in primates by Kostović and Molliver (1974) and Rakic (1982) has been attributed more and more functional importance in the last 10 years. This special layer of the developing cerebral cortex, located below the cortical plate, is made up of neurons generated between E38 and E48 in monkeys (Kostović and Rakic, 1980), before those that will form the six-layered cerebral cortex. It is a transient cellular compartment that contains a heterogeneous population of neurons with a variety of connections and transmitters (Kostović and Rakic, 1990; Schatz et al., 1988) most of which gradually disappear before birth. The subplate has been hypothesized to assume many functions in fetal life, especially as a temporary synaptic target of waiting axonal systems, prior to their entry in the cortical plate and as pioneering the first cortico-subcortical and perhaps cortico-cortical pathways; it might be the site where the processes leading to the parcelling of the cytoarchitectonic areas are regulated (Rakic, 1988). The highest density of monoaminergic terminals in the subplate and thereafter in the deep cortical plate could modulate the effect of incoming excitatory afferents on the output or local circuit neurons. For instance, in the primate prefrontal cortex, the thalamo-cortical and cortico-cortical afferents that have reached the developing cortex earlier, remain in the subplate until the last quarter of gestation when the different cortical layers become more differentiated (Goldman-Rakic, 1987; Kostović and Goldman-Rakic, 1983). Interestingly, the penetration of the cortico-cortical afferents in the cortical plate coincides with the increase of aminergic afferents to the upper cortical layers including the molecular layer. However, the hypothesis that the earliest corticipetal fibers might be monoaminergic in nature (Kostović and Rakic, 1990; Marin-Padilla, 1984) is questionable at least with respect to the visual cortex. Our data clearly show that a very few aminergic fibers have reached the occipital cortex by the time (E83) that cortical afferents originating in the lateral geniculate and the pulvinar geniculate are already present in the subplate and extend sparsely into the cortical plate (Rakic, 1976).

Other puzzling questions arise. What determines the regional variation in the cortical distribution of the NA, DA, and 5-HT systems? Does their time-dependent variable asynchronous development in some areas (motor areas, for instance) imply some sequential effect on the different compartments of the cortical anlage? Finally, the early arrival of monoaminergic projections in the fetal cerebral cortex and their localization in layers where the first synaptic contacts occur and the dendrites begin to mature (marginal and intermediate zones) are consistent with a possible trophic role on cortical neurogenesis. This hypothesis has been widely debated. Contradictory results have been obtained in rodents concerning the influence of the aminergic systems on cortical development. Some authors, for instance, observed no effect of norepinephrine depletion in rat fetuses (Lidov and Molliver, 1982b), whereas others described significant modifications either in that condition (Felten et al., 1982) or after lesioning the dopaminergic and the serotonergic afferents in neonates (Kalsbeek et al., 1988a). These discrepancies could be partly due to the fact that in rodents, as aforementioned, the different steps of corticogenesis and cell differentiation and the development of the cortical monoaminergic innervation overlap within a relatively short period of time, encompassing both pre- and postnatal events. In rodents, brainstem monoaminergic cell groups and cortical neurons are generated during a second half of embryonic life; monoaminergic afferents reach the cerebral cortex during the last third of gestation and some of them arrive postnatally; for instance, the bulk of the dopaminergic projections to the anterior cingulate cortex. During the protracted gestational period of primates, multiple neurotransmitter interactions may occur in the developing cortex, particularly in relation to a transitory expression of peptides, for instance (Huntley et al., 1988), and these matters have to be further investigated.

A major reason to pursue investigations on the cerebral cortex of primates is the considerable difference that is observed in both the regional and laminar distribution of the cortical catecholaminergic afferents, especially the dopaminergic ones, among the rat, monkey, and man. In contrast to the more homogeneous distribution in the rodent cortex, in primate the highest density of noradrenergic fibers is observed in the primary motor and somatosensory cortices and the density is substantially greater in the deeper layers, especially in layer V (Gaspar et al., 1989; Morrison et al., 1982). The dopaminergic innervation extends to the whole cerebral cortex, especially the motor and postcentral cortices; these extended cortical targets contrast with the restricted topography of cortical dopaminergic terminal fields in the rodent, where most of the primary somatomotor and association areas are devoid of dopaminergic afferents (Berger et al., 1985a). Moreover, there is a region-dependent laminar-specific distribution of cortical DA afferents in primates (Berger et al., 1988; Gaspar et al., 1989; Lewis et al 1988). The bilaminar type of distribution in the somatosensory and association areas, predominating in layers I and V–VI, and the highest density of innervation in layer I throughout the cerebral cortex in adults, are also characteristic features of the primate fetal cortex that have no equivalent in the rodent. From an ontogenetic perspective, in primates, limbic areas such as the cingulate cortex, the insula, and the entorhinal cortex are innervated earlier and more densely than other cortical areas. This might be put in line with the presumed atrophy/aplasia process of the entorhinal cortex, which has been hypothesized in some forms of psychoses (Roberts and Crow, 1987).

Finally, the very preliminary comparative results concerning the development of monoaminergic projections to the cerebral cortex in human and nonhuman primates suggest that it might proceed faster in human: for instance, the dopaminergic innervation of the motor cortex seems to be more developed at the beginning of the second half of gestation in the human than in the monkey. The early appearance in fetal life of TH-LIR neurons in the cerebral cortex seems also to be a characteristic feature of the human brain. Such neurons are observed transitorily in the rodent cortex during the first month of postnatal life (Berger et al., 1985), whereas they form a permanent population in human cortex (Gaspar et al., 1987). Also, when comparing corticogenesis in the rat, cat, monkey, and man, it seems that the more protracted the gestation length is, the earlier corticogenesis starts. This underlines the importance of investigating the events occurring in the primate fetal cortex in order to understand the establishment of the functional circuitry in the human cerebral cortex.

Acknowledgments. This research was supported by Center grant MH 44866 for Basic Neuroscience Research in Schizophrenia to P. S. Goldman-Rakic, and a subvention of Singer Polignac Foundation and INSERM to B. Berger.

References

Altman J, Bayer SA (1981): Development of the brain stem in the rat. V. Thymidine radioautographic study of the time of origin of neurons in the midbrain tegmentum. *J Comp Neurol* 198:677–716

Berger B, Thierry AM, Tassin JP, Moyne MA (1976): Dopaminergic innervation of the rat prefrontal cortex: a fluorescence histochemical study. *Brain Res* 106:133–145

Berger B, Febvret A, Greengard P, Goldman-Rakic PS (1990): DARPP-32, a phosphoprotein enriched in dopaminergic neurons bearing D1 receptors: distribution in the cerebral cortex of the newborn and adult monkey. *J Comp Neurol* 299:1–22

Berger B, Verney C (1984): Development of the catecholaminergic innervation in rat neocortex: morphological features. In: *Neurology and Neurobiology*, Descarries L, Reader TR, Jasper HH, eds. "Monoamine innervation of cerebral cortex," Vol. 10, New York: Alan R. Liss, pp. 95–124

Berger B, Verney C, Alvarez C, Vigny A, Helle KB (1985a): New dopaminergic terminal fields in the motor, visual (area18b) and retrosplenial cortex in the young and adult rat. Immunocytochemical and catecholamine histochemical analyses. *Neuroscience* 15:983–998

Berger B, Verney C, Febvret A, Vigny A, Helle KB: (1985b): Postnatal ontogenesis of the dopaminergic innervation in the rat anterior cingulate cortex (area 24). Immunocytochemical and catecholamine fluorescence histochemical analysis. *Dev Brain Res* 21:31–47

Berger B, Verney C, Gaspar P, Febvret A (1985): Transient expression of tyrosine hydroxylase immunoreactivity in some neurons of the rat neocortex during postnatal development. *Dev Brain Res* 23:141–144

Berger B, Trottier S, Verney C, Gaspar P, Alvarez C (1988): Regional and laminar distribution of the dopamine and serotonin innervation in the Macaque cerebral cortex: a radioautographic study. *J Comp Neurol* 273:99–119

Björklund A, Lindvall O (1984): Dopamine containing systems in the central nervous system. In: *Handbook of Neuroanatomy* Vol. 2, Björklund A, Hokfelt T, eds. "Classical transmitters in the central nervous system," Part 1, Elsevier, pp. 55–122

D'Amato RJ, Blue ME, Argent B, Lynch DR, Lesbetter DL, Molliver ME, Snider SH (1987): Ontogeny of the serotoninergic projection to rat neocortex: transient expression of a dense innervation to primary sensory areas. *Proc Nat Acad Sci USA* 84:4322–4326

Dekker JS, Rakic P (1980): Genesis of the motor cortex and ventroanterior and ventrolateral thalamic complex in Rhesus Monkey. *Soc Neurosci Abst* 6:205

Descarries L, Lemay B, Doucet G, Berger B (1987): Regional and laminar density of the dopamine innervation in adult rat cerebral cortex. *Neuroscience* 21:807–824

Felten DL, Hallman H, Jonsson G (1982): Evidence for a neurotrophic role of noradrenaline neurons in the postnatal development of the cerebral cortex. *J Neurocytol* 11:119–135

Gaspar P, Berger B, Febvret A, Vigny A, Henry JP (1989): Catecholamine innervation of the human cerebral cortex revealed by comparative immunohistochemistry of tyrosine-hydroxylase and dopamine-β-hydroxylase. *J Comp Neurol* 279:249–271

Gaspar P, Berger B, Febvret A, Vigny A, Krieger-Poulet M, Borri-Voltattorni C (1987): Tyrosine hydroxylase immunoreactive neurons in the human cerebral cortex: a novel catecholaminergic group? *Neurosci Lett* 80:257–262

Goldman-Rakic PS (1981): Prenatal formation of cortical input and development of cytoarchitectonic compartments in the neostriatum of the Rhesus monkey. *J Neurosci* 7:721–735

Goldman-Rakic PS (1987): Development of cortical circuitry and cognitive function. *Child Dev* 58:601–622

Goldman-Rakic PS, Isseroff A, Schwartz ML, Bugbee NM (1983): The neurobiology of cognitive development. In: *Handbook of Child Psychology: Biology and Infancy Development*, Mussen P, ed. New York: Wiley, pp. 281–344

Goldman-Rakic PS, Leranth C, Williams SM (1989): Dopamine synaptic complex on pyramidal neurons in primate cerebral cortex. *Proc Nat Acad Sci USA* 86:9015–9019

Hicks SP, D'Amato CJ (1968): Cell migrations to the isocortex in the rat. *Anat Rec* 160:619–634

Huntley GW, Hendry SHC, Killackey HP, Chalupa LM, Jones EG (1988): Temporal sequence of neurotransmitter expression by developing neurons of fetal monkey visual cortex. *Dev Brain Res* 43:69–96

Kalsbeek A, De Bruin JPC, Feenstra MGP, Matthijssen MAH, Uylings HBM (1988a): Neonatal thermal lesions of the mesolimbocortical dopaminergic projection decrease foodhoarding behavior. *Brain Res* 475:80–90

Kalsbeek A, Voorn P, Buijs RM, Pool CW, Uylings HBM (1988b): Development of the dopaminergic innervation in the prefrontal cortex of the rat. *J Comp Neurol* 269:58–72

Kosofsky BE, Molliver ME, Morrison JH, Foote SL (1984): The serotonin and norepinephrine innervation of the primary visual cortex in the cynomologus monkey (*Macaca fascicularis*). *J Comp Neurol* 230:168–178

Kostović I, Goldman-Rakic PS (1983): Transient cholinesterase staining in the mediodorsal nucleus of the thalamus and its connections in the developing human and monkey brain. *J Comp Neurol* 219:431–447

Kostović I, Molliver ME (1974): A new interpretation of the laminar development of cerebral cortex: synaptogenesis in different layers of neopallium in the human fetus. *Anat Rec* 395

Kostović I, Rakic P (1980): Cytology and time of origin of interstitial neurons in the white matter in infant and adult human and monkey telencephalon. *J Neurocytol* 9:219–242

Kostović I, Rakic P (1990): Developmental history of the transient subplate zone in the visual and somatosensory cortex of the Macaque monkey and human brain. *J Comp Neurol* 297:441–470

Larroche JC (1981): The marginal layer in the neocortex of a 7 week-old human embryo. A light and electron microscopic study. *Anat Embryol* 162:301–312

Lauder JM, Bloom FE (1974): Ontogeny of monoamine neurons in the locus ceruleus, raphe nucleus and substantia nigra of the rat. I. Cell differentiation. *J Comp Neurol* 155:469–482

Levitt P, Moore RY (1979): Development of the noradrenergic innervation of neocortex. *Brain Res* 162:243–260

Levitt P, Rakic P (1982): The time of genesis, embryonic origin and differentiation of the brainstem monoamine neurons in the rhesus monkey. *Dev brain Res* 4:35–58

Lewis DA, Campbell MJ, Foote SL, Goldstein M, Morrison JH (1987): The distribution of tyrosine hydroxylase immunoreactive fibers in primate neocortex is widespread but regionally specific. *J Neurosc* 7:279–290

Lewis, DA, Foote, SL, Goldstein, M, Morrison, JH (1988): The dopaminergic innervation of monkey prefrontal cortex: A tyrosine hydroxylase immuno histochemical study. *Brain Res.* 449:225–243

Lewis DA, Morrison JH (1989): Noradrenergic innervation of monkey-prefrontal cortex: a dopamine-β-hydroxylase immunohistochemical study. *J Comp Neurol* 282:317–330

Lidov HGW, Molliver ME (1982a): An immunohistochemical study of serotonin neuron development in the rat: ascending pathways and terminal fields. *Brain Res Bull* 8:389–430

Lidov HGW, Molliver ME (1982b): The structure of cerebral cortex in the rat following prenatal administration of 6-hydroxydopamine. *Dev Brain Res* 3:81–108

Lidov HGW, Molliver ME, Zecevic NR (1978): Characterization of the monoaminergic innervation of immature rat neocortex: a histofluorescence analysis. *J Comp Neurol* 181:663–680

Marin-Padilla M (1970): Prenatal and early postnatal ontogenesis of the human motor cortex: a golgi study. I. The sequential development of the cortical layers. *Brain Res* 23:167–183

Marin-Padilla M (1978): Dual origin of the mammalian neocortex and evolution of the cortical plate. *Anat Embryol* 152:109–126

Marin-Padilla M (1984): Neurons of layer I. A developmental analysis. In: *Cerebral Cortex*, Vol. 1, Peters A, Jones EG eds. New York: Plenum Press, pp. 447–478

Morrison JH, Foote SL, O'Connor D, Bloom FE (1982): Laminar, tangential and regional organization of the noradrenergic innervation of monkey cortex: dopamine-β-hydroxylase immunohistochemistry. *Brain Res Bull* 9:309–319

Mrzljak L, Uylings HBM, Kostović I, Van Eden C (1988): Prenatal development of neurons in the human prefrontal cortex: I. A qualitative golgi study. *J Comp Neurol* 271:355–386

Noack HJ, Lewis DA (1989): Antibodies directed against tyrosine hydroxylase differentially recognize noradrenergic axons in monkey neocortex. *Brain Res* 500:313–324

Nobin A, Björklund A (1973): Topography of the monoamine neuron systems in the human brain as revealed in fetuses. *Acta Physiol Scand Suppl* 388:1–40

Olson L, Seiger A (1972): Early prenatal ontogeny of central monoamine neurons in the rat: fluorescence histochemical observations. *Z Anat Entwickl Gesch* 137:301–316

Olson L, Boreus LO, Seiger A (1973): Histochemical demonstration and mapping of 5-hydroxytryptamine- and catecholamine-containing neuron systems in the human fetal brain. *Z Anat Entwickl Gesch* 139:259–282

Papadopoulos GC, Parnavelas JG, Buijs RM (1989): Light and electron microscopic immunocytochemical analyses of the dopamine innervation of the rat visual cortex. *J Neurocytol* 18:303–310

Pearson J, Brandeis L, Goldstein M (1980): Appearance of tyrosine hydroxylase immunoreactivity in the human embryo. *Dev Neuroscience* 3:140–150

Phillipson OT, Kilpatrick IC, Jones MW (1986): Dopaminergic innervation of the primary visual cortex in the rat and some correlations with human cortex. *Brain Res Bull* 18:621–633

Raedler E, Raedler A, Feldhaus S (1980): Dynamical aspects in neocortical histogenesis in the rat. *Anat Embryol* 158:253–270

Rakic P (1974): Neurons in rhesus monkey visual cortex: systematic relation between time of origin and eventual disposition. *Science* 183:425–427

Rakic P (1976): Prenatal genesis of connections subserving ocular dominance in the rhesus monkey. *Nature* 261:467–471

Rakic P (1982): Early developmental events: cell lineages, acquisition of neuronal positions, areal and laminar development. *Neurosci Res Program Bull* 20:439–451

Rakic P (1988): Specification of the cerebral cortical areas. *Science* 241:170–177

Rakic P, Bourgeois JP, Eckenhoff MF, Zecevic N, Goldman-Rakic (1986): Concurrent overproduction of synapses in diverse regions of the primate cerebral cortex. *Science* 232:232–235

Rakic P, Nowakowski RS (1981): The time of origin of neurons in the hippocampal region of the rhesus monkey. *J Comp Neurol* 196:99–128

Rice FL, Gomez C, Barstow C, Burnet A, Sands P (1985): A comparative analysis of the development of the primary somatosensory cortex: interspecies similarities during barrel and laminar development. *J Comp Neurol* 236:477–495

Roberts GW, Crow TJ (1987): The neuropathology of schizophrenia: a progress report. *Brit Med Bull* 43: 599–615

Schatz CJ, Chun JJM, Luskin MB (1988): The role of the subplate in the development of the mammalian telencephalon. In: *Cerebral Cortex*, Vol. 7, Peters A, Jones EG, eds. New York: Plenum Press, pp. 35–58

Schlumpf M, Shoemaker WJ, Bloom FE (1980): Innervation of embryonic rat cerebral cortex by catecholamine-containing fibers. *J Comp Neurol* 192: 361–377

Schmidt RH, Björklund A, Lindvall O, Loren I (1982): Prefrontal cortex: dense dopaminergic input in the newborn rat. *Dev Brain Res* 5:222–228

Seiger A, Olson L (1973): Late prenatal ontogeny of central monoamine neurons in the rat. Fluorescence histochemical observations. *Z Anat Entwickl* 140:281–318

Specht LA, Pickel VM, Joh TH, Reis DJ (1981a): Lightmicroscopic immunocytochemical localization of tyrosine hydroxylase in prenatal brain. I. Early ontogeny. *J Comp Neurol* 199:233–254

Specht LA, Pickel VM, Joh TH, Reis DJ (1981b): Lightmicroscopic immunocytochemical localization of tyrosine hydroxylase in prenatal brain. II. Late ontogeny. *J Comp Neurol* 199:255–276

Van Eden CG, Hoorneman EMD, Buijs RM, Matthissen MAH, Geffard M, Uylings (1987): Immunocytochemical localization of dopamine in the prefrontal cortex of the rat at the light and electron microscopical level. *Neuroscience* 22:849–862

Van Eden CG, Uylings HBM (1985): Cytoarchitectonic development of the prefrontal cortex in the rat. *J Comp Neurol* 242:253–267

Verney C, Alvarez C, Geffard M, Berger B (1991): Ultrastructural double labeling study of dopamine terminals and GABA-containing neurons in rat anteromedial cerebral cortex. *Europ J Neuroscience* 2:960–972

Verney C, Baulac M, Berger B, Alvarez C, Vigny A, Helle KB (1985): Morphological evidence for a dopaminergic field in the hippocampal formation of the young and adult rat. *Neuroscience* 14:1039–1052

Verney C, Berger B, Adrien J, Vigny A, Gay M (1982): Development of the dopaminergic innervation of the rat cerebral cortex. A light microscopic immunocytochemical study using antityrosine hydroxylase antibodies. *Dev Brain Res* 5:41–52

Verney C, Berger B, Baulac M, Helle KB, Alvarez C (1984): Dopamine-β-hydroxylase-like immunoreactivity in the fetal cerebral cortex of the rat: noradrenergic ascending pathways and terminal fields. *Int J Devl Neuroscience* 5:491–503

Verney C, Zecevic N, Nikoloc B, Alvarez C, Berger B (1991): Early evidence of catecholaminergic cell groups and pathways in 6-week-old human human embryos using tyrosine hydroxylase and dopamine beta hydroxlase immunocytochemistry. *Neurosci Letters* (in press).

Zecevic N, Bourgeois JP, Rakic P (1989): Changes in synaptic density in motor cortex of rhesus monkey during fetal and postnatal life. *Dev Brain Res* 50: 11–32

Ontogenetic Development and Decline of Memory Functions in Nonhuman Primates

Jocelyne Bachevalier and Mortimer Mishkin

An emerging parallel view in research on ontogeny and decline of memory abilities is that mnemonic processes are differentially susceptible to age. Some behavioral memory tasks are performed normally by newborn monkeys, whereas other memory tasks are performed normally only if the animal is mature. Similarly, some memory tasks are performed normally by aged subjects while other tasks are not. Thus, it appears that the presence and absence of age-related memory abilities depend on the specific information-processing demands of the behavioral tasks used to measure memory functions. One important outcome of characterizing impaired versus preserved memory function in ontogeny or aging is the insight this information can provide concerning the neural systems that mediate age-dependent cognitive functions. Neuropsychological findings in amnesic patients and in monkeys with experimental brain damage have indicated that different brain regions appear to make rather specialized contributions to learning and memory. Accordingly, specific types of cognitive abilities in infant and aged subjects may reflect the functional status of those brain regions that mediate the component processes of memory. Based on this neuropsychological perspective, this chapter will review actual knowledge on the ability of both infant, mature, and aged monkeys to form habits and memories, two learning processes that are known to depend on separate neural circuits.

Inference from Human Amnesia

Since the classic reports of amnesic syndromes by Ribot (1881) and Korsakoff (1887), it has been recognized that damage in certain subcortical regions of the human brain can result in an amnesic syndrome, and that this syndrome can sometimes appear as a relatively circumscribed deficit. Indeed, amnesic patients may exhibit no discernible impairment of higher cognitive function except an inability to accomplish new learning and a loss of memory for some events that occurred prior to the onset of the disorders. They are not affected in other cognitive functions such as attention, perception, or general intellectual ability. In addition, despite their profound memory deficit, amnesic patients may still learn new habits and skills in an entirely normal manner.

Much of our knowledge concerning amnesia derived from the well-studied case H.M. In 1953 this individual sustained a bilateral resection of the medial temporal lobe in an attempt to relieve otherwise intractable epilepsy. Following surgery, although his seizures were relieved, H.M. could no longer retain any experience. Neuropsychological investigation of other human cases with similar neuropathology to that described in H.M. has revealed that bilateral damage to the medial temporal lobe in humans results in a profound anterograde amnesia. The amnesia is also referred to as "global" in that it extends to many types of information derived from all sensory modalities (Corkin, 1965; Milner, 1972), though amnesic patients remember much more about the past than they do of the present. Paradoxically, despite their severe impairment in the formation of new memories, amnesic patients may still learn new habits and skills in an entirely normal manner. The first piece of evidence for the existence of preserved ability in profoundly amnesic subjects came from Milner's (1962) dramatic demonstration of a nearly normal rate of improvement of mirror-drawing skill in the severely amnesic patient, H.M. Since the report on H.M., there have been numerous studies showing the ability of amnesic patients to learn despite their inability to remember (Brooks and Baddeley, 1976; Cohen and Squire, 1980; Corkin, 1968; Warrington and Weiskrantz, 1968). All these studies demonstrate that gradual learning of skills and habits can proceed independently of the ability to recollect specific events or episodes.

Experimental studies in nonhuman primates converge upon a similar conclusion (Mishkin et al., 1984; Mishkin and Petri, 1984; Zola-Morgan and Squire, 1985a). In our laboratory, Mishkin and his colleagues had found corresponding losses and hidden learning capacities in monkeys with brain damage like H.M.'s. For example, when tested on a visual recognition task, such as delayed nonmatching-to-sample or object-reward association, both with trial-unique objects, monkeys with combined ablation of the medial temporal lobe (including the amygdala, hippocampus, and adjacent cortex) seemed unable either to recognize an object they had seen just a minute or two earlier (Mishkin, 1978), or to remember for even a few seconds whether or not the object had been baited (Spiegler and Mishkin, 1981). Also, their memory deficits appear to be global since monkeys with medial temporal lobe lesions are severely impaired on a nonmatching-to-sample test of tactile recognition (Murray and Mishkin, 1983). Interestingly, the same monkeys are able to learn a long series of different object pairs, each containing one baited object, when the series is presented only once a day (Malamut et al., 1984). Thus, in this object discrimination test, with time they learn to choose the object carrying the reward by forming an automatic connection between a stimulus and a response.

These paradoxical results have led most of the investigators to conclude that there are at least two different forms of retention, only one of which is affected in amnesia. One learning and retention system supports gradual and incremental learning and is involved in the acquisition of habits and skills (such as mirror-drawing in humans and object discrimination in monkeys), while the other supports rapid one-trial learning and is necessary for forming memories that represent specific situations and episodes (such as object recognition). Furthermore, the fact that one of the systems is dependent upon the integrity of the limbic structures whereas the other is not has led Mishkin and his collaborators (1984, 1987) to propose that these two retention systems are in fact subserved by two independent neural circuits.

The Neural Circuits for Memories and Habits

The detailed knowledge now available concerning the neuroanatomy of the medial temporal lobe and its interconnections with diencephalic, prefrontal, and basal forebrain areas, as well as the memory loss described following lesion or neuropathology in these neural structures, has guided the formulation of a functional neural system for the formation of cognitive or representational memory.

The neuroanatomical studies have shown that the amygdala directly and the hippocampus via the perirhinal and entorhinal cortex receive highly processed sensory information from sensory specific cortical areas (Amaral, 1987; Turner et al., 1980; Van Hoesen and Pandya, 1975). The amygdala, hippocampus, and rhinal cortex are connected in turn reciprocally with various medial thalamic nuclei and the mamillary bodies (Aggleton et al., 1986; Aggleton and Mishkin, 1984; Murray, 1991). In addition, these diencephalic regions are connected not only with medial temporal lobe structures, but also with the ventral and medial portions of the prefrontal cortex (Baleydier and Mauguiere, 1980; Nauta, 1962; Tobias, 1975). Finally, both the limbic, diencephalic, and prefrontal areas are interconnected with the basal forebrain, which sends widespread projections back to the entire cortical mantle (Amaral and Cowan, 1980; Mesulam et al., 1983). A diagrammatic summary of the anatomical circuit that has been implicated in visual recognition memory is shown in Fig. 1. Severe lesion-induced memory losses have been demonstrated in monkeys at different loci along this pathway: (1) area TE, the visual higher association area in the anterior portion of the inferior temporal cortex (Mishkin, 1982); (2) the connections between area TE and the amygdalo-hippocampal complex (Zola-Morgan et al., 1982); (3) the amygdala, hippocampus, and rhinal cortex (Meunier et al., 1990; Mishkin, 1978; Murray et al., 1989; Zola-Morgan et al., 1989); (4) the limbo-diencephalic pathway (Bachevalier et al.,1985); (5) the limbic targets in the medial thalamus, including the medial parts of the medial dorsal nucleus, the anterior nuclei, and mamillary bodies (Aggleton and Mishkin, 1983a, b, 1985; Zola-Morgan and Squire, 1985b); (6) sub-, pre-, and supracallosal, and orbital prefrontal cortex (Bachevalier and Mishkin, 1986); and (7) the basal forebrain (Aigner et al., 1984). As for the neurological evidence, these different regions comprising the proposed circuit are also the major sites of neuropathology associated with amnesia in humans. These sites include the medial temporal area, in patients with temporal lobe resections or herpes encephalitis (Lilly et al., 1983; Penfield and Milner, 1958; Scoville and Milner, 1957), the medial diencephalic area in patients with Korsakoff's disease (Butters and Cermak, 1980; Mair et al., 1979; Squire, 1987; Talland, 1965; Victor et al., 1971), the ventromedial aspect of the prefrontal cortex in patients with aneurysms of the anterior communicating artery (Alexander and Freedman, 1984; Gade, 1982; Talland et al., 1967), frontal lobe neoplasms (Angelergues et al., 1955; Provensal, 1951) or prefrontal surgical removals (Whitty and Levin, 1960), and basal forebrain in patients with Alzheimer's disease (Arendt et al., 1983; Candy et al., 1983; Whitehouse et al., 1982).

As shown for visual recognition memory in Fig. 1, the neural model proposed by Mishkin and his colleagues (e.g., Mishkin and Appenzeller, 1987) postulates that inspection of a new stimulus activates a chain of cortical areas (V1–V4), which extract physical qualities of the stimulus, until the final stations of the inferior temporal cortex (areas TEO and TE) synthesize a complete representation of the stimulus, constituting a novel perception. Storage of the new perception is realized each time the

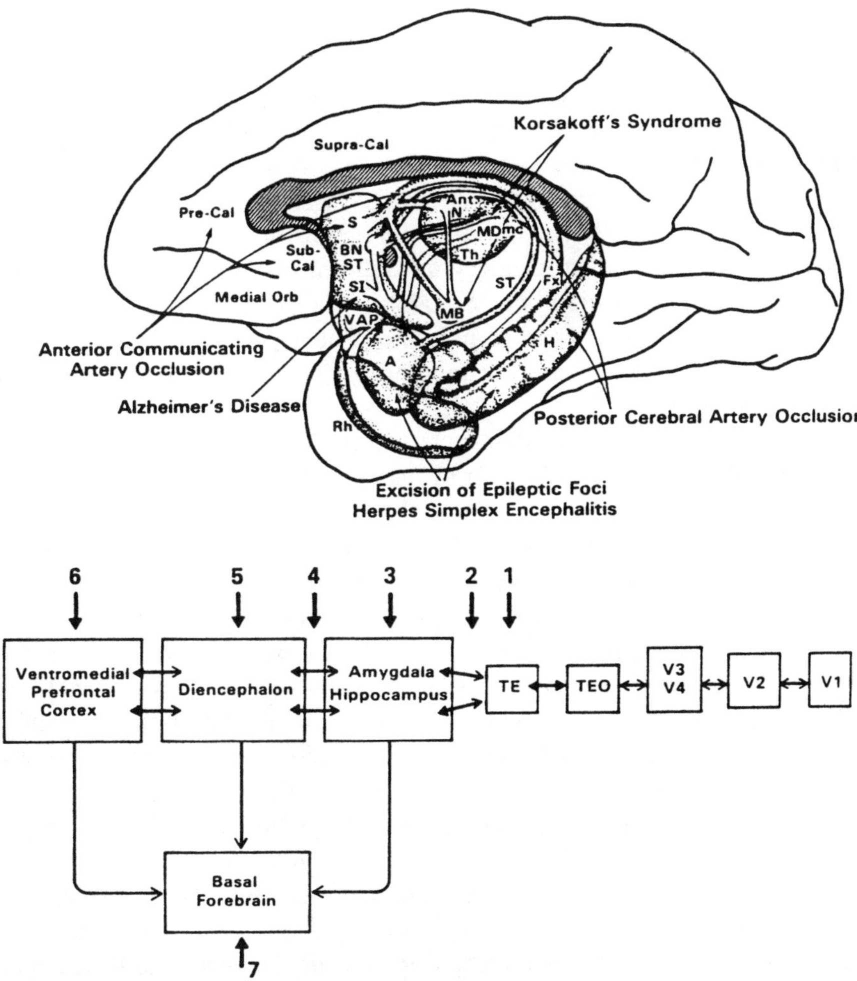

Figure 1. The neural circuit for visual memory. The diagram at the top indicates the brain structures crucial for memory functions and the sites of brain damage by diseases that can cause memory loss in human beings. The chart at the bottom indicates the flow of visual information from striatal cortex (V1), to prestriate cortex (V2, V3, V4), to inferior temporal cortex (areas TEO and TE), to the limbic structures (amygdala, A; hippocampus, H; rhinal cortex, Rh), and via the limbo-diencephalic pathways (fornix, Fx; stria terminalis, ST; ventroamygdalofugal pathway, VAP) to the diencephalon (anterior, Ant. N, and medial dorsal, MD, nuclei of the thalamus; mamillary bodies, MB), to the ventromedial prefrontal cortex (supra-, pre-, and subcallosal cortex, and orbital cortex), and to the basal forebrain (substantia innominata, SI; bed nucleus of the stria terminalis, BNST; septum, S). The numbers indicate lesion-induced visual recognition loss in monkeys. (From Bachevalier et al., 1991b.)

neurons in area TE that have been activated by the stimulus perception trigger a cortico-limbo-thalamo-cortical circuit. It is postulated that the feedback action of these neural stations on sensory areas stores the neural representation of the sensory event that had just taken place. Once stored, this central representation can be reactivated whenever the same stimulus reappears in the field, yielding visual recognition. Furthermore, since all sensory modalities appear to be represented centrally by a hierarchical arrangement from primary to higher processing areas, and since the highest cortical levels in each modality appears to project to both the amygdala, hippocampus, and rhinal cortex, it is proposed that the same circuit that mediates recognition memory in vision does so in the other sensory modalities.

As for the neural circuit that underlies the formation of visual habits, here the evidence is only beginning to emerge. Both neuroanatomical and lesion studies, however, indicate that the striatum is the most likely candidate. As shown in Fig. 2, the striatum receives projections from sensory processing areas of the cerebral cortex and sends projections to the globus pallidus, substantia nigra, and associated structures within the extrapyramidal system that control movements. Behavioral evidence also supports the view that the corticostriatal system serves noncognitive habit formation. For example, impairment in visual discrimination has now been demonstrated following lesions of either (1) the inferior temporal cortex (Mishkin, 1954; Phillips et al., 1988); (2) the projections from the visual areas to the neostriatum (Horel, 1978; Zola-Morgan et al., 1982); or (3) the striatal regions (Divac et al., 1967; Buerger et al., 1974; Wang et al., 1990). The neurological evidence also indicates that patients with structural damage to the neostriatum, such as patients with Parkinson's disease, are selectively impaired in the task of skill learning while remaining intact in recognition tests (Saint-Cyr et al., 1988). Thus, for visual habit formation, the neural model proposed by Mishkin and his colleagues (Mishkin and Appenzeller, 1987; Phillips et al., 1988) postulates that, through a series of connections from the striatum to the globus pallidus and substantia nigra, to the ventral thalamus, and, finally, to the premotor cortex, cortically processed visual inputs could become associated with extrapyramidally generated motor outputs and so yield the stimulus-response bonds that constitute visual habits.

The Ontogeny of Habits and Memories

On the evidence that, in adult monkeys, there may be two relatively independent systems for retention of information, we sought to compare the

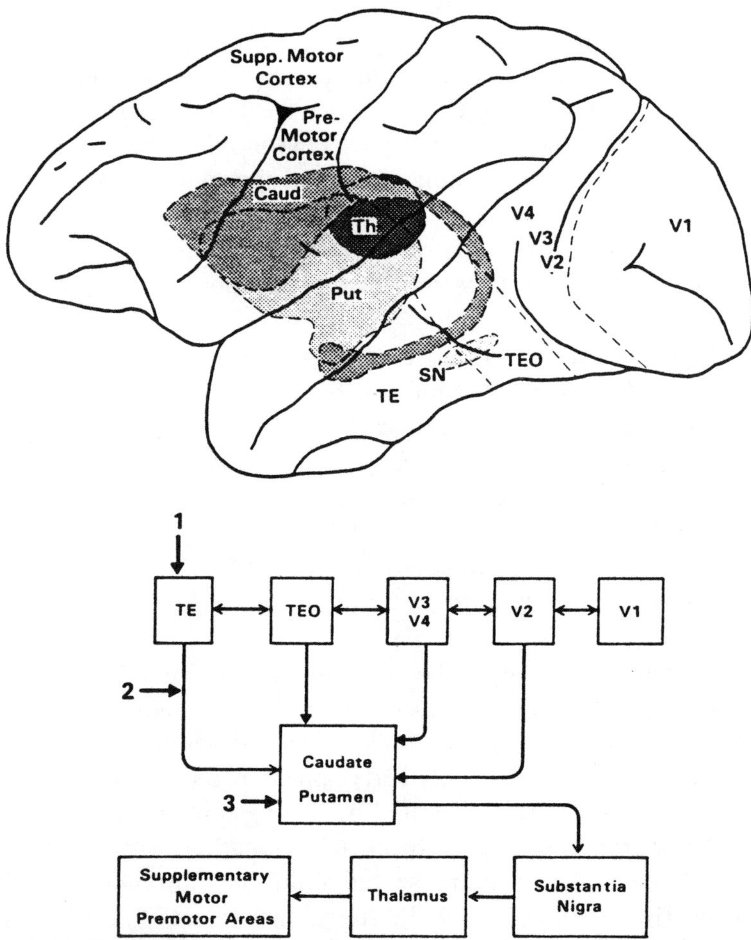

Figure 2. The neural circuit for visual habit. The diagram at the top indicates the brain structures crucial for the formation of visual habits. The chart at the bottom indicates the flow of visual information in the postulated neural circuit for visual habit formation. The numbers indicate lesion-induced loss in visual habit formation. Abbreviations: Caud: caudate; Put: putamen; SN: substantia nigra; Th: thalamus. (From Bachevalier et al., 1991b.)

functional development of the two systems ontogenetically (Bachevalier and Mishkin, 1984).

First, the development of cognitive memory in infant monkeys was investigated by testing monkeys between three months and four years of age in the visual recognition task (delayed nonmatching-to-sample, DNMS) that was used to demonstrate anterograde amnesia in adult monkeys with limbic lesions (Mishkin, 1978). In this task, the animal has to remember on the basis of a single trial whether or not an object had been seen before. Having learned the principle of consistently selecting the novel object in the pair, the animals were given a performance test in which the delay between sample presentation and choice was increased in steps from 10 sec to 2 min; after this, the number of objects to be remembered was also increased in steps from a single object to a list of ten objects. The results indicated that three-month-old monkeys failed to solve the recognition task until they were approximately four months of age.

With further maturation, there was a gradual improvement in learning ability, yet it did not reach adult levels of proficiency even at one year. Only at about two years of age did monkeys master the recognition task as efficiently as adult animals (Bachevalier, 1990). This slow ontogenetic development in learning the principle was accompanied by a slow improvement in performance in the memory recognition task. With the longer delays and lists, average scores declined systematically in inverse relation to the group's age, from 96% in the adults to 86% in the youngest animals.

Like amnesic monkeys, the three-month-old infants that could not remember an object they had just seen a few moments ago were nevertheless able to learn in as few trials as normal adults to pick the baited object out of two. The relatively early appearance in life of the noncognitive habit system was demonstrated by testing three-month-old monkeys in a multiple-trial concurrent object discrimination task (24-hr ITI) that was used to demonstrate the preserved ability in adult monkeys with limbic lesions (Malamut et al., 1984). In this task, a set of 20 pairs of easily discriminable objects was presented for concurrent learning; but these 20 different pairs were presented just once a day, on successive days, until the animals attained a criterion of 90 correct choices in 100 trials. Despite the 24-hour intertrial intervals, the infant monkeys performed as well as adult monkeys (Bachevalier and Mishkin, 1984).

Together these results strongly suggest that the two learning systems that were found to be separable in the adult monkeys are also developmentally dissociable. Moreover, a similar developmental dissociation in the performance on the visual recognition (DNMS) and visual discrimination tasks was recently demonstrated in human infants (Overman, 1991). In short, in primates the system for noncognitive association (habit system), which is preserved in adult monkeys rendered amnesic by limbic lesions, seems to develop early in infancy. By contrast, the system for cognitive associations (memory system), which is impaired in adult monkeys with limbic lesions, appears to develop late in infancy, presumably because the neural circuit on which it depends has a slow ontogenetic maturation. Thus, which is the station of the neural circuit underlying recognition memory requiring a prolonged maturation? A possible answer to this question emerges from a comparison between the effects of early versus late lesions of higher-order areas of the cortex on the two visual tasks.

Delayed Maturation of Association Areas of the Cortex

Newborn monkeys received damage to either the limbic system, i.e., amygdalo-hippocampal complex, or the inferior temporal cortical area TE. The bilateral lesions were performed in two unilateral stages at approximately one week and three weeks of age, respectively. Each operated animal was age-matched with an unoperated control.

The formation of visual discrimination habits in operated and unoperated monkeys was measured by the 24-hr ITI task when the animals were three months of age (Bachevalier et al., 1990). Performance of infant monkeys was compared to that of monkeys that had been tested in the same way after receiving identical brain lesions when they were adults. As shown in Fig. 3A, damage to the limbic system appeared to leave habit formation intact whether the lesions were early or late. By contrast, lesions of area TE yielded a severe and long-lasting deficit in visual habit formation if the damage occurred late, but they produced only a mild and transitory deficit of this function if the damage occurred early. Indeed, by the second set of visual discriminations, infant monkeys with area TE damage required no more sessions to learn than normal adult monkeys (averaging 11 and 10 sessions, respectively).

With regard to cognitive memory (as measured by the DNMS task), there is again evidence of a greater compensatory potential after neonatal cortical than after neonatal limbic damage (Bachevalier, 1990). Thus, whereas infants with neonatal limbic lesions were severely impaired in visual recognition at 10 months of age, those with neonatal ablation of the visual association cortex (area TE) showed significant functional sparing (Fig. 3B). Both early and late damage to the limbic structures yielded severe impairment in visual recognition, reflected in drops in performance of 22% for the infants and 32% for the adults (as compared with their normal controls). By contrast, although late damage to area TE resulted in a drop in performance of 29%, early lesions of this cortical area resulted in a drop of only 8%.

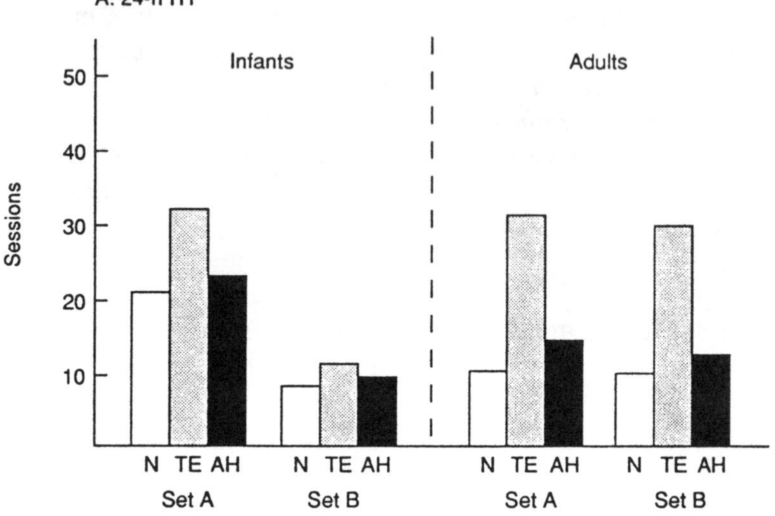

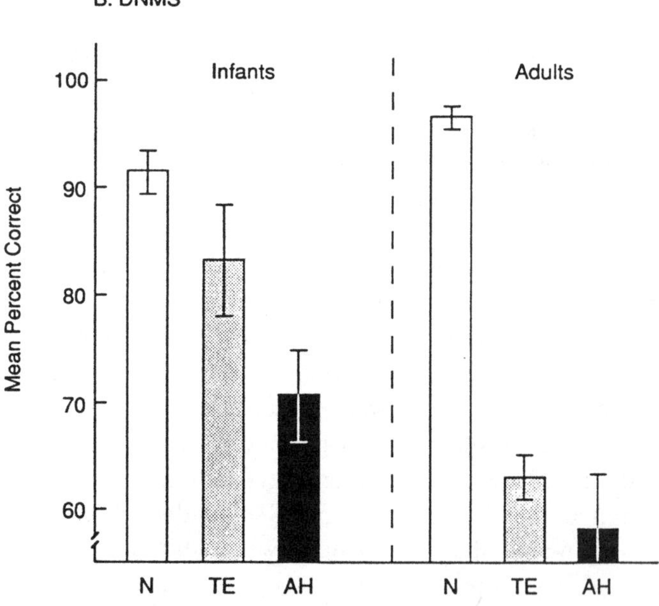

Figure 3. (A) Average number of sessions to attain criterion on two consecutive sets of discriminations (sets A and B) by three-month-old monkeys (operated on neonatally) and adult monkeys and (B) average scores on the recognition performance test by 10-month-old monkeys (operated on neonatally) and adult monkeys. N, normal controls; TE, animals with bilateral area TE lesions; AH, animals with bilateral amygdalohippocampal lesions. (Adapted from Bachevalier et al., 1991b.)

Some of the first infants to receive these specific neonatal lesions are now four-year-olds and have been retested to assess the long-term effects of these early lesions on the formation of visual habits and visual memories (Bachevalier and Mishkin, 1988). Their performance on the 24-hr ITI and DNMS tasks was compared to that of four-year-olds that had received the same lesions in adulthood. At this age, the monkeys with neonatal damage to area TE exhibited sparing of both habit and memory formation of the same magnitude as that found earlier when they were 3 and 10 months of age. By contrast, the monkeys with neonatal limbic lesions resembled those with late limbic lesions in that they were unimpaired in habit formation but severely impaired in memory.

In summary, these behavioral data are consistent with the notion that association areas of the cortex possess considerable plasticity at birth, implying a degree of functional immaturity. Indeed, compensatory mechanisms appear to operate early to promote permanent recovery from neonatal temporal cortical lesions but not from limbic lesions. In support of this proposal, we have recently obtained direct evidence of neocortical immaturity in the macaque. First, the distribution of both opiate and muscarinic cholinergic receptors appears to be adult-like at birth in subcortical structures and allocortical areas but is not fully developed in any neocortical areas, including area TE (Bachevalier et al., 1986; O'Neill et al., 1986). Second, adult levels of metabolic activity, as measured by the 2-deoxyglucose technique, are not reached in visual association cortex until the animals are about four months of age (Bachevalier et al., 1991). These behavioral and neurobiological findings suggest that the relatively poor recognition ability of normal neonates is probably due to slow maturation of the cortical association areas, rather than to neonatal immaturity of the limbic system. Immaturity of the neocortex could also underlie the absence of a fully functional memory system in human infants and, by implication, could help explain the phenomenon of infantile amnesia, i.e., the lack of memory for the events of infancy and earliest childhood.

Decline of Habits and Memories

Elderly persons commonly report that their memory is not as it used to be. For example, they may not recall someone's name on a particular occasion, even though that person is well known to them, or they may forget a particular word when speaking, even though the word will seem to be on the tip of their tongue. The elderly also perform more poorly than younger subjects on a wide variety of memory tests. Thus, are the changes in memory that accompany normal aging caused by a malfunctions of the same neural system that is damaged in amnesia? One important point to emerge from clinical research is that normal aging involves a spectrum of cognitive changes, which include memory but are not limited to it. Accordingly, one might suppose that the neuropathological correlates of aging might correspond to those of amnesia, insofar as the defects in memory resemble those in amnesia, but can be expected to involve other brain regions, insofar as there are other cognitive defects.

To gain better understanding of age-associated changes in mnemonic and behavioral functions, we initiated a series of studies designed to delineate the learning and memory abilities of aged rhesus monkeys, with the ultimate goal of directly correlating age related behavioral impairments with measures of neural alterations in the same monkeys. First, as for the infant monkeys, we chose tasks to measure visual object recognition (DNMS) and visual habit formation (24-hr ITI and standard two-objects discrimination) in order to compare with the deficit seen in amnesic monkeys with medial temporal lobe lesions. In addition, we also included tasks to measure spatial memory (delayed response), visuospatial orientation, visually guided reaching, motor skill learning, and reaction time (Bachevalier et al., 1991). These additional tasks were selected because they are known to be sensitive to damage of cerebral areas outside the limbic system and, thus, provide a way to investigate whether the behavioral deficit seen in aged monkeys was widespread or restricted to specific brain areas.

At the time that testing began, monkeys used in this study were assigned to four different age groups of four or six monkeys each, ranging from 3–6, 14–17, 20–24, and 26–30 years of age. Initially, the animals were tested in the DNMS task (Presty et al., 1987). The results indicate that, on average, visual recognition ability in monkeys deteriorates progressively with age. This gradual decline in visual recognition memory was reflected in a progressive decrease in scores on the performance test across the four different age groups, the oldest monkeys of 26–30 years of age having significantly poorer scores than the youngest animals (Fig. 4A). When tested in the 24-hr ITI task to measure their ability to form visual habits, the oldest monkeys (26–30 years of age) required again significantly more sessions to learn the three sets of discriminations than the youngest ones (Fig. 4B). Thus, in contrast to monkeys with damage to the medial temporal lobe that showed excellent performance on tasks measuring visual habit formation, the oldest monkeys were significantly impaired in the ability to form new visual discrimination habits. Nonetheless, the oldest monkeys learned standard two-object discrimi-

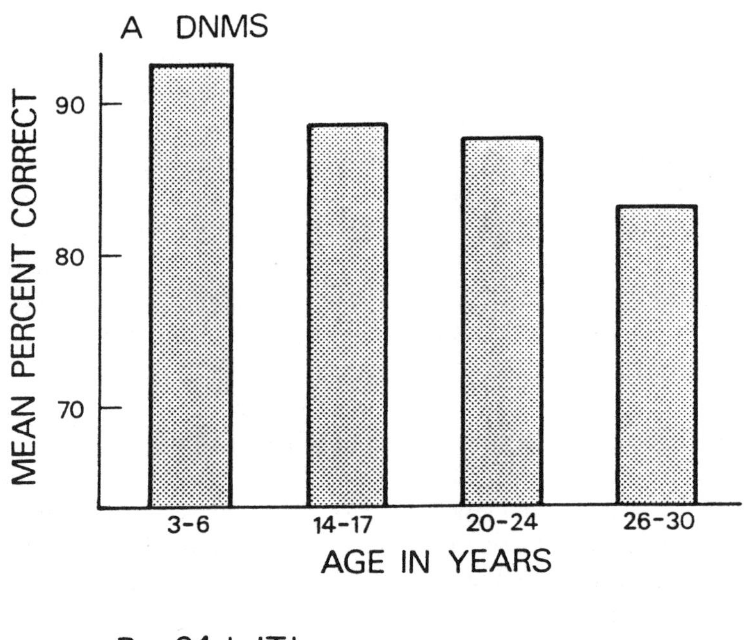

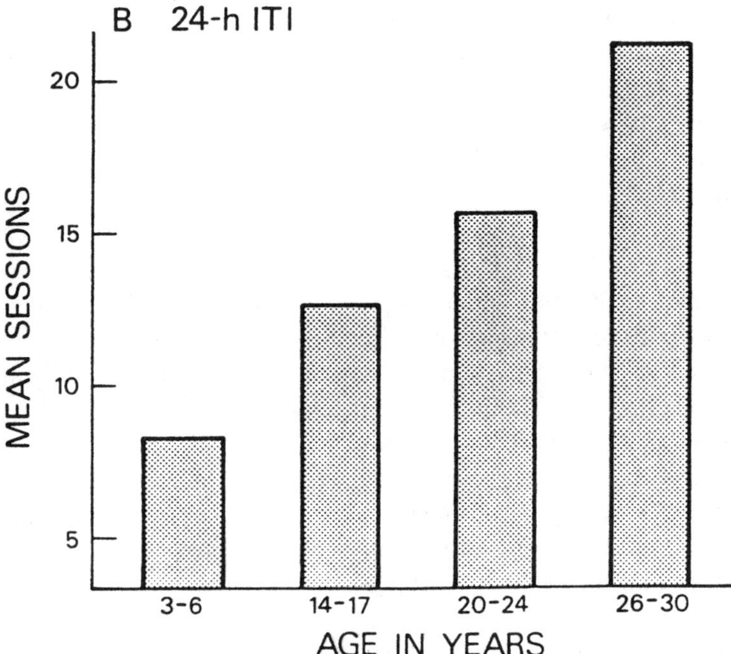

Figure 4. Average scores on the recognition performance task (A) and average number of sessions to attain criterion on three consecutive sets of discriminations (B) in aged monkeys.

nation as readily as the young controls, suggesting that simple forms of habits can be acquired readily by aged monkeys.

In addition, aged monkeys were significantly impaired relative to young adult animals in delayed response, a task measuring spatial memory. In this task, the monkey is trained to remember in which food well, the left or the right, the food reward had been concealed. Monkeys of 21 years of age and older were performing relatively well when the delays were short (0 and 1 sec) but were significantly worse than the youngest animals in delays above 2 sec. Thus, aged monkeys were also impaired in spatial memory and this decline appears to become manifest at an earlier age than the decline in visual recognition and object discrimination reported above.

Interestingly, abilities to perform on a visuospatial task are affected at even earlier age. In this task, monkeys must free a ring-shape "Lifesaver" candy from metal rods that have been fashioned into routes of varying complexity (Petrides and Iversen, 1979). There was no significant difference in the performance of the four groups of animals on the simpler patterns (0 to 2 bends), but there was a progressive increase in the time to solve more difficult patterns (3 to 5 bends) with increasing age. Further, the percent of failure in solving these routes was significantly higher in animals of 16 years of age or older than in young control animals. Since there was no group difference on the easy routes, and also none in a visually guided finger orientation task, which measures fine motor dexterity, the age-related deficits found on the difficult routes probably reflect impairments in visuospatial orientation rather than simply in motor dexterity.

The overall findings in our study in aged monkeys are summarized in Table 1. The results indicate that behavioral decline with age occurs in a variety of abilities, although performance on some tasks, often on the easy versions of these tasks, remains intact even in the oldest animals. This pattern of results corroborates earlier reports of age-related behavioral deficits in aged monkeys (Bartus, 1979; Davis, 1985) and humans (Ellis and Oscar-Berman, 1989) and suggests that the losses observed are in the particular ability being measured and not simply the reflection of an amotivational state or of a total deterioration of behavior. Second, the data indicate that the decline in certain spatial abilities begins relatively early, in the late teens, and does not become widespread until the late 20s, suggesting that some cerebral systems are compromised by aging earlier than others. Third, the presence of correlations between scores in aged animals primarily within tests as opposed to across tests suggests that different animals have different patterns of cerebral involvement. This tentative conclusion is further supported by the finding that, even on tests in which differences between old and young animals were statistically significant, the performance of individual older animals often matched that of younger animals. These variations among tasks and among animals present a unique opportunity to determine the types and locations of abnormalities in the brain that may underlie behavioral decline with aging.

Table 1

Age (years)	3–6	14–17	20–24	26–30
Visuospatial Abilities				
Easy routes				
Difficult routes		***	***	***
Delayed Response				
Short delays				
Long delays			***	***
Object Discrimination				
Standard				
Concurrent				***
Object Recognition				
Acquisition				
Performance test				***

Widespread behavioral deficits in aged rhesus monkeys. *** indicates significant differences from the youngest animals of age 3–6 years.

Widespread Cerebral Dysfunction in Aging

The eventual goal of this series of studies in aged monkeys is to relate behavioral impairments to specific changes in the brains of individual animals. Previous studies of functional localization in nonhuman primates have helped to identify candidate regions to be scrutinized for the neuropathological basis of age-related behavioral decline. To the extent that we can extrapolate from these localization studies, our earlier finding of impairment in visual recognition memory in old monkeys suggests dysfunction of the limbic system, i.e., medial temporal, medial diencephalic, or medial prefrontal regions, or some combination of these areas (Presty et al., 1987). These may occur relatively late, however. Similarly, impaired visual habit formation, implicating abnormalities in the inferior temporal cortex (Phillips et al., 1988) and possibly the caudoventral portion of the neostriatum (Wang et al., 1990) also seem to occur relatively late.

In corroboration of reports by others (Rapp and Amaral, 1989), an earlier onset of deficit was found in spatial memory, implying that the dorsolateral prefrontal cortex (Arnsten and Goldman-Rakic, 1985; Brozovski et al., 1979; Goldman et al., 1971; Passingham, 1985; Rosenkilde, 1979), or dorsal portion of the caudate nucleus (Battig et al., 1960; Divac et al., 1967; Goldman and Rosvold, 1970; Pribram et al., 1952), or both, may be functionally impaired earlier in the aging process than either limbic or temporo-striatal systems. Finally, the earliest deficits of all to appear were those in visuospatial orientation and visually guided reaching, implying that still earlier dysfunction occurs in the parietal cortex (Petrides and Iversen, 1979).

Of course, the age-associated abnormalities that disrupt function in these (and perhaps other) areas remain to be identified, and different neuropathological factors may be at play in different structures. Anatomical, histochemical, and neurochemical analyses of the brains of these and other behaviorally tested animals are now being conducted and should help lead to the identification of the degenerative events that underlie behavioral decline in elderly humans.

Summary

Both clinical investigation of amnesic patients and lesion studies in adult monkeys have suggested that an experience enters into memory in two ways: as cognitive information stored in a cortico-limbo-thalamo-cortical system and as a habit stored in a cortico-striatal system. Our studies of behavioral development in infant monkeys as well as those in human infants provide complementary evidence by suggesting that these two systems are developmentally dissociable, in that the limbic-dependent cognitive memory system appears to mature considerably more slowly than the cortico-striatal habit system. The effects of early lesions on the development of cognitive behavior in primates has indicated that early damage to the limbic system (like damage in adulthood) leaves habit formation intact but severely impairs the formation of new memories. By contrast, neonatal damage to area TE yields relative sparing of both habit and memory formation (as compared with the severe deficit in both functions seen in adults with the same lesions). Thus, association areas of the cortex appear to possess considerable plasticity at birth, implying a degree of functional immaturity. These behavioral data are supported by direct neurobiological evidence of neonatal cortical immaturity, reflected in the homogeneous distribution of opiatergic and cholinergic receptors as well as in the low level of metabolic activity.

By contrast, the dissociation between the two retention systems was not apparent with aging. That is, both the formation of habits and memories decline at the same tempo and relatively late in aging. These findings indicate that neuropathological processes may develop concurrently in the neural circuits underlying visual habits and visual memories, unless the neuropathological processes affect a neural structure subserving the two functions, such as the inferior temporal cortex. Finally, the behavioral decline with aging began in the teens for visuospatial abilities but did not affect other spatial abilities until the late 20s, suggesting that although the cerebral dysfunction eventually becomes widespread, the cerebral systems underlying spatial abilities are compromised earlier than others by aging. Interestingly, these spatial abilities, which appear to decline very early with aging, are known to attain functional maturity relatively late in development (Goldman, 1975; Corinne Hagger, personal communication).

References

Aggleton JP, Desimone R, Mishkin M (1986): The origin, course, and termination of the hippocampothalamic projections in the macaque. *Journal of Comparative Neurology* 243:409–421

Aggleton JP, Mishkin M (1983a): Visual recognition memory impairment following medial thalamic lesions in monkeys. *Neuropsychologia* 21:189–197

Aggleton JP, Mishkin M (1983b): Memory impairments following restricted medial thalamic lesions in monkeys. *Experimental Brain Research* 52: 199–209

Aggleton JP, Mishkin M (1984): Projections of the amygdala to the thalamus in the cynomolgus monkey. *Journal of Comparative Neurology* 222:56–68

Aggleton JP, Mishkin M (1985): Mammillary-body lesions and visual recognition in the monkey. *Experimental Brain Research* 58:190–197

Aigner T, Mitchell S, Aggleton JP, DeLong M, Struble R, Wenk G, Price D, Mishkin M (1984): Recognition deficits in monkeys following neurotoxic lesions of the basal forebrain. *Society for Neuroscience Abstracts* 10:386

Alexander MP, Freedman M (1984): Amnesia after anterior communicating artery aneurysm rupture. *Neurology* 34:752–757

Amaral DG (1987): Memory: anatomical organization of candidate brain regions. In: *Handbook of Physiology. Section 1. The Nervous System*, Mountcastle VB, Plum F, Geiser SR, eds. Bethesda: American Physiological Association, pp. 211–294

Amaral DG, Cowan WM (1980): Subcortical afferents to the hippocampal formation in the monkey. *Journal of Comparative Neurology* 189:573–591

Angelergues R, Hecaen H, De Ajuriaguerra J (1955): Les troubles mentaux au cours des tumeurs du lobe frontal. *Annals of Medical Psychology* 113:577–642

Arendt T, Bigl V, Arendt A, Tennstedt A (1983): Loss of neurons in the nucleus basalis of Meynert in Alzheimer's disease, paralysis agitans, and Korsakoff's disease. *Acta Neuropathologica* 61:101–108

Arnsten AFT, Goldman-Rakic PS (1985): Alpha-adrenergic mechanisms in prefrontal cortex associated with cognitive decline in aged nonhuman primates. *Science* 230:1273–1276

Bachevalier J (1990): Ontogenetic development of habit and memory formation in primates. In: *The Development and Neural Bases of Higher Cognitive Functions*, Diamond A, ed. *Annals of New York Academy of Sciences* 608:457–486

Bachevalier J (1991) Cortical vs limbic immaturity: Relationship to infantile amnesia. In: *Minnesota Symposium on Child Psychology: Developmental Neuroscience*, Gunnar M, Nelson CA, eds. New York: Erlbaum, 24:129–153

Bachevalier J, Brickson M, Hagger C, Mishkin N (1990): Age and sex differences in the effects of selective temporal lobe lesion on the formation of visual discrimination habits in rhesus monkeys (*Macaca mulatta*). *Behavioral Neuroscience* 104:885–899

Bachevalier J, Hagger C, Mishkin M (1991a): Functional maturation of the occipitotemporal pathway in infant rhesus monkeys. In: *Alfred Benzon Symposium 31: Brain Work and Mental Activity*, Copenhagen: Munksgaard, pp. 231–240

Bachevalier J, Landis LS, Walker LC, Brickson N, Mishkin M, Price DL, Cork LC (1991b): Aged monkeys exhibit behavioral deficits indicative of widespread cerebral dysfunction. *Neurobiology of Aging* 12:99–111

Bachevalier J, Mishkin M (1984): An early and a late developing system for learning and retention in infant monkeys. *Behavioral Neuroscience* 98:770–778

Bachevalier J, Mishkin M (1986): Visual recognition impairment follows ventromedial but not dorsolateral prefrontal lesions in monkeys. *Behavioral Brain Research* 20:249–261

Bachevalier J, Mishkin M (1988): Long-term effects of neonatal temporal cortical and limbic lesions on habit and memory formation in rhesus monkeys. *Society for Neuroscience Abstracts* 14:1

Bachevalier J, Parkinson JK, Mishkin M (1985): Visual recognition: effects of separate vs. combined transection of fornix and amygdalofugal pathways. *Experimental Brain Research* 57:554–561

Bachevalier J, Ungerleider LG, O'Neill B, Friedman DP (1986): Regional distribution of [3H]naloxone binding in the brain of a newborn rhesus monkey. *Developmental Brain Research* 25:302–308

Baleydier C, Mauguiere F (1980): The duality of the cingulate gyrus in monkey: neuroanatomical study and functional hypothesis. *Brain* 103:525–554

Bartus RT (1979): Effects of aging on visual memory, sensory processing and discrimination learning in nonhuman primate. In: *Sensory Systems and Communication in the Elderly*, Ordy JM, Brizzee K, eds. New York: Raven, pp. 85–144

Battig K, Rosvold HE, Mishkin M (1960): Comparison of the effects of frontal and caudate lesions on delayed response and alternation in monkeys. *Journal of Comparative Physiology and Psychology* 53:400–404

Brooks DN, Baddeley AD (1976): What can amnesic patients learn? *Neuropsychologia* 14:111–122

Brozovski TJ, Brown RM, Rosvold HE, Goldman PS (1979): Cognitive deficit caused by regional depletion of dopamine in prefrontal cortex of rhesus monkey. *Science* 205:929–932

Buerger AA, Gross CG, Rocha-Miranda CE (1974): Effects of ventral putamen lesions on discrimination learning by monkeys. *Journal of Comparative Physiology and Psychology* 86:440–446

Butters SN, Cermak LS (1980): *Alcoholic Korsakoff's Syndrome—An Information-Processing Approach to Amnesia.* New York: Academic Press

Candy J, Perry R, Perry E, Irving D, Blessed G, Fairnairn A and Tomlinson B (1983): Pathological changes in the nucleus of Meynert in Alzheimer's and Parkinson's diseases. *Journal of Neurological Science* 54:277–289

Cohen NJ, Squire LR (1980): Preserved learning and retention of pattern-analyzing skill in amnesia: dissociation of "knowing how" and "knowing that." *Science* 210:207–209

Corkin S (1965): Tactually-guided maze learning in man: Effects of unilateral cortical excisions and bilateral hippocampal lesions. *Neuropsychologia* 3:339–351

Corkin S (1968): Acquisition of motor skill after bilateral medial temporal-lobe excision. *Neuropsychologia* 6: 255–266

Davis RT (1985): The effects of aging on the behavior of rhesus monkeys. In: *Behavior and Pathology of Aging in Rhesus Monkeys. Monographs in Primatology Vol. 8*, Davis RT, Leathers CW, eds. New York: Alan R. Liss, pp. 57–82

Divac I, Rosvold HE, Szwarcbart MK (1967): Behavioral effects of selective ablation of the caudate nucleus. *Journal of Comparative Physiology and Psychology* 63:184–190

Ellis RJ, Oscar-Berman M (1989): Alcoholism, aging, and functional cerebral asymmetries. *Psychological Bulletin* 106:128–147

Gade A (1982): Amnesia after operations on aneurysms of the anterior communicating artery. *Surgical Neurology* 18:46–49

Goldman PS (1975): Age, sex, and experience as related to the neural basis of cognitive development. In: *Brain Mechanisms in Mental Retardation*, Buchwald NA, Brazier MAB, eds. New York: Academic Press, pp. 379–392

Goldman PS, Rosvold HE (1970): Localization of function within the dorsolateral prefrontal cortex of the rhesus. *Experimental Neurology* 27: 291–304

Goldman PS, Rosvold HE, Vest B, Galkin TW (1971): Analysis of the delayed-alternation deficit produced by dorsolateral prefrontal lesions in the rhesus monkey. *Journal of Comparative Physiology and Psychology* 77:212–220

Horel JA (1978): The neuroanatomy of amnesia: A critique of the hippocampal memory hypothesis. *Brain* 101:403–445

Korsakoff SS (1887): Disturbance of psychic function in alcoholic paralysis and its relation to the disturbance of the psychic sphere in multiple neuritis of nonalcoholic origin. *Psychiatrii* 4: fascicle 2

Lilly R, Cummings JL, Benson F, Frankel M (1983): The human Klüver-Bucy syndrome. *Neurology* 83:1141–1145

Mair WGP, Warrington E, Weiskrantz L (1979): Memory disorders in Korsakoff's psychosis. A neuropathological and neuropsychological investigation of two cases. *Brain* 102:749–783

Malamut BL, Saunders RC, Mishkin M (1984): Monkeys with combined amygdalo-hippocampal lesions succeed in object discrimination learning despite 24-hour intertrial intervals. *Behavioral Neuroscience* 98:759–769

Mesulam M, Mufson E, Levey A, Wainer B (1983): Cholinergic innervation of cortex by the basal forebrain: Cytochemistry and cortical connections of the septal area, diagonal band nuclei, nucleus basalis (substantia innominata): and hypothalamus in the rhesus monkey. *Journal of Comparative Neurology* 214:170–197

Meunier M, Murray EA, Bachevalier J, Mishkin M (1990) Effects of perirhinal cortical lesions in visual recognition memory in rhesus monkeys. *Society for Neuroscience Abstracts* 16:616

Milner B (1962): Les troubles de la mémoire accompagnant des lésions hippocampiques bilatérales. In: *Physiologie de l'hippocampe*, Passouant P, ed. Paris: Centre de la Recherche Scientifique, pp. 257–272

Milner B (1972): Disorders of learning and memory after temporal lobe lesions in man. *Clinical Neurosurgery* 19:421–446

Mishkin M (1954): Visual discrimination performance following partial ablations of the temporal lobes Ventral surface vs. hippocampus. *Journal of Comparative Physiology and Psychology* 47:187–193

Mishkin M (1978): Memory in monkeys severely impaired by combined but not separate removal of amygdala and hippocampus. *Nature* 273:297–298

Mishkin M (1982): A memory system in the monkey. *Philosophical Transactions of the Royal Society of London* Series B, 298:85–95

Mishkin M, Appenzeller T (1987): The anatomy of memory. *Scientific American* 256:80–89

Mishkin M, Malamut BL, Bachevalier J (1984): Memories and habits: Two neural systems. In: *Neurobiology of Learning and Memory*, Lynch G, McGaugh L, Weinberger NM, eds. New York: Guilford, pp. 65–77

Mishkin M, Petri HL (1984): Memories and habits: Some implications for the analysis of learning and retention. In: *Neuropsychology of Memory*, Butters N, Squire L, eds. New York: Guilford, pp. 287–296

Murray EA (1991): Medial temporal lobe structures contributing of recognition memory: The amygdaloid complex versus the perirhinal cortex. In: *Neurobiology of the Amygdala*, Aggleton JP, ed. New York: Wiley-Liss, pp. 453–470

Murray EA, Bachevalier J, Mishkin M (1989): Effects of rhinal cortical lesions on visual recognition memory in rhesus monkeys. *Society for Neuroscience Abstracts* 15:342

Murray EA, Mishkin M (1983): Severe tactual memory deficits in monkeys after combined removal of the amygdala and hippocampus. *Brain Research* 270:340–344

Nauta WJH (1962): Neural associations of the amygdaloid complex in the monkey. *Brain* 85:505–520

O'Neill B, Friedman DP, Bachevalier J, Ungerleider LG (1986): Distribution of muscarinic cholinergic receptors in the brain of a newborn monkey. *Society for Neuroscience Abstracts* 12:809

Overman WH (1991): Performance on traditional match-to-sample, nonmatch-to-sample, and object discrimination tasks by 12 to 32 month-old children: A developmental progression. In: *Development and Neural Bases of Higher Cognitive Functions*, Diamond A, ed. *Annals of New York Academy of Sciences* 608:365–390

Passingham RE (1985): The memory of monkeys (*Macaca mulatta*) with lesions in prefrontal cortex. *Behavioral Neuroscience* 99:3–21

Penfield W, Milner B (1958): Memory deficit produced by bilateral lesions in the hippocampal zone. *American Medical Association Archives of Neurology and Psychiatry* 79:475–497

Petrides M, Iversen SD (1979): Restricted posterior parietal lesions in the rhesus monkey and performance on visuospatial tasks. *Brain Research* 161:63–77

Phillips RR, Malamut BL, Bachevalier J, Mishkin M (1988): Dissociation of the effects of inferior temporal and limbic lesions on object discrimination learning with 24-hr intertrial intervals. *Behavioral Brain Research* 27:99–107

Presty SK, Bachevalier J, Walker LC, Struble RG, Price DL, Mishkin M, Cork LC (1987): Age differences in recognition memory of the rhesus monkey (*Macaca mulatta*). *Neurobiology of Aging* 8:435–440

Pribram KH, Mishkin M, Rosvold HE, Kaplan SJ (1952): Effects on delayed-response performance of lesions of dorsolateral and ventromedial frontal cortex of baboon. *Journal of Comparative Physiology and Psychology* 45:565–575

Provensal JC (1951): *Les tumeurs frontales. Etudes anatomo-cliniques de 82 observations*. Thesis, Marseille

Rapp PR, Amaral DG (1989): Evidence for task-dependent memory dysfunction in the aged monkey. *Journal of Neuroscience* 9:3568–3576

Ribot T (1881): *Les maladies de la mémoire.* Paris: Germer-Baillere

Rosenkilde CE (1979): Functional heterogeneity of the prefrontal cortex in the monkey: A review. *Behavioral and Neural Biology* 25:301–345

Saint-Cyr JA, Taylor AE, Lang AE (1988): Procedural learning and neostriatal dysfunction in man. *Brain* 111:941–959

Scoville WB, Milner B (1957): Loss of recent memory after bilateral hippocampal lesions. *Journal of Neurology, Neurosurgery and Psychiatry* 20:11–21

Spiegler BJ, Mishkin M (1981): Evidence for the sequential participation of inferior temporal cortex and amygdala in the acquisition of stimulus-reward associations. *Behavioral Brain Research* 3:303–317

Squire LR (1987): *Memory and Brain.* New York: Oxford University Press

Talland G (1965): *Deranged Memory.* New York: Academic Press

Talland GA, Sweet WT, Ballantine HT (1967): Amnesic syndrome with anterior communicating artery aneurysm. *Journal of Nervous Mental Disease* 145:179–192

Tobias TJ (1975): Afferents to prefrontal cortex from the thalamic mediodorsal nucleus in the rhesus monkey. *Brain Research* 83:191–212

Turner B, Mishkin M, Knapp M (1980): Organization of the amygdalopetal projections from modality-specific cortical association areas in the monkey. *Journal of Comparative Neurology* 191:515–543

Van Hoesen GW, Pandya DN (1975): Some connections of the entorhinal (area 28) and perirhinal (area 35) cortices of the rhesus monkeys: I. Temporal lobe afferents. *Brain Research* 95:1–24

Victor M, Adams RD, Collins GH (1971): *The Wernicke–Korsakoff Syndrome.* Philadelphia: Davis

Wang J, Aigner T, Mishkin M (1990): Effects of neostriatal lesions on visual habit formation in rhesus monkeys. *Society for Neuroscience Abstracts* 16:617

Warrington E, Weiskrantz L (1968): New method of testing long-term retention with special reference to amnesic patients. *Nature (Lond)* 217:972–974

Whitehouse P, Price DL, Struble R, Clark A, Coyle J, DeLong M (1982): Alzheimer's disease and senile dementia: Loss of neurons in the basal forebrain. *Science* 215:1237–1239

Whitty CWM, Levin W (1960): A Korsakoff's syndrome in the postcingulectomy confusional state. *Brain* 83:648–653

Zola-Morgan S, Squire LR (1985a): Complementary approaches to the study of memory: Human amnesia and animal models. In: *Memory Systems of the Brain: Animal and Human Cognitive Processes*, Weinberger NW, McGaugh JL, Lynch L, eds. New York: Guilford, pp 463–477

Zola-Morgan S, Squire LR (1985b): Amnesia in monkeys after lesions of the mediodorsal nucleus of the thalamus. *Annals of Neurology* 17:558–564

Zola-Morgan S, Squire LR, Amaral DG, Suzuki WA (1989): Lesions of perirhinal and parahippocampal cortex that spare the amygdala and hippocampal formation produce severe memory impairment. *Journal of Neuroscience* 9:4355–4370

Zola-Morgan S, Squire LR, Mishkin M (1982): The neuroanatomy of amnesia: amygdala-hippocampus versus temporal stem. *Science* 218:1337–1339

Anoxia and Memory Processes

K. Krnjević and Y. Z. Xu

Introduction

With few exceptions, mammals cannot function for more than very brief periods in the absence of an adequate supply of O_2 (cf. Hochachka, 1986). Indeed their very survival is jeopardized by anoxia prolonged beyond a few minutes.

Mental function is particularly susceptible to anoxic disruption. Higher cognitive processes are known to be depressed even by moderate hypoxia such as is experienced at high altitudes (Bert, 1878; Ernsting, 1986; Haldane and Priestley, 1935). In experiments on humans or dogs, the first signs of dysfunction are detectable within a few seconds (Rossen et al., 1943). These observations are readily explained if certain brain cells or processes are highly sensitive to even brief O_2 lack.

In studies on animals, some brain areas do in fact show a particularly early and marked depression of electrical activity (Sugar and Gerard, 1938). Among the very first is the hippocampus. This is of special interest because the hippocampus appears to play a vital role in learning and memory (O'Keefe and Nadel, 1978; Olton et al., 1979; Rawlins, 1985; Squire, 1987). The aim of the present chapter is to review briefly some relevant evidence: first, on the cellular mechanisms by which anoxia disrupts hippocampal activity; second, on the effects of anoxia on learning and memory retrieval in whole animals; and third, on the effects of anoxia on a possible hippocampal manifestation of learning, the sustained enhancement of synaptic responses evoked by repetitive activation, generally known as long-term potentiation (LTP) (Bliss and Lømo, 1973; Teyler and DiScenna, 1987).

Neuronal Function in Hippocampus during Anoxia

Changes in intrinsic cellular properties. Several intracellular studies have been made on hippocampal pyramidal cells (mainly in the area CA1), using the techniques of current clamp (Fujiwara et al., 1987; Hansen et al., 1982; Leblond and Krnjević, 1989; Misgeld and Frotscher, 1982) and voltage clamp (Krnjević and Leblond, 1989). In most respects they are in agreement: brief periods of anoxia (2–4 min) lead to a hyperpolarization, associated with a marked drop in resistance and in excitability (Fig. 1). This effect, which is readily explained by an increase in K conductance (G_K) is probably sufficient to account for the early loss of cell firing, whether spontaneous or evoked (synaptically or electrically). The mechanism of G_K increase is more controversial. A variety of observations indicate that it is probably not caused by activation of ATP-sensitive K channels (Ben-Ari, 1990; Ben-Ari et al., 1990; Krnjević and Leblond, 1988; Leblond and Krnjević, 1989), but rather is generated by Ca-sensitive K channels, following the release of Ca^{2+} from intracellular Ca^{2+} stores (Duchen et al., 1990; Krnjević and Xu, 1989; Leblond and Krnjević, 1989). There may be a significant contribution of adenosine-sensitive G_K activated by adenosine liberation during anoxia (Berne et al., 1974).

At about the same time anoxia also causes a marked, albeit reversible, reduction of postsynaptic voltage-dependent Ca current (I_{Ca}) (Fig. 2), especially the high-threshold, more slowly inactivating I_{Ca} (Krnjević and Leblond, 1988, 1989). How I_{Ca} is suppressed is at present not known with certainty. Ca currents are recorded during nearly complete pharmacological suppression of G_K so the block of I_{Ca} is probably not simply due to the G_K increase normally evoked by anoxia. Ca currents are known to be sensitive to the cytosolic Ca^{2+} concentration (Eckert and Chad, 1984); therefore the same mechanism, that is, the anoxic rise in cytosolic $[Ca^{2+}]$, could explain the two principal manifestations of anoxia, the GK increase *and* the depression of I_{Ca}. But some other mechanisms, perhaps more directly related to protein phosphorylation and the depletion of ATP, have not been excluded. A further point of interest is that NMDA-evoked inward currents can also be depressed by anoxia (Krnjević et al., 1989).

A late effect of anoxia is a second, *postanoxic* phase of hyperpolarization, often seen when O_2 is reintroduced (Figs. 1 and 3). It is believed to be principally due to accelerated Na,K-pumping and its corresponding electrogenic action (Fujiwara et al., 1987; Leblond and Krnjević, 1989).

Extrinsic changes. Processes that are not intrinsic to the recorded cells are of two kinds. The first is a variable—sometimes pronounced—

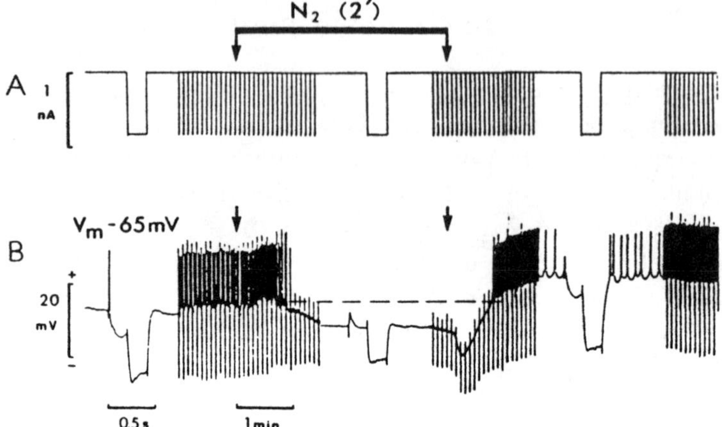

Figure 1. Hyperpolarizing effect of brief anoxia (N_2) on CA1 pyramidal neuron. Intracellular recording was with 3M KCI microelectrode from conventional hippocampal slice (Sprague-Dawley rat) in the interface chamber, at 330. Input resistance was estimated at regular intervals by 200 ms hyperpolarizing current pulses, monitored in upper trace (A). Shortly preceding these pulses, stratum radiatum stimulation evoked a synaptic response, including an EPSP, a spike and a prolonged hyperpolarizing IPSP (see accelerated trace at left). Anoxia was produced by substituting 95% N_2 for 95% O_2 (keeping 5% CO_2 constant throughout) for 2 min, as indicated by arrows. Note during anoxia the marked reduction in resistance, and disappearance of both spontaneous and evoked firing, as well as of IPSP. Return of oxygen is associated with a further sharp postanoxic hyperpolarization (just after second arrow).

depolarization that is superimposed on, or supersedes the early hyperpolarization. If anoxia is maintained beyond a few minutes, a progressive depolarization is always seen, culminating in a gross depolarizing shift that may or may not be reversible after the return of O_2. In some preparations, a reversible paroxysmal depolarization is evoked quite early by anoxia. Its characteristics are very similar to those of spreading depression (SD) (Bures et al., 1974; Leao, 1944).

Depolarizing effects of anoxia are especially common in the CA3 area, where they appear to be principally due to glutamate release. There is good reason to believe that this glutamate release is to a significant extent "modulated" by presynaptic ATP-sensitive K channels: they are activated by falling ATP-levels and thus oppose the anoxic depolarization of glutamatergic terminals (such as the mossy fibers) and the consequent release of glutamate (Ben-Ari, 1990; Ben-Ari et al., 1990; Krnjević and Ben-Ari, 1989; Miller, 1990).

The other main type of extrinsic change is a depression of synaptic potentials. Inhibitory synaptic potentials (IPSPs) are especially suscep-

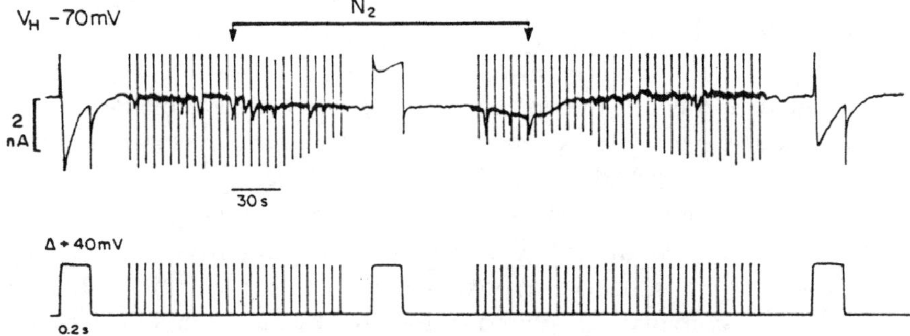

Figure 2. Anoxia also suppresses voltage-dependent Ca current. Upper trace: membrane current of CA1 neuron, recorded by single-electrode (3M KCI) voltage clamp, at 33° and at holding potential of –70 mV. Brief depolarizing commands (+40 mV), applied at regular 5 sec intervals are monitored on lower traces. To eliminate inward Na and outward K currents the following were added to the superfusate: 1 μM tetrodotoxin, 4 mM Cs, 10 mM tetraethylammonium and 0.5 mM 4-aminopyridine. Currents are also displayed on faster trace initially, during anoxia and at end. Note gradual disappearance of Ca inward current (downward deflections) during anoxia, and its later substantial recovery.

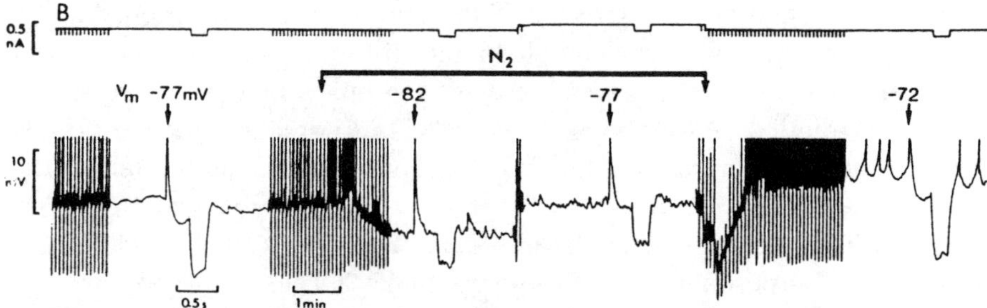

Figure 3. Selective suppression of IPSPs during anoxia. As in Fig. 1, recording from CA1 neurons shows usual hyperpolarization (to –82 mV) and marked reduction in resistance (monitored with regular hyperpolarizing current-pulses, recorded on upper trace). IPSP has disappeared, but not the EPSP and the evoked spike. When membrane potential was returned to initial value (–77 mV), by injection of depolarizing current, IPSP was not restored. This indicates a true suppression of inhibitory transmission.

tible to anoxia (Fig. 3). Typically they disappear well before EPSPs do and they return later (and perhaps less completely) after reoxygenation. The mechanism of synaptic block probably includes suppression of voltage-dependent Ca currents in nerve terminals (comparable to the above-described suppression of postsynaptic I_{Ca}) an excessive rise in extracellular K^+, and perhaps presynaptic effects of glutamate and/or GABA released during anoxia (Nicholls, 1989).

Anoxia and Memory

In animal experiments—those on rats are particularly relevant in the present context—brief anoxia can produce a marked retroactive amnesia. For example, D'Andrea and Kesner (1973) trained rats *not* to press a reward-delivering lever by giving them foot shocks (passive avoidance). Once trained, the rats performed reliably for over 24 hr. However, a 15 sec exposure to hypoxia—sufficient for the rat to become inanimate—or convulsions elicited by electrical stimulation (electroshock or ECS), resulted in a delayed, temporary loss of the newly-learned behavior, reaching a peak at 3 hr posthypoxia, with complete recovery by 24 hr. Exhaustive controls showed that neither convulsions—which could be prevented by a paralysing agent—nor the ECS were essential. The crucial factor in all cases was O_2 lack.

Similarly, Frieder and Allweis (1982) trained rats to flee from an electrified cage (active avoidance). When brief anoxia was applied at various times after the end of the training period, there was a delayed disruption of the active avoidance. As illustrated graphically in Fig. 4, the loss of retention was greatest if the anoxia occurred "immediately" at the end of training; it was much smaller if the anoxia followed 15 min later; and was no longer detectable when the rats were exposed to anoxia 2 hr after training. As in the experiment of D'Andrea and Kesner (1973), the retroactive amnesia was long-lasting, but it disappeared after 4 hr.

In order to define more precisely the critical early post-training period during which anoxia is most disruptive, Allweis et al. (1984) made further comparable experiments on newborn chicks. These learned to avoid pecking at red beads that have a bitter flavor. When fully trained, the birds consistently showed a very marked preference for the non-bitter blue beads (cf. open circles (red beads) and closed circles (blue beads) in Fig. 5). The usual brief (15 sec) anoxic treatment was then applied after various post-training intervals. If the anoxia came immediately (0 min) after the training (Fig. 5A), there was no evidence of any loss of the learned avoidance of red beads. By 2–5 min post-training, however, anoxia elicited a delayed disturbance of specific avoidance (Fig. 5B). This reached a maximum when the anoxia was given at 5 min post-training (Fig. 5C). As before, the loss of retention (or recall) began well after the brief anoxia, in all cases being greatest some 30 min post-training. The critical period for the anoxic effect lasted only until 10 min post-training; after that, anoxia no longer produced any subsequent loss of recall.

The principal conclusion from these results is that anoxia can have a strongly disruptive effect on a recently learned behavior. Though pro-

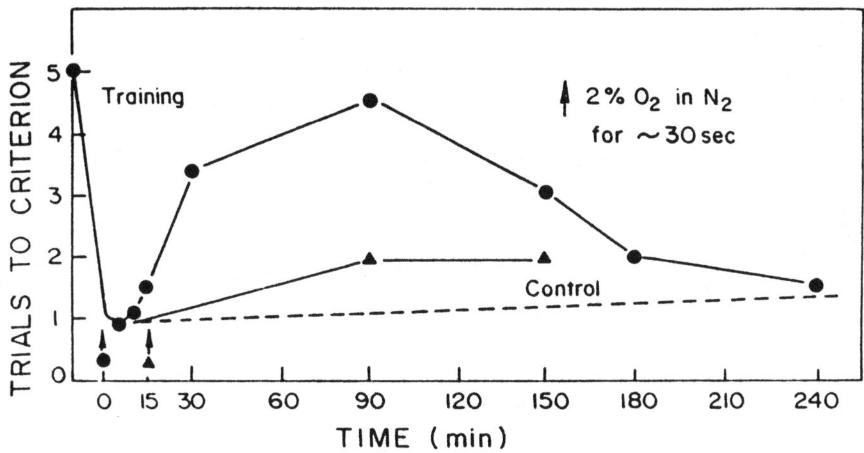

Figure 4. Brief hypoxia causes a delayed, but only temporary, failure of recall (retroactive amnesia) in rats trained to avoid an electrified compartment. The initial training resulted in a sharp drop in number of "trials to criterion" (T.T.C.), from five (in naive rats) to one. The rats were exposed to 2% O_2 for 30 sec (arrows), either immediately (circles) or 15 min (triangles) after training. This caused a delayed but prolonged loss of recall (T.T.C. increased), that was most pronounced when hypoxia was administered "immediately" after the end of training. (Fig. 2 from Frieder and Allweis, 1982).

longed, this apparent loss of memory is not permanent. Therefore it cannot be due to the disappearance of whatever memory trace (engram) the original training "inscribed" in cerebral neurons, synapses and/or circuits. Clearly, what is disturbed must be either the retrieval of the stored information or the animal's ability to use it for the purposes of behavior (which is otherwise not grossly and unspecifically disturbed; cf. in Fig. 5B, C the continued normal pecking at blue beads). One possible interpretation postulates the existence of a "working" memory—intermediate between short-term and long-term memory (Gibbs and Ng, 1976; Olton et al., 1979; Rawlins, 1985)—which is essential for the retrieval of memory traces, and is temporarily susceptible to anoxia (Allweis et al., 1984).

Perhaps the most intriguing aspects of such retroactive amnesia are its slow and delayed post-anoxic onset and long duration. They point to a complex chain of events initiated by anoxia—perhaps the activation of such Ca-sensitive elements as calmodulin-dependent protein kinase or kinase C, followed by phosphorylation of proteins that can modulate the recall mechanism.

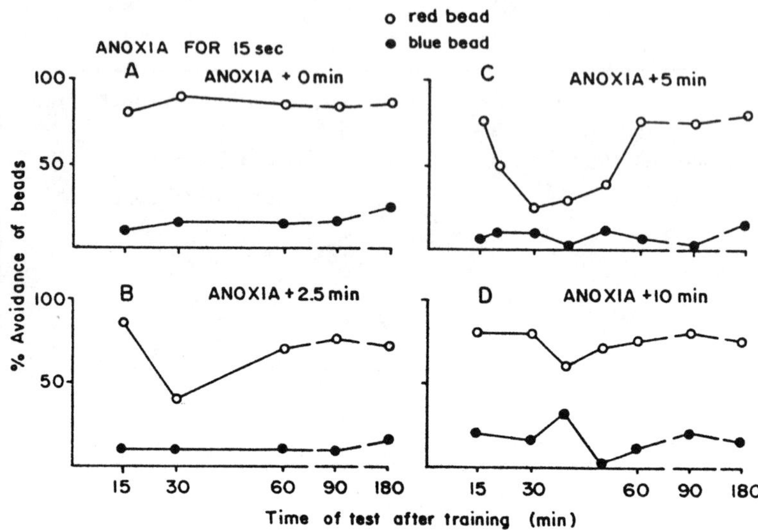

Figure 5. Further evidence of a critical period post-training when anoxia elicits retroactive amnesia. Newborn chicks, trained to avoid red but not blue beads, were briefly exposed to anoxia at various times *after* training: 0, 2.5, 5, and 10 min respectively in A, B, C, and D. The specific avoidance was essentially unaffected when exposure to anoxia was at 0 min or >10 min post-training; but between 2 and 10 min, there was a sharp, but delayed, loss of retention, reaching a peak at 30 min and vanishing by 60 min (Fig. 1 from Allweis et al., 1984).

Anoxia and LTP

What is long-term potentiation, or LTP? Some 20 years ago, Bliss and Lømo (1973) discovered the very prolonged enhancement of synaptic responses that follows brief tetanic stimulation of hippocampal pathways. The realization that LTP could be a cellular manifestation of learning and memory has generated enormous interest in its properties (biophysical and biochemical): hence the intense effort that is being directed towards uncovering the precise underlying mechanisms (Lynch and Baudry, 1984; Malenka et al., 1989; Teyler and DiScenna, 1987). Some 10 unexpected features are now well established: LTP cannot be obtained if the tetanic stimulation does not evoke a critical minimum amount of postsynaptic firing or depolarization; it requires the influx of Ca^{2+}, for which activation of NMDA receptors (by the natural transmitter glutamate) is essential; and once established it is associated with an enhanced glutamate sensitivity, mediated via non-NMDA receptors.

In a plausible scheme (Davies et al., 1989; Lynch and Baudry, 1984; Malenka et al., 1989; Malinow et al., 1989), Ca^{2+} influx is triggered whenever tetanic stimulation liberates enough glutamate to produce the required amount of postsynaptic depolarization (perhaps reinforced by disinhibition that activates NMDA receptors). Acting via calmodulin- and kinase C-dependent kinases, Ca^{2+} stimulates a self-maintained phosphorylation of glutamate (AMPA-type) receptors, thus causing a long-lasting enhancement of synaptic responses. Whether there is also a sustained increase in evoked glutamate *release* during LTP—either during the initial phase, that lasts some 5–20 min, or even during the late sustained phase—remains controversial.

Although a comparable LTP can be induced by repetitive stimulation in the three main regions of the hippocampus (cf. Fig. 6), in CA3 its mechanism may not involve the activation of NMDA receptors (Martin, 1983).

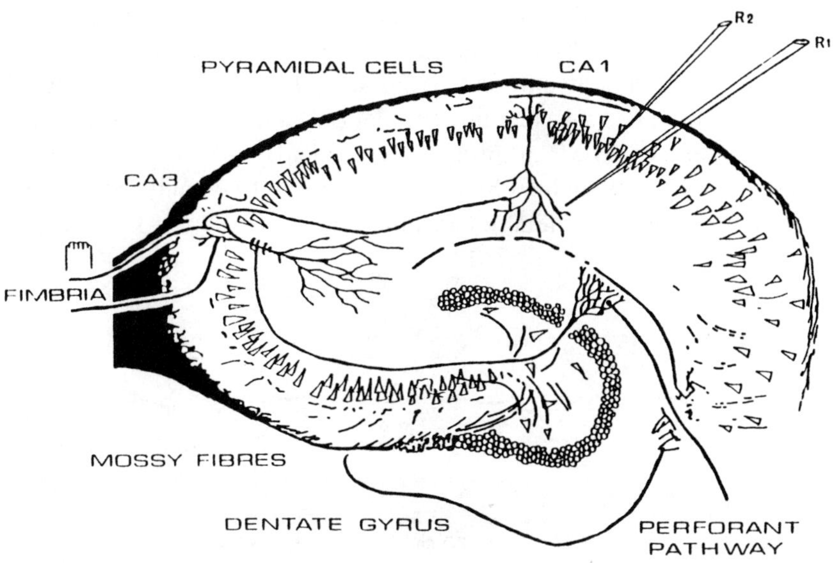

Figure 6. Diagram of hippocampal slice showing the three principal regions and some sites of stimulation and recording.

Anoxia on LTP Hippocampal Slices

The experiments described here have already been reported briefly (Leblond, 1988; Xu and Krnjević, 1990). They were performed on conventional, 500 μM thick slices (Dingledine, 1984), prepared from the brains of Sprague–Dawley rats. The slices were kept in an interface-type chamber, where their lower surface was continuously superfused with an appropriate saline and their upper surface aerated by a humidified gas mixture consisting of 95% O_2 and 5% CO_2, at 33–340. The slices were always allowed to equilibrate in the recording chamber for 1–2 hr before starting electrical recording and stimulation.

Recording was exclusively with low-resistance extracellular microelectrodes. An electrode was placed either in the apical dendritic region (e.g., R_1, in Fig. 6) or in the cell body layer (R_2 in Fig. 6) where stable negative fields (population EPSPs or population spikes) were respectively evoked by stimulation of the afferent pathway (Schaffer collaterals for CA1, the hilus of the dentate gyrus for CA3, or the perforant pathway at the angular bundle for dentate granule cells (Fig. 6)). In all cases, stimulation was at a constant frequency (0.05 Hz) and intensity throughout, except when series of pulses of regularly incremented intensity were applied to assess the strength-response relation over a wide range (as in Fig. 9). LTP was induced by a standard 100 Hz tetanus, lasting for 1 sec, and repeated once after an interval of 20 sec. In CA1, where most of the data were obtained, this tetanic stimulation caused a sustained (>30 min) enhancement of EPSP amplitude, typically by 40–50%. Figure 7 gives some examples of such EPSPs recorded in CA1 (A) and the dentate gyrus (B), before and after tetanic stimulation.

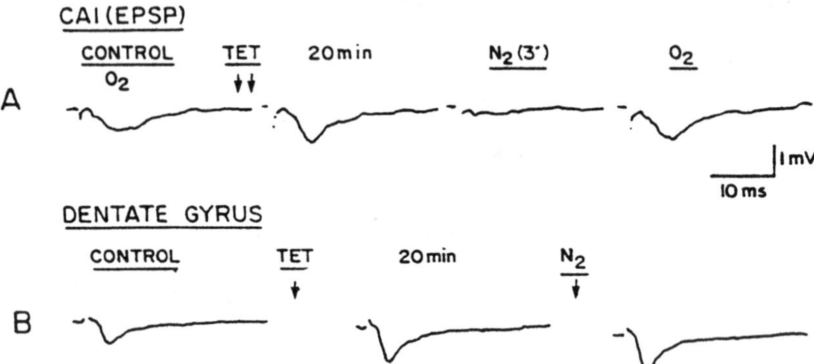

Figure 7. Extracellular recordings of EPSP fields from a hippocampal slice, showing LTP and effects of anoxia (N_2). A: in CA1 apical dendritic region, EPSPs evoked by Schaffer collateral stimulation of constant intensity 26 throughout (at 0.05 Hz). After two tetanic stimulations (100 Hz for 1 sec), EPSP was increased in amplitude and rate of rise, as indicated by trace recorded at 20 min. When O_2 was replaced by N_2 for 3 min, the EPSP virtually vanished (but not the small early response that reflects the arrival of the afferent volley). Within 5 min of reoxygenation the EPSP returned to its potentiated size. Lower traces: comparable data from dentate gyrus also show LTP after tetanus and full post-anoxic return to potentiated amplitude.

The time course of LTP was shown more clearly in "dot plots" of either the peak amplitude or the rate of rise of individual responses (stored on disk and analyzed off-line). An example of such a plot can be seen in Fig. 8, on a very slow time scale. In most cases, the tetanic stimulation evoked an early, short-lasting enhancement, comparable to the relatively brief (1–5 min) post-tetanic potentiation (PTP) observed at more peripheral synapses (Eccles, 1964). It is followed by the very slowly decaying phase of potentiation, that is generally known as LTP. To produce anoxia, the 95% O_2 in the aerating gas mixture was substituted by 95% N_2, keeping CO_2 constant at 5%: it lasted long enough (2–4 min) to cause a major (90%) reduction in EPSP amplitude.

Anoxia before LTP. When anoxia (N_2) was applied (as described above) during the 10–15 min period that preceded the tetani, there was no *obvious* interference with the subsequent LTP (no attempt was made to detect any minor changes by 13 statistical comparisons of LTP in N_2-treated and untreated slices).

Anoxia during tetanic stimulation. In these experiments, the tetani were applied during an anoxic test of sufficient duration to cause a major reduction of EPSPs (by >50%). This procedure quite consistently prevented the development of any subsequent LTP.

Anoxia after the appearance of LTP. In CA1, N_2 was tested systematically after the onset of potentiation, at times ranging from 3–30 min post-tetanus. As can be seen in Fig. 8, the anoxic suppression of the EPSP was always followed (within 2–3 min) by full return of the EPSP amplitude to the *enhanced*, post-tetanic level. A comparison of data from seven such experiments in CA1 showed no significant difference between potentiated EPSPs, recorded pre-N_2 and post-N_2, over a wide range of stimulus intensities (Fig. 9). Similar results were obtained when population spikes enhanced by LTP were compared before and after anoxia.

Two further points are of interest: first, even spreading depression (evoked by anoxia) failed to abolish an established LTP; second, there was no indication of any *delayed* depression of responses over periods of 20–60 min following anoxia.

Anoxia and LTP in other hippocampal areas. Experiments were also made recording in CA3 or the dentate gyrus. As illustrated for the dentate gyrus in Fig. 7B, LTP of EPSP fields was readily evoked by similar tetanic stimulation. The EPSPs were suppressed by N_2, but when they returned, they regained their pre-anoxic enhanced amplitude and rate of rise, as in CA1.

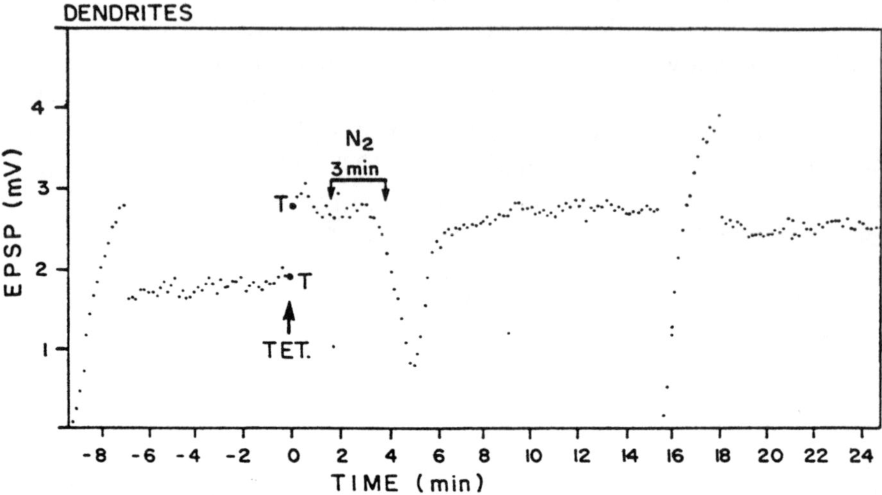

Figure 8. Dot diagram plots time-course of LTP and the effect of anoxia. Each dot represents peak amplitude of EPSP field in the CA1 apical dendritic layer. The intensity of stratum radiatum stimulation remained constant, except when it was increased in regular steps to obtain data for EPSP: stimulus intensity curves (cf. Fig. 9). Two 1 sec 100 Hz tetani elicited a typical LTP. Anoxia for 3 min, applied between 2 and 5 min post-tetanus, caused a temporary suppression of EPSPs but with subsequent full return of LTP.

Discussion

In view of the vast amount of experimentation being carried out on LTP, one needs to know to what degree LTP in slices conforms to various features of learning and memory as expressed in behavioral studies, preferably in the same species. One phenomenon of relevance is the apparent retroactive amnesia produced by anoxia and ECS—the latter, however, seems to be caused mainly by the accompanying anoxia rather than by ECS itself, being absent when anoxia is prevented by appropriate treatment (D'Andrea and Kesner, 1973).

Post-anoxic retroactive amnesia has some curious properties. It appears only several minutes after the end of anoxia; and, though quite long-lasting, it is not permanent being fully reversible after an hour or so. Moreover, it is seen only if anoxia occurs within a short time frame, between 2 and 10 min after the end of a specific learning task. This indicates that "permanent" memory traces are firmly implanted at a very early stage. A retrieval mechanism ("working" or "intermediate" memory) develops after the end of the initial "short-term" memory that does not outlive immediate awareness (Gibbs and Ng, 1976; Olton et al.,

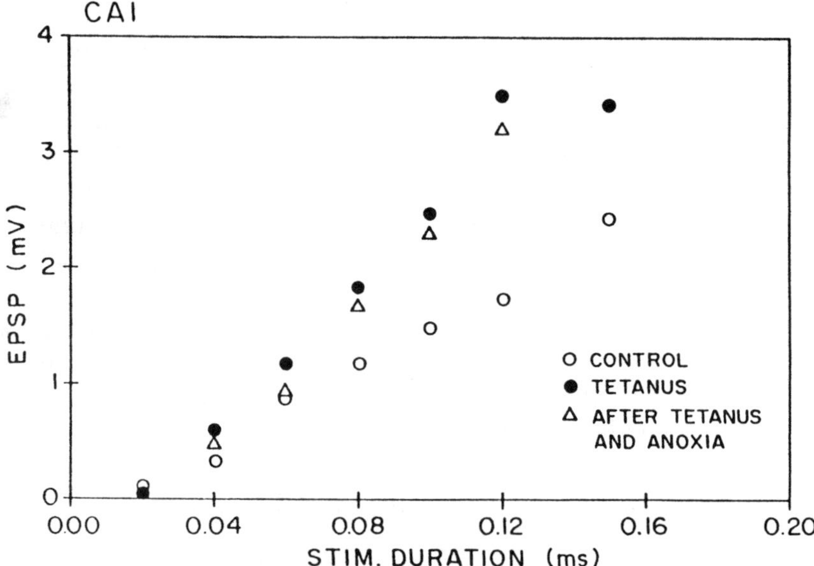

Figure 9. LTP of EPSP field is evident over a wide range of stimulus intensities (cf. open and closed circles, recorded before and after usual tetanic stimulation). Triangles show little significant reduction of EPSPs after recovery from anoxic suppression (as in Fig. 8). Stimulation intensity was varied by changing the duration of stimulating pulses of constant current.

1979; Rawlins, 1985). During its early phase, the "working" memory is evidently sensitive to anoxia.

How this disruption operates can only be guessed at. The block of EPSPs produced by anoxia is quite quickly reversible. This appears *not* to be the case with IPSPs or postsynaptic Ca currents, both of which show very early suppression during anoxia, as well as much slower post-anoxic recovery than do EPSPs or such membrane properties as resting potential, resistance and excitability (Leblond and Krnjević, 1989). A prolonged inactivation of IPSPs and/or postsynaptic I_{Ca} might therefore disrupt the mechanism by which memories are *retrieved*. Inhibition plays a vital role in the handling of specific information in the CNS (Ito, 1976): it is perhaps equally important for the recall of such information from memory. Similarly, Ca-influx may be crucial not only for the induction of LTP—and presumably the laying down of memory—but also for the mechanism of memory read-out.

The *delayed* character of retroactive amnesia, however, requires some additional (or alternative) effect of anoxia. The easiest way of accounting for it would be through a more complex chain of intracellular events, presumably 16 mediated by second messengers activating (or inactivating) protein phosphorylation, capable of changing significantly the permeability or voltage-dependence of ionic channels, the affinity or efficacy of transmitter receptors, etc. An obvious trigger would be a rise in cytosolic [Ca^{2+}], which is likely to be largely responsible for at least some of the cellular manifestations of anoxia (Duchen et al., 1990; Krnjević and Xu, 1989; Leblond and Krnjević, 1989).

At any rate, our observation that anoxia caused neither an immediate nor a delayed disruption of established potentiation suggests that, in so far as LTP in hippocampal slices truly reflects a process involved in learning and memory (McNaughton et al., 1986), this is more likely to be a "permanent" memory, or one that would not be disrupted by anoxia—rather than the retrieval mechanism ("working" memory), which *is* susceptible to temporary post-anoxic inactivation (retroactive amnesia).

Acknowledgments. This research was financially supported by the Medical Research Council of Canada.

References

Allweis C, Gibbs ME, Ng T, Hodge RJ (1984): Effects of hypoxia on memory consolidation: implications for a multistage model of memory. *Behav Brain Res* 11:117–121

Ben-Ari Y (1990): Galanin and glibenclamide modulate the anoxic release of glutamate in rat CA3 hippocampal neurons. *Eur J Neurosci* 2:62–68

Ben-Ari Y, Krnjević K, Crepel V (1990): Activators of ATP-sensitive K^+ channels reduce anoxic depolarization in CA3 hippocampal neurons. *Neurosci* 37:55–60

Berne RM, Rubio R, Curnish R (1974): Release of adenosine from ischemic brain. Effect on cerebral vascular resistance and incorporation into cerebral adenine nucleotides. *Circ Res* 35:262–271

Bert P (1878): *La pression barométrique.* Paris: Masson

Bliss TVP, Lømo T (1973): Long lasting potentiation of synaptic transmission in the dentate area of the anaesthetised rabbit following stimulation of the perforant path. *J Physiol* 232:331–356

Bures J, Buresova O, Krivanek J (1974): *The Mechanism and Applications of Leao's Spreading Depression of Electroencephalographic Activity.* Prague: Academia Publishing House of the Czechoslovak Academy of Sciences

D'Andrea JA, Kesner RP (1973): The effects of ECS and hypoxia on information retrieval. *Physiology and Behavior* 11: 747–752

Davies SN, Lester RAJ, Reymann KG, Collingridge GL (1989): Temporally distinct pre- and postsynaptic mechanisms maintain long-term potentiation. *Nature* 338:500–552

Dingledine R (1984): *Brain Slices.* New York: Plenum Press

Duchen MR, Valdeolmillos M, O'Neill SC, Eisner DA (1990): Effects of metabolic blockade on the regulation of intracellular calcium in dissociated mouse sensory neurones. *J Physiol* 424:411–426

Eccles JC (1964): *The Physiology of Synapses.* Berlin: Springer-Verlag

Eckert R, Chad JE (1984): Inactivation of Ca channels. *Progress Biophys Molec Biol* 44:215–267

Ernsting J (1986): The effects of hypoxia upon human performance and the electroencephalogram. In: *Oxygen Measurements in Blood and Tissues*, Payne JP and Hill DW, eds. London: J and A Churchill Ltd, pp. 245–259

Frieder B, Allweis C (1982): Delayed post hypoxic transient amnesia is not associated with electrical brain seizures. *Physiology and Behavior* 1929:1059–1064

Fujiwara N, Higashi H, Shimoji K, Yoshimura M (1987): Effects of hypoxia on rat hippocampal neurones *in vitro. J Physiol* 384:131–151

Gibbs ME, Ng KT (1976): Memory formation: A new three-phase model. *Neurosci Letts* 2:165–169

Haldane JS, Priestley JG (1935): *Respiration.* New Haven: Yale University Press

Hansen AJ, Hounsgaard J, Jahnsen H (1982): Anoxia increases potassium conductance in hippocampal nerve cells. *Acta Physiol Scand* 115:301–310

Hochachka PW (1986): Defense strategies against hypoxia and hypothermia. *Science* 231:234–241

Ito M (1976): Roles of GABA neurons in integrated functions of the vertebrate CNS. In: *GABA in Nervous System Function*, Roberts E, Chase TN, Tower DB, eds. pp. 427–448

Krnjević K, Ben-Ari Y (1989): Anoxic changes in dentate granule cells. *Neuroscience Letts* 107:89–93

Krnjević K, Cherubini E, Ben-Ari Y, (1989): Anoxia on slow inward currents of immature hippocampal neurons. *J Neurophysiol* 62:896–906.24

Krnjević K, Leblond JE (1987): Anoxia reversibly suppresses neuronal 20 calcium currents in rat hippocampal slices. *Can J Physiol Pharmacol* 65:2157–2161

Krnjević K, Leblond J (1989): Changes in membrane currents of hippocampal neurons evoked by brief anoxia. *J Neurophysiol* 62:15–30

Krnjević K, Leblond J (1988): Are there hippocampal ATP-sensitive K-channels that are activated by anoxia? *Pflugers Archiv Supp* 411:R145.

Krnjević K, Xu Y (1989): Dantrolene suppresses the hyperpolarization or outward current observed during anoxia in hippocampal neurons. *Can J Physiol Pharmacol* 67:1602–1604

Leao AAP (1944): Spreading depression of activity in the cerebral cortex. *J Neurophysiol* : 359–390

Leblond J (1988): Anoxia and long-term potentiation (LTP) in the CA1 field of hippocampal slices of rats. *J Physiol* 406:38P

Leblond J, Krnjević K (1989): Hypoxic changes in hippocampal neurons. *J Neurophysiol* 62:1–14

Lynch G, Baudry M (1984): The biochemistry of memory: a new and specific hypothesis. *Science* 224:1057–1063

Malenka RC, Kauer JA, Perkel DJ, Nicoll RA (1989) The impact of postsynaptic calcium on synaptic transmission—Its role in long-term potentiation. *TINS* 12:444–450

Malinow R, Schulman H, Tsien RW (1989): Inhibition of postsynaptic PKC or CaMKII blocks induction but not expression of LTP *Science* 245:862–866

Martin MR (1983): Naloxone and long term potentiation of hippocampal CA3 field potentials *in vitro. Neuropeptides* 4:45–50

McNaughton BL, Barnes CA, Rao G, Baldwin J, Rasmussen M (1986): Long-term enhancement of hippocampal synaptic transmission and the acquisition of spatial information. *J Neurosci* 6:563–571

Miller RJ (1990): Glucose regulated potassium channels are sweet news for neurobiologists. *TINS* 13:197–199

Misgeld U, Frotscher M (1982): Dependence of the viability of neurons in hippocampal slices on oxygen supply. *Brain Res Bull* 8:95–100

Nicholls DG (1989): Release of glutamate, aspartate, and y-aminobutyric acid from isolated nerve terminals. *J Neurochem* 52:331–341

O'Keefe J, Nadel L (1978): *The Hippocampus as a Cognitive Map.* Oxford University Press

Olton DS, Becker JT, Handelmann GE (1979): Hippocampus, space and memory. *The Behavioral and Brain Sciences* 2:313–365

Rawlins JNP (1985): Associations across time: The hippocampus as a temporary memory store. *The Behavioral and Brain Sciences* 8:479–496

Rossen R, Kabat H, Anderson JP (1943): Acute arrest of cerebral circulation in man. *Arch Neurol Psychiat* 50:510–528

Squire LR (1987): *Memory and Brain.* New York: Oxford University Press

Sugar O, Gerard RW (1938) Anoxia and brain potentials. *J Neurophysiol* 1:558–572

Teyler TJ, DiScenna (1987): Long-term potentiation. *Ann Rev Neurosci* 10: 131–161

Xu Y, Krnjević K (1990) Anoxia does not abolish LTP. *Soc Neurosci Abstr* 16:980

Brain before Cognition: EEG Maturation in Preterm Infants

Giovanni Cioni, Enrico Biagioni, and Carlo Cipolloni

Spontaneous electrical activity of the brain in humans can be easily recorded from the scalp. The recording reflects the neurophysiological activity of many neurons located under the electrodes and they change according to neural level of the activity.

Electroencephalography (EEG) has been widely applied in the assessment of brain functioning. Berger (1929) first noted that an 8–12 Hz "alpha" rhythm can be recorded at rest in the occipital regions. This rhythm disappears or is greatly suppressed while the subject opens the eyes or performs a mental task. Since then, EEG, and in particular changes in alpha rhythm, assessed visually or more recently by means of computer analysis (Nuwer, 1988), has been intensively used as a measure of attention in adults and in children.

Alpha rhythm has become also a measure of brain maturation. In fact, a linear correlation has been found between the EEG frequency obtained from the occipital region, while closing the eyes, and age: it appears slower in infancy and reaches adult frequency at around adolescence.

In adults, specific EEG patterns are highly correlated to wakefulness and sleep states. This correlation develops gradually in the ontogeny of the brain; the differentiation of sleep-related background EEG activity and the appearance of specific EEG features, as sleep spindles (at 11–15 Hz) in quiet sleep, are other EEG indexes of brain maturation often used in research. In the neonatal and early postnatal period rapid changes are observable both in the brain and in behavior and, from the first recording, neurologists were very interested in looking at the EEG during this early period of human life.

Berger himself (1932) recorded an EEG from a 35-day-old infant. However, because of the high number of artifacts in neonatal EEG, the lack of an alpha rhythm at that age, and the short duration (15–20 min) of routine recordings, the EEG of the neonate was considered for years a chaotic phenomenon, with no changes between wakefulness and sleep. In the 1950s, thanks to the French school in Paris, systematic investigations of EEG in preterm and full-term infants began (Dreyfus-Brisac et al., 1955). Since then, several studies have described the changes on EEG patterns from 24–26 weeks of postmenstrual age (limit of viability of a premature) to term. Long duration EEGs and simultaneous recording of other physiological variables, such as heart rate, eye and body movements, and respiration have permitted the recognition of the organization of behavioral states in the preterm period and of EEG configuration related to these states.

The scenario that emerges from these studies is very impressive. In a very short time, such as the last 2–3 months of gestation, EEG undergoes faster and more consistent changes than in any other period of life. These changes are relatively independent of environmental stimuli, and EEG seems to develop in parallel in the fetus and in preterm infants. In fact, no systematic differences have been found in the EEG of preterm infants of the same postmenstrual (though of different gestational and postnatal) age. Therefore, prematurely born infants provide a unique opportunity for studying early ontogeny of brain electrical activity. In the last few years more strict selection of preterm infants showing no brain lesion, now detectable by means of brain imaging techniques (ultrasound, CT, MRI), and new methods of EEG recording and analyzing, have provided new findings on EEG maturation in low-risk preterm infants.

In order to illustrate suitable data for comparison with anatomical and behavioral findings in the same period of life, in this chapter literature and personal studies on the main maturational features of EEG in preterm infants will be reviewed.

EEG and Behavioral State Organization

EEG and sleep are highly correlated and, therefore, the interpretation of an EEG during infancy, and particularly in the neonatal period, requires a good knowledge of the ontogeny of sleep behavior. A full description of the development of sleep and wakefulness in infancy is beyond the aim of this chapter, but a brief review on the characteristics of behavioral states in the newborn is necessary before describing the maturation of EEG.

In full-term infants five distinct behavioral states—quiet sleep, active sleep, quiet wakefulness, active wakefulness, and crying (Prechtl, 1974)—are observable. They are easily detectable by direct observation or polygraphic recordings. Besides eye and body movements, respiration, and other variables, the type of EEG is one of the vectors for behavioral state definition, and states of wakefulness and sleep are strong determinants of EEG patterns. Polygraphic recordings, including EEG, EKG, respiration, eye movements, and EMG are mandatory for a proper evaluation of the EEG.

Behavioral states gradually emerge in the last three months of gestation, very similarly in the fetus and in preterm infants. The first ultradian rhythmic activities are distinguishable after 30 weeks of postmenstrual age (brain rest–activity cycles of Kleitman, 1963). In both extra- and intrauterine conditions, periods of activity become gradually synchronized with the other physiological variables. After 36 weeks, fetuses (Nijhuis et al., 1982) and preterm infants (Prechtl et al., 1979) show distinct behavioral states similar to those of full-term infants.

Again, a tremendous change is observable in sleep organization and in its relationship with the EEG in a very short period of time. From 28 to 40 weeks not only an increasing number of EEG patterns are observable (see below), but also the association between specific EEG patterns and the emergent behavioral states gradually increases. At 28 weeks the infant is permanently in an indeterminate state and only one type of EEG can be recorded. At term, 3 months later, the full range of behavioral states is present: two distinct EEG pattern are associated with quiet sleep, two with active sleep, and one with wakefulness.

In a series of studies (Curzi-Dascalova et al., 1985; Karch et al., 1982; Parmelee et al., 1968; Prechtl et al., 1979), the time of rising of behavioral states, the relative proportion of each state at a given postmenstrual age, and the degree of concordance between state, EEG, and other state criteria have been investigated. The data obtained were often contradictory. The criteria selected by the authors and the different techniques of electrophysiological recording and analyzing probably account for the different results.

Similarity predominates in the development of states and state-related EEG patterns in the fetus and in preterm infants. However, some authors have pointed out small differences between full-term infants and prematures reaching term age. But these differences, i.e., more mature or immature patterns in preterm infants, are not univocal. Dreyfus-Brisac (1975) states that sleep cycles are better organized in full-term infants. On the contrary, slightly more mature EEG patterns than would be expected from their postmenstrual age (PMA) have been reported in preterm infants with a long extrauterine development (Nolte and Haas, 1978).

EEG Maturation: An Overview

Spontaneous brain electrical activity largely precedes any organization of ultradian cyclicity in the neural activity. First electric waves in human embryos are detectable in the pons at 60 days of postmenstrual age. Around the fourth month of gestation some discontinuous activity appears in the hippocampus and in the diencephalon (Bergstrom, 1969). At 19 weeks of PMA it has been possible to register from the cerebral cortex of a fetus some very low deflections of the isoelectric line (Engel, 1975). Dreyfus-Brisac (1968) recorded a discontinuous high-voltage pattern characterized by attenuated periods lasting between 10 sec and 3 min in preterm infants of 24–27 weeks surviving for a few days. In the near future, improvement of neonatal intensive care will probably provide very low gestational age infants showing no cerebral damage and a normal neurological outcome. When their number is sufficiently high, a full description of "normal" brain electric activity in very early prematurity will be possible. After 27 weeks of PMA normal EEG features are well known. Examples of tracings from 28 to 40 weeks of low-risk preterm infants from our sample are produced in Fig. 1.

At 28 weeks, cerebral electric activity is characterized by a constantly discontinuous tracing: "bursts," often occurring synchronously on the two hemispheres and constituted by high voltage slow waves, alternate with attenuated periods lasting up to 40–60 sec. Within the bursts, in temporal regions, a rhythmic high voltage 4–6 Hz activity generally known as "temporal sawtooth" is detectable. Medium amplitude faster activities (8–20 Hz) are also present.

Two weeks later (30 weeks), temporal sawtooth is well represented but rapid activity (8–20 Hz) increases in incidence and amplitude. It generally occurs superimposed on slow waves: this pattern, more evident on temporal and occipital regions, is called "delta brush." EEG is still constantly discontinuous but, sometimes, periods characterized by longer bursts and shorter intervals are observable.

At 32 weeks, a continuous tracing first appears although the activity is sometimes interrupted by short low-voltage intervals ("bursting tendency"). Background activity is constituted by medium- to high-voltage delta brushes. This continuous pattern is more frequently detectable during phases in which spontaneous motility is abundant; periods of coincidence of sleep state parameters begin to be present. During rest phases

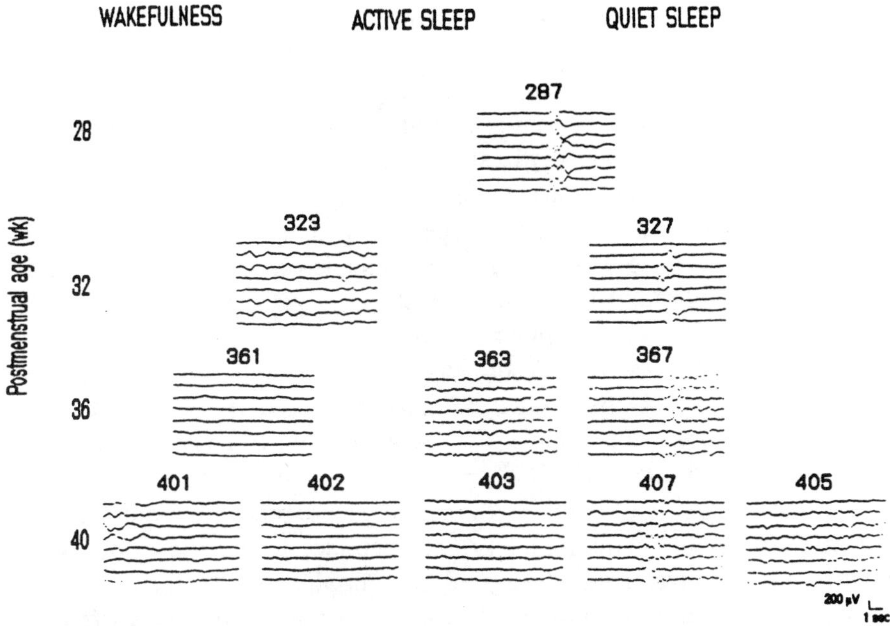

Figure 1. Examples of EEG patterns typical of the different behavioral states for the different postmenstrual ages (PMA) from 28 to 40 weeks. The code number of each EEG indicates: the first two digits the PMA in weeks, the third one the basic features of the tracing. Bipolar leads: Fp2–C4, C4–O2, O2–T4, T4–Fp2, Fp1–C3, C3–O1, O1–T3, T3–Fp1.

tracing is often discontinuous; bursts, characterized by high voltage fast activity delta brushes, are not rarely asynchronous. Temporal theta is still present but appears reduced in incidence and amplitude.

At 34 weeks, continuous tracing is well organized and bursting tendency is no longer observable. Sleep states are more differentiated: periods of concordance between EEG and other parameters (body and eye movements, regularity of respiration, etc.) are more often present. Usually, continuous tracing is detectable in active sleep and discontinuous EEG in quiet sleep. In discontinuous tracing, bursts are generally synchronous and constituted by delta brushes including high-amplitude sharp rapid activities. Temporal sawtooth almost completely disappears. In active sleep either slow waves or fast rhythms are lower in amplitude.

At 36 weeks, as previously mentioned, behavioral states are well differentiated. Temporal sawtooth is generally no longer observable. Both in active and in quiet sleep, slow waves, but also superimposed rapid activity, decrease in amplitude and are more evident on occipital areas. Often, bilaterally synchronous sharp transients, known as "encoches frontales," appear in frontal regions. During wakefulness and the phase of active sleep following quiet sleep, a medium- to low-voltage pattern constituted by delta and theta activity superimposed by very low amplitude rapid rhythms is observable. A certain reactivity of EEG (generally a transient flattening) to sudden nonspecific stimuli is evident.

At 38 weeks, in the phase of active sleep preceding quiet sleep, a continuous tracing characterized by delta and theta activities, rarely superimposed by very low voltage fast rhythms, is observable. In wakefulness and in the phase of active sleep, following quiet sleep rapid activity almost disappears. In this phase of active sleep theta frequencies are predominant, whereas in wakefulness theta and delta are equally represented. In quiet sleep, tracing is discontinuous and characterized by generally synchronous bursts formed by slow waves. They are superimposed by fast activities, which appear less sharp and lower in amplitude.

At 40 weeks, in the phase of active sleep preceding quiet sleep, fast rhythms are almost completely undetectable. In the phase following quiet sleep, theta frequencies are predominant and some trains of 4–5 Hz rhythmic waves are observable. In quiet sleep a discontinuous tracing is still present, but also a high voltage slow continuous pattern appears. Wakefulness phases are characterized by a medium- to low-voltage delta-theta mixed-activity tracing.

During the first month after term, discontinuous tracing disappears. Rapid rhythms superimposed on slow waves (delta brushes), only present in quiet sleep at term age, are no longer detectable. Also "encoches frontales" and 4–5 Hz rhythmic wave trains disappear before 44 weeks of postmenstrual age. At that time, in wakefulness and in the phase of active sleep following quiet sleep, a predominant theta low-voltage tracing is observable. The phase of active sleep preceding quiet sleep is characterized by medium-voltage theta-delta mixed activity: this pattern becomes rarer because infants fall asleep more frequently in quiet sleep. This last state is characterized by a high voltage slow pattern, which precedes deep sleep tracing of older children and adults. In the same period, low voltage often asynchronous 11–15 Hz activity, antecedent of sleep spindles (called "prespindles"), appear on rolandic regions.

Data reported above show the complexity and the rapidity of change of neonatal brain electric activity. Trained observers can rather easily assess the maturational age of an EEG by means of a "gestaltic" observation of the tracing. This global appreciation is due to the detection of various specific EEG features, which gradually change in relation to postmenstrual age. The developmental trend of the main maturational features, already mentioned in the previous description, will be reported in more detail in the following paragraph. In particular, discontinuity of tracing, synchrony of bursts during discontinuous activity, appearance and disappearance of waveforms as temporal theta, delta brushes, and "encoches frontales," will be reported.

Developmental Trends of Specific EEG Features

Discontinuity. In early prematurity, as already reported, EEG is characterized by constantly discontinuous activity. Bursts alternate with long intervals (even more than 30 sec) of attenuated activity. Approaching term age, discontinuous tracing is detectable only during quiet sleep, and intervals appear much shorter (generally less than 10 sec). Then both percentage of discontinuous activity and interval length decrease with increasing postmenstrual age.

Some authors have thoroughly studied the limits in which these parameters vary at different ages. Connell et al. (1987) carried out very long (24 hours) two-lead cassette recordings within the first days of life in low-risk preterm infants. The authors divided recordings into five-minute epochs, which were allocated to one of the following five categories of continuity or discontinuity: (1) continuous: complete continuity of activity throughout the epoch; (2) mixed, predominantly continuous: continuous activity for at least 80% of the epoch but with at least one interval of activity attenuated to 10–20 μV for at least five seconds; (3) mixed: both types of activity with no clear predominance of one over the other; (4) mixed, predominantly discontinuous: mainly discontinuous with at least 20% but less than 50% of continuous activity; (5) discontinuous: activity for less than 20% of the epoch.

Continuous activity represents 10% of total recording time at 26 weeks and it reaches over 70% at 37 weeks (Fig. 2A). Discontinuous activity decreases from 32% (at 27 weeks) to 11% (at 39 weeks) (Fig. 2B). Both mixed continuous and mixed discontinuous activities decrease from around 30% (at 27 weeks) to less than 10% (at 39 weeks). Mixed activity incidence is always low but it generally decreases with increasing postmenstrual age.

The authors described also the length of uninterrupted runs of completely continuous and completely discontinuous activities. The duration of runs of completely continuous activity increases from less than 10 min to 1 hr. The length of runs of completely discontinuous activity remains constant between 5 and 15 min. Mean interval duration falls from 14 to 8 sec between 26 and 37 weeks (Fig. 2C); maximum interval duration, in the same period, falls from 75 to 15 sec (Fig. 2D).

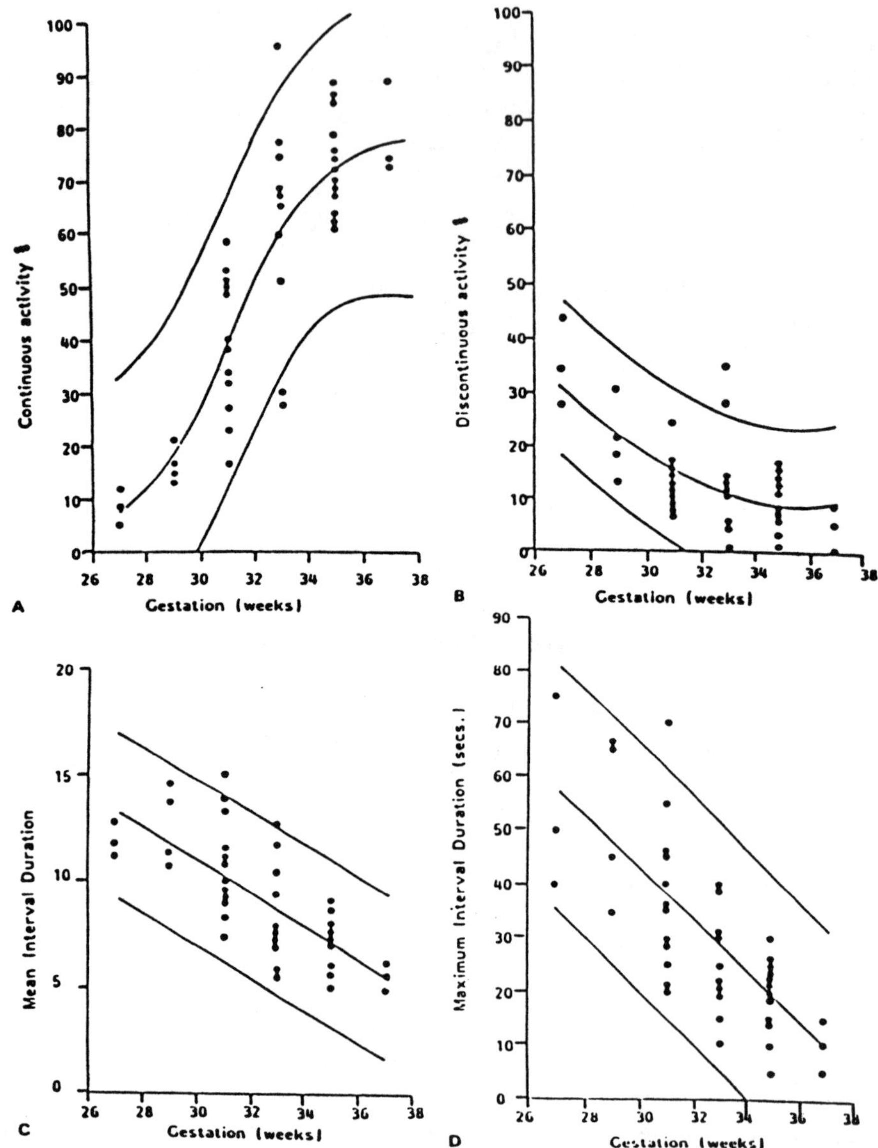

Figure 2. Quantification of EEG maturation from 26 to 38 weeks gestation derived from 24 hour Medilog recordings in 44 low-risk infants: (A) increases in total continuous activity; (B) decrease in total discontinuous activity; (C) decrease in mean interburst interval duration; (D) decrease in maximum interburst interval duration (mean +/– 250) (reproduced with permission from Connell and Oozeer, 1988).

Hahn et al. (1989) measured interburst interval duration in one-hour recordings of infants with normal neurological outcome. They defined two types of intervals: type 1 (intervals lasting more than 2 sec containing no cerebral activity >15 μV) and type 2 (intervals lasting more than 2 sec containing a single transient >15 μV at 1–2 electrodes or continuous cerebral activity >15 μV in a single electrode). The restrictive criteria used to label type 1 intervals justify the short durations reported for this pattern (less than 14 sec at 26–27 weeks and less than 4 sec at 38–40 weeks). Values recorded for type 2 interval are higher (less than 20 sec at 26–29 weeks and less than 8 at 38–40). According to these authors, continuous tracing appears after 30 weeks; its percentage increases from 20% (30–31 weeks) to 70% (38–40 weeks).

Synchrony. During discontinuous activity bursts may occur synchronously on the two hemispheres or there may be a delay on one side. Lombroso (1979) measured the incidence of synchronous bursts during 5 min of discontinuous tracing in 10 healthy newborns at different postmenstrual ages. Bursts were considered synchronous if one hemisphere preceded the burst of the opposite one by less than 1.5 sec and their duration was approximately equal (discrepancies shorter than 2 sec were ignored). Percentage of bursts satisfying these criteria was calculated at different postmenstrual ages. The study indicated that synchrony increases from values of 50–60% (at 31–32 weeks) to 100% (at term age).

Anderson et al. (1985) stressed the presence of a high incidence of synchronous bursts (more than 90%) even in early prematurity. These authors defined the onset of a burst asynchronous when it precedes that on the opposite hemisphere by 2 sec or more. Other authors (Ferrari et al., 1990) applied Lombroso's criteria in another group of low-risk preterm infants. Synchrony range was 57–85% at 31–32 weeks and 69–88% at term. These changes were not significant.

Temporal theta. As already reported, this EEG pattern appears on temporal regions in early prematurity. It consists of high amplitude bursts of rhythmic theta activity, frequently synchronous on the two hemispheres (Fig. 3). It is generally known as "temporal sawtooth" (Werner et al., 1977).

Hughes et al. (1987), studying a large population of low- and high-risk infants, stated that this pattern occurs very early and reaches its maximum representation at 29–31 weeks of PMA. Afterwards it gradually diminishes and disappears before term. Theta frequency tends to increase and amplitude to decrease with increasing postmenstrual age. A slight right predominance is also reported.

Delta brushes. This pattern consists of 8 to 20 Hz activity superimposed on slow waves; it is detectable even in early prematurity but becomes

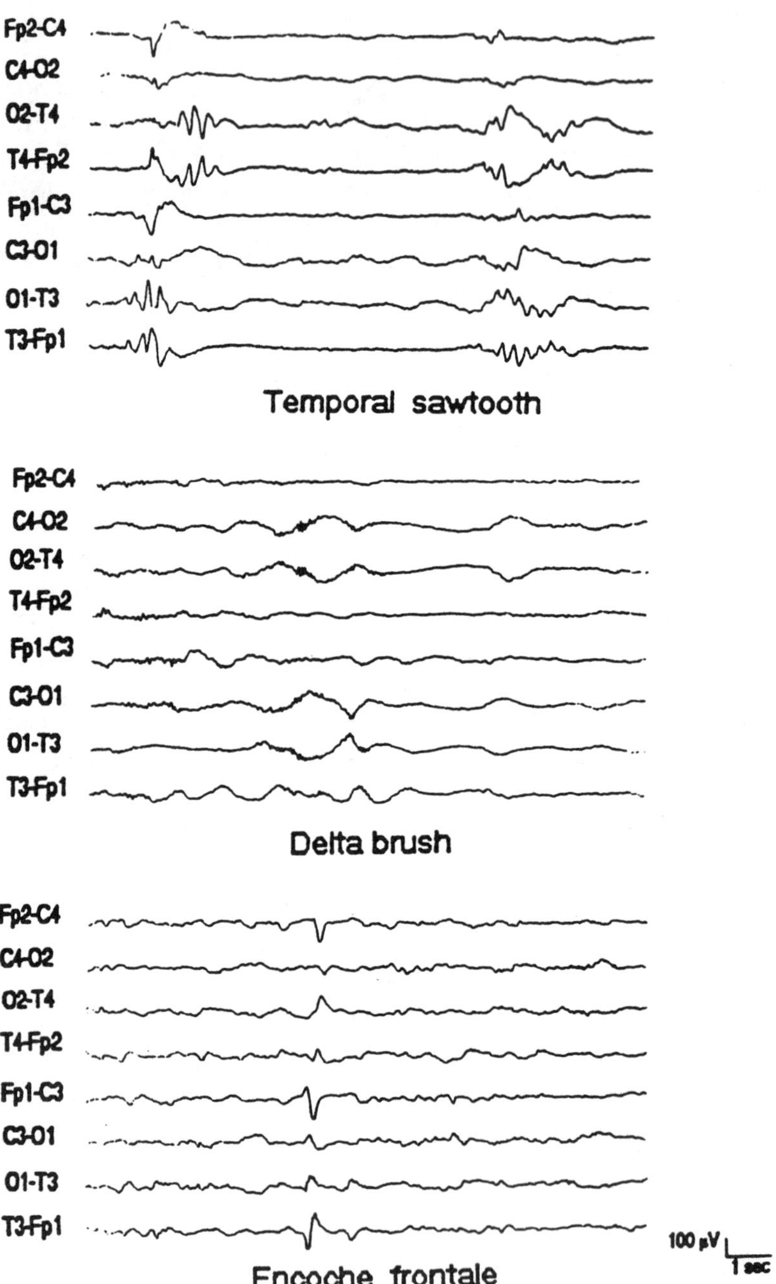

Figure 3. Examples of waveforms typical of neonatal EEG: temporal sawtooth, delta brush, *encoche frontale*. They were recorded in low-risk newborn infants of 28, 34, and 40 weeks of PMA, respectively.

more evident after 29–30 weeks of PMA (Fig. 3). Approaching term age it diminishes first in active sleep and then in quiet sleep. Around 40 weeks rapid rhythms are observable only during bursts of discontinuous tracing.

Lombroso (1979) counted the number of delta brushes during 5 min of quiet sleep and 5 min of active sleep in 10 healthy preterm infants at different postmenstrual ages. This author defined "delta brush" as a 0.8–1.5 Hz, 50–200 μV-high wave lasting 0.2–1.2 sec and superimposed by 8–22 Hz, 20–150 μV activities. The incidence of this finding decreases in quiet sleep from 19–49/5 min at 31–32 weeks to 2–6/5 min at term age and disappears at around 42 weeks. In active sleep it is 30–57/5 min at 31–32 weeks and 1–3 at 37–38. Delta brushes are almost completely undetectable in active sleep after 38 weeks. Ferrari et al. (1990) replicated Lombroso's research but with quite different results. According to these authors, the number of detectable delta brushes did not change significantly until 42 weeks of PMA. In quiet sleep, a rate of 16–20 delta brushes per 5 min at 31–32 weeks and a rate of 13–25/5 min at 40–41 weeks were reported. This waveform was undetectable at 42–43 weeks.

We studied the development of delta brushes by measuring maximum amplitude of both delta and rapid activities at different postmenstrual ages during continuous and discontinuous tracing in 67 low-risk newborns. To be included in this low-risk group, infants had to fulfil the following criteria: (1) gestational age reliably obtained from mother's last menstrual date; (2) no cerebral lesions detected by ultrasound scan except for intraventicular hemorrhage below grade three of Papile et al. (1978) or periventricular increased echogenicity lasting less than 10 days; (3) normal neurological follow-up until an age of at least six months after term.

Recordings (lasting 90–120 min) have been executed by means of 8-lead cassette electroencephalographs ("Brain-Spy" by Micromed Company–Italy) within the first two weeks of life. Electrodes were placed on Fp1–2, C3–4, O1–2, and T3–4 positions of the International 10–20 System. Results are shown in Fig. 4 and Fig. 5.

During discontinuous pattern maximum delta amplitude is high (near 400 μV) before 31 weeks and then remains around values of 200–250 μV till term age (Fig. 4, above). Rapid activities appear low (around 100 μV) at 27–30 weeks, then increase (up to 200 μV) at 31–34 weeks and finally decrease again approaching term age (60–90 μV) (Fig. 4, below).

Data concerning continuous patterns are available only after 30 weeks of PMA. Among the different types of continuous EEG, medium-high voltage pattern detectable during the phase of active sleep preceding

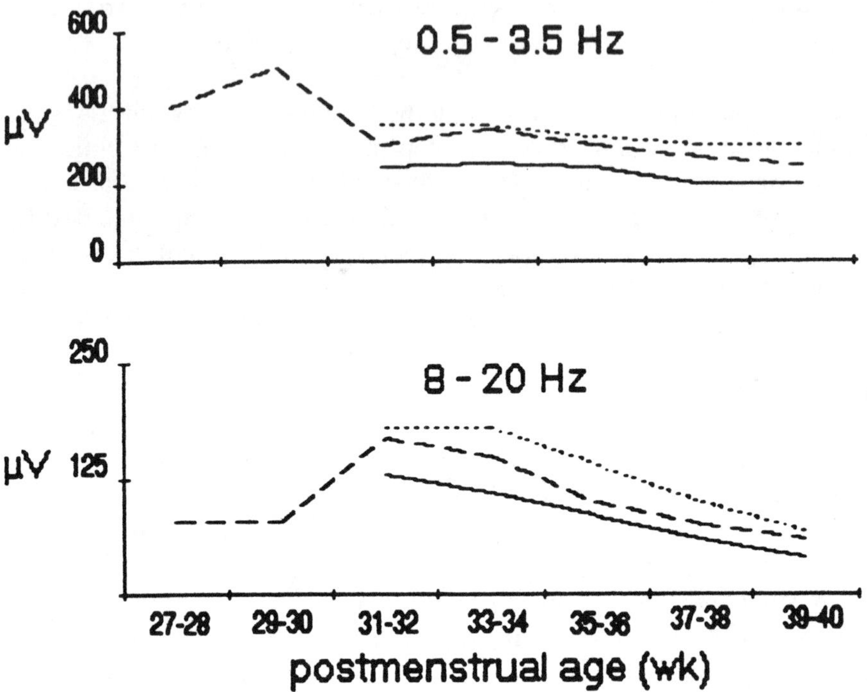

Figure 4. Maximum delta and rapid activity amplitude during discontinuous tracing from 28 to 40 weeks of PMA in low-risk preterm infants. Median values and, when allowed by the number of subjects, 25 and 75 percentiles are shown.

quiet sleep was selected. In fact this EEG pattern is always present from 31 to 40 weeks. Within this tracing both delta and rapid maximum amplitudes decrease from 31 weeks to term age. Delta values are around 200–250 µV at 31–32 weeks and under 150 µV at term (Fig. 5, above). Rapid activity falls from around 80 µV to around 40 µV at 37–38 (Fig. 5, below). After 38 weeks fast rhythms in active sleep appear rare and low in amplitude (less than 40 µV).

These data permit a more refined description of delta brush pattern from early prematurity to term age. Changes of this waveform appear not only related to the time of its appearance and disappearance, but also to a modification of its features and, in particular, to the amplitude and the sharpness of superimposed rapid activities.

Encoches frontales. This finding is also known as "frontal sharp transient" (Kellaway and Crawley, 1964), but the French diction is usually conserved in homage to the Parisian School of Dreyfus-Brisac and Monod

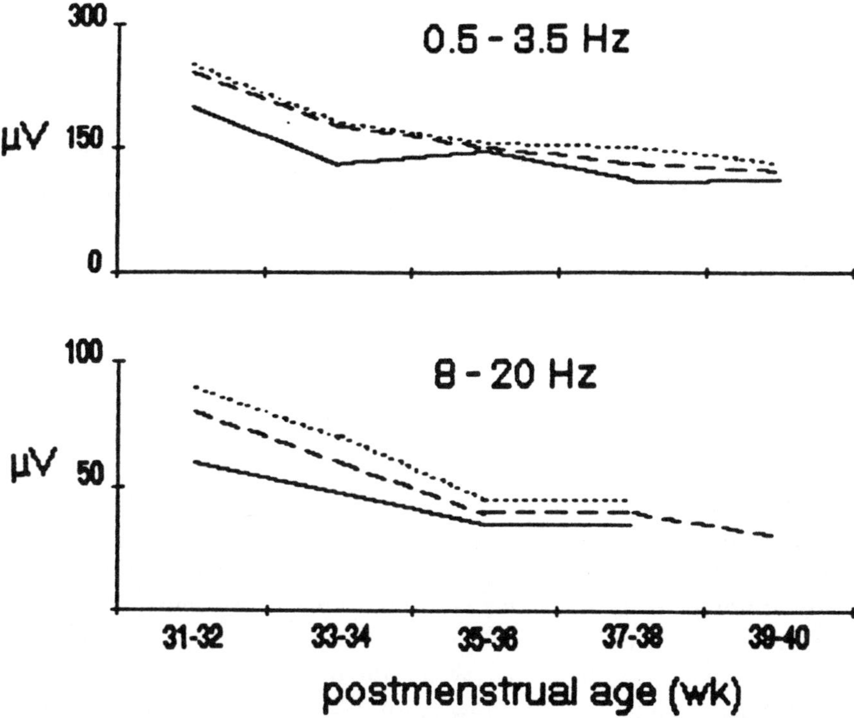

Figure 5. Maximum delta and rapid activity amplitude during continuous tracing from 31 to 40 weeks of PMA in low-risk preterm infants. Median values and, when allowed by the number of subjects, 25 and 75 percentiles are shown.

who described it more than 30 years ago (Monod et al., 1960). It consists of a sharp, often bilaterally synchronous, biphasic graphoelement, lasting up to 250 msec. It is localized in frontal regions but often spreads posteriorly (Statz and Dumermuth, 1982) (Fig. 3). Frequently it looks like a brief and low-in-amplitude initial negative phase followed by a higher-in-amplitude and longer secondary positive phase (Arfel et al., 1977). A degraded form, characterized by a blunt aspect with a more or less pronounced loss of the biphasic configuration, is also observable. In their typical aspect, *encoches frontales* appear around 34–35 weeks of PMA and disappear within the first month after term. Nevertheless, the occurrence of a blunt form of frontal sharp transient is reported also in early prematurity (Anderson et al., 1985). The range of incidence of this transient is generally wide (from 1 per 50 sec to 1 per 20 min in full-term babies in quiet sleep). It is less frequent in active sleep, particularly in that phase preceding wakefulness (Statz and Dumermuth, 1982).

EEG Maturational Codes

None of the neonatal EEG features described above seems to be able, per se, to provide sufficient information for determining postmenstrual age of an infant. In other words, no single quantitative parameter (e.g., interval duration, percentage of synchronous bursts, number of delta brushes, etc.) allows us to define the electroencephalographic age of a neonatal tracing. This is probably due to the high range of interindividual variability of some parameters and to the small maturational changes of others. On the contrary, a global assessment of all parameters by means of a "gestaltic" perception of EEG maturational features permits a precise evaluation of electroencephalographic age.

Bioelectrical activities can also be quantitatively assessed by visual pattern analysis in consecutive epochs, using a system of EEG-type coding firstly described by Parmelee et al. (1968). This method is based on the definition of three digit "maturational codes" describing EEG patterns. The third digit defines basic features of the tracing. In particular "nn1" code indicates a medium–low amplitude tracing detectable during wakefulness, "nn2" code is a low amplitude tracing typical of active sleep, "nn3" is medium–high continuous tracing observable in active sleep, "nn5" is a high voltage slow pattern detectable during quiet sleep only at term age, and "nn7" represents discontinuous tracing (Nolte and Haas, 1978). The first two digits of the code indicate the maturational variations of the basic pattern type, expressed as the approximate postmenstrual age in weeks. For example, code "367" means a discontinuous tracing typical of 36 weeks of PMA. Parmelee et al. (1968) and, later, Nolte and Haas (1978) provided a description of the codes of 28, 32, 36, 40, and 44 weeks. At 28 weeks only the discontinuous pattern ("nn7") is present, at 32 weeks the pattern "nn3" appears; at 36 weeks the pattern "nn1" is also described, whereas at 40 and 44 weeks every code is represented. Consecutive epochs (generally of three minutes) can be assessed by this method. This approach allows a quantification for the age-specific patterns of the whole EEG record and of the EEG type in relation to behavioral states.

Reliability of this pattern classification has been tested (Parmelee et al., 1968). Serial polygraphic recordings of low-risk preterm infants have shown a high correlation between postmenstrual age and EEG maturation, assessed by means of EEG codes (Nolte and Haas, 1978). When sleep organization is well defined (from 36 weeks), basic types of EEG patterns are closely related to behavioral states. In particular, "nn5" and "nn7" patterns are observable in quiet sleep, "nn2" and "nn3" in active sleep, and "nn1" in wakefulness (or in the phase of active sleep following quiet sleep). EEG maturation (indicated by the first two digits of the code) is not state-dependent.

Computer Analysis of EEG Maturation

Visual analysis of neonatal EEG is very complex and very subjective, although reliability tests and trials of more quantitative assessment, such as the maturational codes described in the previous paragraph, have confirmed the value of the results provided by this approach. With the introduction of computers and the development of algorithms for the analysis of EEG signals new frontiers in assessing neonatal brain electric activity are now opening.

In the literature some reports on computer analysis of newborn EEG are already available. Neonatal tracings have been submitted to coherence function study (Kuks et al., 1988; Prechtl et al., 1974), to spectral analysis (Richards et al., 1986), to Berg-Fourier transformation (Aziz et al., 1986), and also total EEG power have been calculated (Ioffe and Chernick, 1990). In general, some difficulties have been encountered. First of all, it is often impossible to apply on a newborn's scalp as high a number of electrodes as required for computer EEG analysis in the adult, i.e., not less than 16 (Nuwer, 1988). Moreover, neonatal tracings are always affected with many artifacts due to spontaneous motility, to cardiac or respiratory activities, but also to the electric apparatus of neonatal intensive care units; therefore computer analysis, which needs long periods of artifact-free activity, may become impossible.

In our Institute, spectral analysis (by means of fast Fourier transformation, FFT) on low-risk preterm infants' EEGs at different postmenstrual ages has been performed. Tracings have been recorded using the same methodology (cassette EEG) described in a previous paragraph. They were affected with many artifacts due to movements, respiration, muscular activity, etc., but with no artifact due to electric apparatus (incubators, ventilators, etc.) because of battery power. The signal was transferred in a computerized EEG analyzer ("Brain Quick" by Micromed Company—Italy) where it was digitalized and mathematically transformed into common averaged-reference EEG.

Afterwards at least 40–50 artifact-free epochs (every epoch lasts two seconds) were visually selected. If a discontinuous tracing was examined, only "bursts" were submitted to analysis. After epoch selection, FFT was performed. In every tracing, each EEG pattern was studied separately; therefore a specific power spectrum was achieved from each sleep state.

The FFT spectrum represents the power on each electrode of selected frequency bands. Three bands concerning slow (0.5–3.5 Hz), medium (4–7.5 Hz), and rapid (8–20 Hz) frequencies have been defined. A computerized EEG analyzer produces a spectrum for each epoch submitted to FFT and then, by means of an averaging process, a medium spectrum of all epochs. An example of a power spectrum (absolute value) on the left temporal region from an EEG of a 28 week infant is shown in Fig. 6: frequency values are represented on the horizontal axis and the power on the vertical one. A high and large peak is observable on the delta frequency band, followed by a smaller peak on the theta band.

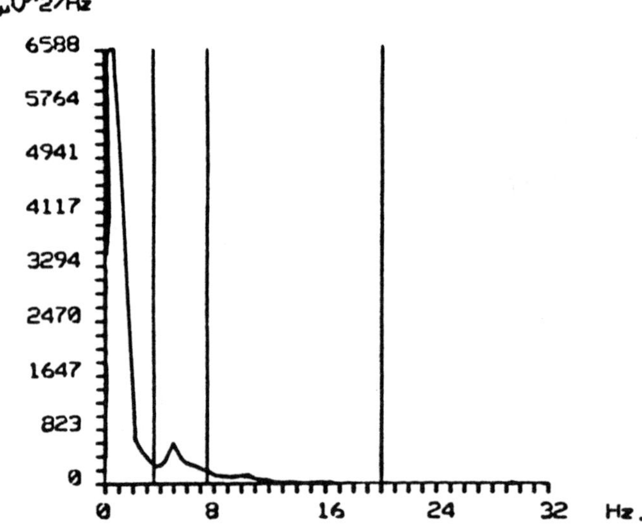

Figure 6. Power spectrum (absolute value) from left temporal region achieved in a low-risk preterm infant recorded at 28 weeks of PMA Data were obtained by fast Fourier transform of 40 two-second epochs of bursts in discontinuous tracing.

An EEG analyzer can also provide a topographic representation of the power of selected bands on the scalp. This analysis is known as Brain Electric Activity Mapping (BEAM). Several examples of BEAMs are shown in Fig. 7 and Fig. 8. Different values of power are represented by means of different colors: our apparatus indicates low values as dark green and high values as red, passing through light green, yellow, and light red. On the left side of the "map" power values are represented;

it is important to stress that the lowest value in each frequency band is represented by dark green and the highest one by red, whether the range is very wide (e.g., 120–4150 μV2/Hz) or whether it is quite narrow (61–67).

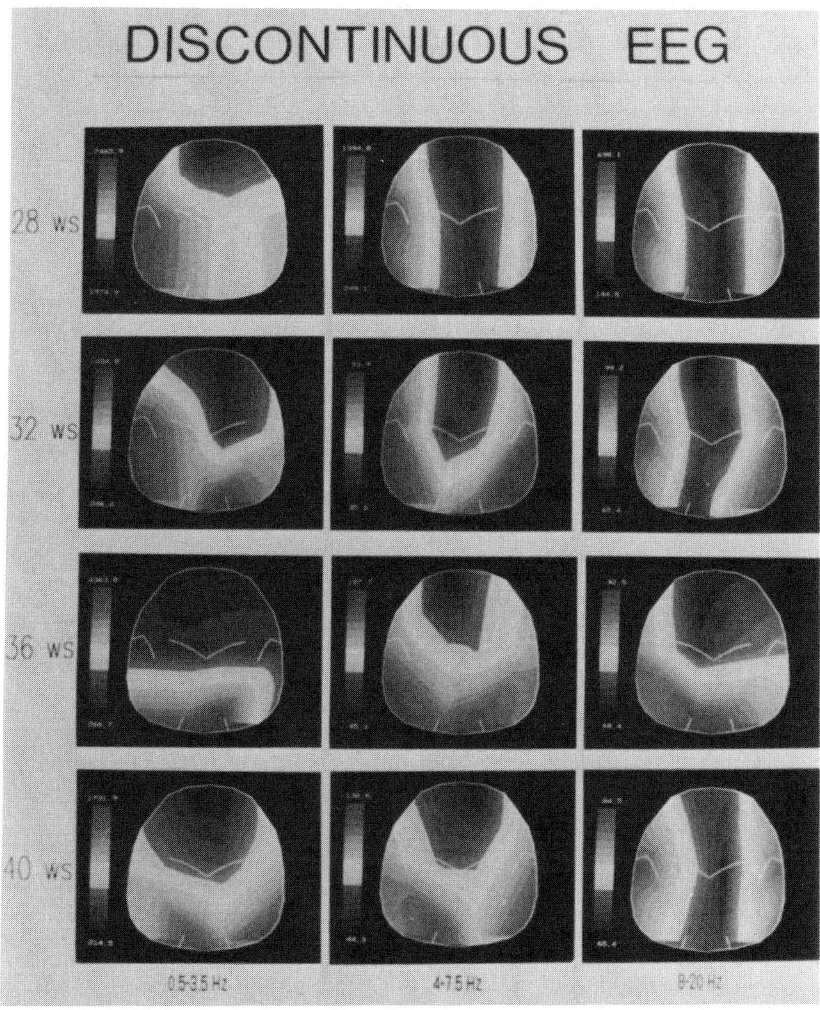

Figure 7. Examples of Brain Electrical Activity Mappings (BEAMs) from discontinuous tracings in four low-risk newborn infants recorded at 28, 32, 36, and 40 weeks of PMA. Three maps are presented for the different frequency bands (see text).

An EEG analyzer only "knows" data relative to the electrodes applied on the scalp. Every other pixel on the screen is obtained by means of an interpolation process. It is evident that the reliability of a map increases with the number of electrodes. As previously indicated, in the adult this number is generally at least 16–19. The use of such a large number of electrodes on low-weight infants, sometimes submitted to intensive care, appears very difficult. Maps shown in Fig. 7 and Fig. 8, achieved from 8-lead tracings, provide only a gross topographic representation of the power of selected bands on the scalp, but give interesting information about regional organization of brain electric activities at different postmenstrual ages.

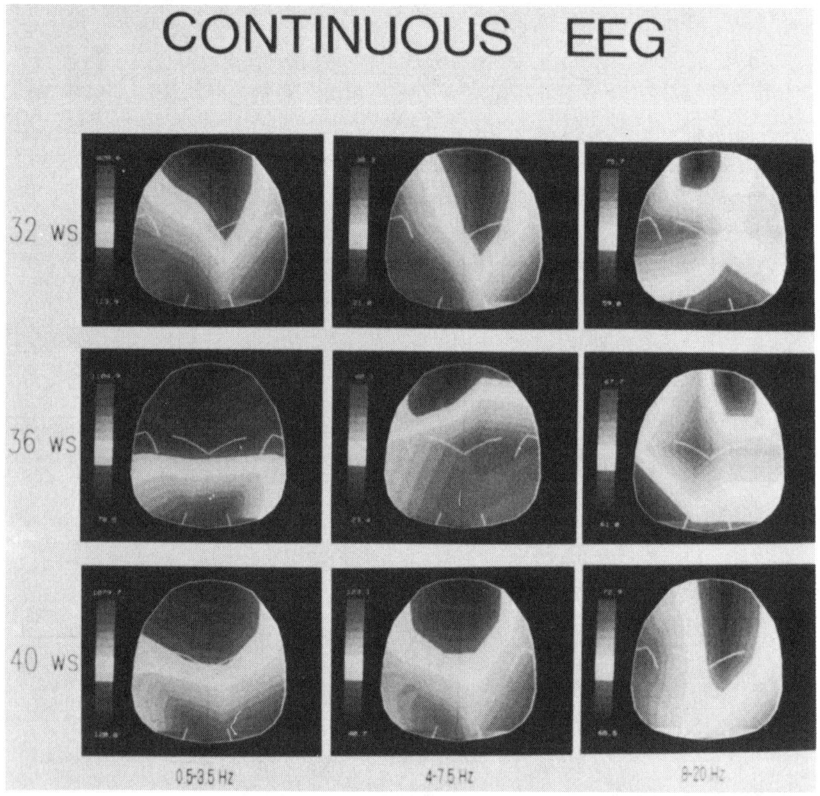

Figure 8. Examples of Brain Electrical Activity Mappings (BEAMs) from continuous tracings recorded at 32, 36, and 40 weeks of PMA in three infants (see text).

In these figures seven groups of maps are shown: every group is formed by one map (on the left) producing the topographic distribution of low frequencies (0.5–3.5 Hz), one map (in the middle) concerning theta activity (4–7.5 Hz) and one (on the right) concerning rapid rhythms (8–20 Hz). Maps have been obtained from the tracings recorded within the first two weeks of life in four low-risk newborns at different PMA

(28, 32, 36, and 40 weeks). In each EEG different maps were produced during discontinuous and continuous tracings. In Fig. 8 only maps from 32 weeks are shown; continuous tracing is not available before that age. At 36 and 40 weeks, among different types of continuous tracings, medium–high amplitude patterns recorded during the phase of active sleep preceding quiet sleep were selected. Hence, Nolte and Haas's 287, 327, 323, 367, 363, 407, and 403 codes have been submitted to computer analysis and to BEAM.

At 28 weeks frequency maps show a temporal predominance of each band. Slow frequencies have a high power (reaching more than 7000 μV2/Hz) but also the theta band is well represented (see "temporal sawtooth"). Rapid activity power reaches one half of that of theta frequencies. In the first two maps a slight but evident left predominance is observable; in our experience this finding is common in low-risk infants at this postmenstrual age (Fig. 7).

At 32 weeks, during discontinuous tracing, delta appears more represented in occipital regions; left predominance now is no longer significant. Theta/delta ratio has decreased (disappearance of temporal sawtooth) and theta power is similar to that of fast rhythms (represented by the rapid component of "delta brushes") (Fig. 7). In active sleep delta values are lower but the topographic distribution is analogous; rapid activities exceed theta (but both values are low) and appear diffuse (as shown by the narrow range of the power) (Fig. 8).

At 36 weeks, in quiet sleep, delta and rapid activity predominate on occipital regions (posterior predominance of delta brushes). Theta appears more diffuse (Fig. 8). In active sleep delta is less powerful but keeps the same distribution. Theta and rapid bands are diffuse (the range of 8–20 Hz band appears very narrow: 61.0–67.7 μV2/Hz, and therefore the respective map is almost meaningless) (Fig. 8).

At term age (40 weeks) slow frequencies appear more diffuse but are still prevalent on posterior areas. Fast activities are more concentrated on temporal regions (Fig. 8). In active sleep each frequency band keeps the same spatial organization. Delta power decreases whereas theta and rapid bands do not show any significant change (therefore rapid/slow ratio is higher in active sleep) (Fig. 8).

In conclusion, spectral analysis and BEAM allow us to confirm data obtained by means of visual examination about the occurrence of particular findings (i.e., temporal sawtooth or delta brushes) and their spatial organization. A wider application of this technique on a large number of low-risk newborns will provide more precise information about its utility in clinical practice and in research.

Conclusions

Literature and personal findings on EEG maturation, here briefly reported, have documented the rapidity and the abruptness of EEG developmental changes in the late fetal period. As shown in the previous paragraphs, these changes concern a number of characteristics:

- temporal organization (from discontinuous to more continuous activity)
- topographic organization (prominence of the activity in a specific region of the scalp at a given PMA; e.g., occipital predominance after 32 weeks of PMA)
- synchrony of the activity in the two hemispheres
- reactivity of the EEG to external stimuli
- specific waveforms typical of a given PMA, as temporal theta at 29–31 weeks of PMA or delta brush after 31 weeks
- increasing correlation between EEG and behavioral states.

Structural, connectional, and histochemical changes in the brain account for developmental changes of these EEG features. Nevertheless, the structural correlates of brain electric activity are still very poorly understood and the relationship between anatomical and EEG changes during early development remains largely unknown.

The sixth month of gestation begins a period of constant organization which lasts for several years postnatally. These organizational events include elaboration of dendritic and axonal ramifications, alignment, orientation, and layering of cortical neurons, establishment of new synaptic contacts, cell death, and selective elimination of neural processes, glial proliferation and differentiation (Volpe, 1987). Electrical potentials recorded from the scalp mainly result from extracellular current flow associated with summed postsynaptic dendritic potentials in synchronously activated, vertically oriented pyramidal cells (Martin, 1985). Therefore, maturation of spontaneous EEG probably reflects the rapid development of dendrites, dendritic spines, and axodendritic synapses. It has been documented that major changes in the organization of the visual cortex during the last trimester (appearance and elaboration of basilar dendrites and of dendritic spines, tangential spread of apical dendrites) match with a parallel maturation of visual evoked potentials (Purpura, 1975). Maturation of intracortical and cortical-subcortical connections permits a more synchronous and continuous activation of cortical neurons and, thus, a more organized and continuous electrical activity recorded from the scalp. Behavioral state maturation and the increasing state–EEG relationship may reflect the differentiation of subcortical sleep-related pacemakers and of their connections with the other neural structures.

In the late fetal period a common feature is noticeable both in the anatomical and the electrophysiological organization. Transient patterns are observable in the structural, histochemical features, which may account for physiological (and behavioral) transient patterns. We have shown several examples of transitional EEG phenomena, sometimes lasting only few weeks, like specific waveforms (e.g., *encoches frontales* between 35 and 43 weeks) or predominance of electrical activity in a region of the scalp (e.g., in temporal regions between 28 and 30 weeks). Such a rapid emergence and disappearance of an EEG feature cannot be observed in other periods of the human life. As stated by Kostović (1992), the late fetal period can be defined as a stage of transient brain organization. Regressive organizational events, such as cell death and selective elimination of neural processes, may be the morphological substrate for transient features on the EEG.

Interesting contributions to identify the neural substrates of EEG development could be provided in the future by the study of the temporal relationship between anatomical and EEG changes in specific parts of the cortex. Transient features in EEG and anatomical structures are some of the best candidates for this type of research. Another approach, which may be promising in revealing casual relationships between aspects of the EEG maturation and structural changes, is represented by clinical studies on preterm infants with clearly identified brain lesions.

Acknowledgments. The authors are most grateful to Ada Lombardi, who performed the EEG recordings, and to Paolo Magni for the skillfully produced figures. We acknowledge gladly the encouragement and advice of Prof. Pietro Pfanner. This research was partially supported by grants from the Italian Ministries of Health and of Scientific.

References

Anderson CM, Torres F, Faoro A (1985): The EEG of the early premature. *Electroencephalogr Clin Neurophysiol* 60:95–105

Arfel G, Leonardon N, Moussalli F (1977): Densité et dynamique des encoches pointues frontales dans le sommeil du nouveau-né et nourrisson. *Rev Electroencephalogr Neurophysiol* 7:351–360

Aziz SS, Wallace SJ, Murphy JF, Sainsbury CPQ, Gray OP (1986): Cotside EEG monitoring using computerised spectral analysis. *Arch Dis Childhood* 61:242–246

Berger H (1929): Uber das Elektrenkephalogramm des Menschen. *Arch fur Psychiatrie und Nervenkrankheiten* 87:527–570

Berger H (1932): Uber das Elektrenkephalogramm des Menschen, funfte Mitteilung. *Archiv fur Psychiatrie und Nervenkrankheiten* 99:213–254

Bergstrom RM (1969): Electrical parameters of the brain during ontogeny. In: *Brain and Early Behaviour*, Robinson RJ, ed. New York: Academic Press

Connell J, Oozeer R (1988): Neonatal electro-encephalography. In: *Fetal and Neonatal Neurology and Neurosurgery*, Levene MI, Bennett MJ, Punt J, eds. Edinburgh: Churchill Livingstone

Connell J, Oozeer R, Dubowitz V (1987): Continuous 4-channel EEG monitoring: a guide to interpretation, with normal values, in preterm infants. *Neuropediatrics* 18:138–145

Curzi-Dascalova L, Peirano P, Silvestri L, Korn G (1985): Organisation du sommeil des nouveau-nés prématurés normaux. Etude polygraphique. *Rev EEG Neurophysiol* 15:237–242

Dreyfus-Brisac C (1968): Sleep ontogenesis in early human prematurity from 24 to 27 weeks of conceptional age. *Dev Psychobiol* 1:162–169

Dreyfus-Brisac C (1975): Neurophysiological studies in human premature and full-term newborns. *Biological Psychiatry* 10:485–496

Dreyfus-Brisac C, Samson-Dreyfus D, Fishgold H (1955): Activité électrique cérebrale du prématuré et du nouveau-né. *Ann Pediatr* 31:1–7

Engel RCH (1975): *Abnormal Electroencephalograms in the Neonatal Period.* Springfield: Charles C. Thomas Publisher

Ferrari F, Cioni G, Biagioni E, Ori L, Incerti E, Sferrazza-Papa A (1990): Semeiotica elettroencefalografica nel neonato pretermine. *Atti del Convegno Internazionale "Dal nascere al divenire nella realtá e nella fantasia"*. Torino: Minerva Medica, pp. 29–38

Hahn JS, Monyer H, Tharp BR (1989): Interburst interval measurements in the EEGs of premature infants with normal neurological outcome. *Electroencephalogr Clin Neurophysiol* 73:410–418

Hughes JR, Fino JJ, Hart LA (1987): Premature temporal theta. *Electroencephalogr Clin Neurophysiol* 67:7–15

Ioffe S, Chernick V (1990): Prediction of subsequent motor and mental retardation in newborn infants exposed to alcohol in utero by computerized EEG analysis. *Neuropediatrics* 21:11–17

Karch D, Rothe R, Jurisch R, Heldt-Hildebrandt R, Lubbesmeier A, Lemburg P (1982): Behavioural changes and bioelectric brain maturation of preterm and full-term newborn infants: a polygraphic study. *Develop Med Child Neurol* 24:30–47

Kellaway P, Crawley J (1964): *A Primer of Electroencephalography of Infants.* Sections I and II of *Methodology and Criteria of Normality*, Bethesda: NIH

Kleitman N (1963): *Sleep and wakefulness.* 2nd edition. Chicago: University of Chicago Press

Kostović I, Petanjek Z, Delalle I, Judaš M (1992): Developmental reorganization of the human association cortex during perinatal and postnatal life. In: *Neurodevelopment, Aging, and Cognition*, Kostović I, Knežević S, Wisniewski H, Spilich G, eds. Boston: Birkhäuser

Kuks JBM, Vos JE, O'Brien MJ (1988): EEG coherence functions for normal newborns in relation to their sleep state. *Electroencephalogr Clin Neurophysiol* 69:295–302

Lombroso CT (1979): Quantified electrographic scales on 10 pre- term healthy newborns followed up to 40–43 weeks of conceptional age by serial poligraphic recordings. *Electroencephalogr Clin Neurophysiol* 46:460–474

Martin JH (1985): Cortical neurons, the EEG, and the mechanism of epilepsy. In: *Principles of Neural Sciences*, 2nd edition. Kandel ER, Schwartz JH eds., New York: Elsevier

Monod N, Dreyfus-Brisac C, Ducas P, Mayer M (1960): L'EEG du nouveau-né aterme: étude comparative chez le nouveau-né en présentation céphalique et en présentation de siege. *Rev Neurol* (Paris) 102:375–379

Nijhuis JG, Prechtl HFR, Martin CB, Bots RSGM (1982): Are there behavioural states in the human fetuses? *Early Hum Develop* 6:177–195

Nolte R, Haas G (1978): A polygraphic study of bioelectrical brain maturation in preterm infants. *Develop Med Child Neurol* 20:167–82

Nuwer MR(1988): Quantitative EEG: I Techniques and problems of frequency analysis and topographic mapping. *J Clinic Neurophysiol* 5:1–43

Papile LA, Burstein J, Burstein R (1978): Incidence and evolution of subependymal and intraventricular hemorrhage: a study of infants with birth-weight less than 1,500 gr. *Pediatrics* 92:529–34

Parmelee AH, Schulte FJ, Akiyama Y, Schultz M, Stern E (1968): Maturation of EEG activities during sleep in premature infants. *Electroencephalogr Clin Neurophysiol* 24:319–29

Prechtl HFR (1974): The behavioural states of the newborn infant (a review). *Brain Research* 76:185–212

Prechtl HFR, Fargel JW, Weinmann HM, Bakker HH (1979): Posture, motility and respiration of low-risk preterm infants. *Develop Med Child Neurol* 21:3–27

Prechtl HFR, Vos JE, Akiyama Y, Casaer P (1974): Neonatal EEG: age codes and coherence functions in relation to intra-uterine growth and oestrogen levels. In: *Ontogenesis of the Brain*, Jilek L, Trojan S eds., Vol. 2, Proc. of Int. Symposium Neuroontogeneticum Secundum, Universita Karlova, Prague

Purpura DP (1975): Dendritic differentiation in human cerebral cortex: normal and aberrant developmentl patterns. In: *Advances in Neurology*, Kreutzberg GW ed., New York: Raven Press

Richards JE, Parmelee AH Jr, Beckwith L (1986): Spectral analysis of infant EEG and behavioural outcome at age five. *Electroencephalogr Clin Neurophysiol* 64:1–11

Statz A, Dumermuth G (1982): Transient EEG patterns during sleep in healthy newborns. *Neuropediatrics* 13:115–22

Volpe JJ (1987): *Neurology of the Newborn.* Philadelphia: WB Saunders Company

Werner BS, Stockard JE, Bickford RG (1977): *Atlas of Neonatal Electroencephalography.* New York: Raven Press

Mechanisms of Asymmetrical Development of the Human CNS

Richard S. Nowakowski

Does the asymmetrical development of the CNS have a genetic basis? Arguments for and against this idea have been made frequently in the past (e.g., Annett, 1985a; Bianki et al., 1979; Collins, 1977; Corballis and Morgan, 1978; Levy, 1977; Morgan, 1977) without resolving the basic question. Any theory that attempts to provide a genetic basis for the occurrence of asymmetries in the human nervous system has to accommodate a number of facts and overcome a variety of conceptual difficulties. In addition, a plausible genetic hypothesis needs accompanying plausible developmental mechanisms, complete with cell biological and molecular facets, through which the gene(s) could act to produce an asymmetric organization in the adult structure. It is the purpose of this chapter to speculate on the possible nature of genetic and developmental mechanisms that could produce structural asymmetries in the adult CNS. It is hoped that these speculations will provide both stimulus and a foundation for future experiments.

Most of the major factual and conceptual difficulties with linking genes to CNS asymmetries can be grouped into three major categories. First, it is not known exactly where in the brain or when an asymmetry first develops; thus, resolution of this issue ultimately devolves to identification of the developmental cause of the original asymmetry. Indeed, there is some evidence that angulation of the Sylvian fissure is asymmetric as early as 16 weeks of gestation (LeMay and Culebras, 1972) and that sulcal asymmetries appear before 31 weeks of gestation (Chi et al., 1977a,b), but there are no realistic suggestions as to what developmental processes (e.g., cell proliferation and/or neuronal migration (see review by Nowakowski, 1987) lead to these earliest asymmetries. Second, for most asymmetries of the CNS most people show the same "sidedness."

For example, speech lateralization is 90–95% on the left (Milner, 1974); right-handedness ranges from 65–90%, depending on how it is measured (Annett, 1985a); and the "sidedness" of a variety of neuroanatomical asymmetries ranges from 60–80% (Witelson and Kigar, 1988). Third, it is essential to define meaningfully the relevant phenotype of each structural and functional asymmetry if a coherent link from genetic locus to mechanism to asymmetry is to be established. For example, much of the research on functional lateralization and asymmetry has been concerned with handedness, as evaluated by performance of a varied array of complex motor behaviors, and with lateralization of speech. Clearly the behaviors involved in such performances incorporate numerous skills and anatomical structures (e.g., Petersen et al., 1988). Thus, it would be difficult (and perhaps unwise) to attribute a functional asymmetry with a wide and varied nature to a single structural asymmetry, but if this is not done how does one make a link to a genetic mechanism? In this context it should be noted that not all alleles at all genetic loci produce single, simple phenotypes. Since complex, multigenic systems are known or believed to underlie a variety of quantifiable traits and even behavior (e.g., Plomin, 1990), it is conceivable that such a complex system could produce asymmetries in the CNS. In this case, the use of linkage maps constructed using restriction fragment length polymorphisms (Lander and Bottstein, 1989; Paterson et al., 1988) may be necessary to unravel the effects of multiple genes on asymmetrical development of the CNS, thus pointing out the necessity of interaction by scientists in a variety of disciplines to resolve this complicated issue.

How Could Genes Produce Asymmetries?

The issue of the possible contribution (and lack of contribution) of genes to asymmetries in the CNS has been extensively reviewed and debated (Annett, 1985a; Collins, 1977; Corballis and Morgan, 1978; Levy, 1977; McManus, 1985a,c; Morgan, 1977). The proffered views range from "genes are left–right agnosic" (Morgan, 1977) to comprehensive theories about the mode of action of hypothetical genes and their roles in producing particular functional asymmetries (Annett, 1985a; McManus, 1985a,c).

Several authors have argued cogently that genes can produce "sidedness" (Annett, 1985a; Levy, 1977; McManus, 1985c). It must, however, be kept in mind that designing a hypothetical genetic system that provides a plausible basis for asymmetric development is not equivalent to proving that such a system is the underlying basis of asymmetric development. As the intent of this chapter is to speculate on possible mechanistic links between genes and structural/functional asymmetries, we will assume that there is an underlying genetic basis. Given this caveat, what is left is to determine how many genetic loci are needed and what mode of inheritance they might have. Perhaps the easiest way to understand this issue is to examine one of the weak links in this field—namely discordance in identical twins, which is frequently used as an argument that asymmetries of the CNS (as well as other diseases, disorders, behaviors, or traits) are not inherited (Corballis and Morgan, 1978; Plomin, 1990).

Various environmental influences and cytoplasmic factors (Corballis and Morgan, 1978; Plomin, 1990) have been suggested as necessary to account for discordant twin data. However, there is a simple genetic mechanism that could account for discordance in the studies of identical twins. Simply stated, this suggestion is that there exist some alleles at some genetic loci whose action in producing a given phenotype is stochastic. In its simplest form the stochastic action of the allele would be to produce an equal proportion of each of two possible phenotypes. A well-documented precedent in mammalian genetics for such a stochastic action by a gene can be found at the *situs inversus* locus in the mouse (Layton, 1976). There exist two alleles at this locus: *Iv*, the dominant form, and *iv*, the recessive form (Brueckner et al., 1989; Layton, 1976). Mice that are homozygous (*Iv/Iv*) or heterozygous (*Iv/iv*) for the dominant allele have typical left-sided asymmetry of the viscera. Mice that are homozygous for the recessive allele (*iv/iv*) have a 50% chance of developing a reversed asymmetry (Layton, 1976). In addition, the *iv/iv* offspring of both reversed asymmetric and typically asymmetric mice also have a 50% chance of developing reversed asymmetry even after many generations of inbreeding (Layton, 1976).

The simplest interpretation of these findings is that the *Iv* allele "biases" the development of the viscera to promote the establishment of their typical asymmetry and that the *iv* allele is a "loss of function" mutation that, therefore, removes this bias. In the absence of bias the "sidedness" of the asymmetry is decided at random, and, hence, a 50% incidence of each type results. It is useful to note that the apparent random direction of the asymmetry in the *iv/iv* mice is not likely to be due to the influence of other genetic loci, because the experiments have been performed on inbred strains (Brueckner et al., 1989; Layton, 1976) that are, in essence, genetically homogeneous (Green, 1981). The *iv* locus has recently been localized to chromosome 12 (Brueckner et al., 1989).

A stochastically acting "biasing" gene (like *iv*) could account for dis-

INHERITANCE OF ATYPICAL ASYMMETRY HAS A STOCHASTIC COMPONENT

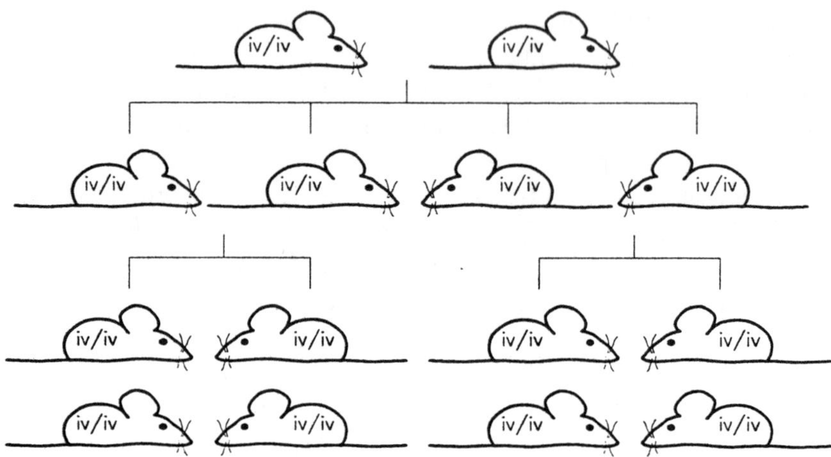

Figure 1. The determination of the direction of asymmetry in mice which are homozygous for the *iv* allele at the *situs inversus* locus is stochastic. This means that an equal proportion of *iv/iv* mice have each type of asymmetry regardless of the direction of asymmetry of their parents.

cordance in CNS asymmetry in identical twins. A hypothetical loss of function allele at a biasing gene for CNS asymmetry would, by analogy with *iv*, produce a 50% rate of reversed asymmetry in homozygotes. Thus, in the case of identical twins homozygous for the hypothetical allele, 25% of the pairs would share a typical asymmetry, 50% of the pairs would be discordant, and 25% would share an atypical asymmetry. Thus, in identical twins that lack the "biasing" gene, one would expect to find a 1:2 ratio of pairs of identical twins concordant for atypical asymmetry to pairs of identical twins discordant for atypical asymmetry. This is a concordancy rate of only 33%. This rate is clearly much lower than the 100% concordancy rate expected for genes that lack a stochastic component, but still higher than most previous studies have reported for handedness (for summary see McManus, 1985c), perhaps because of an association of left-handedness with prenatal pathological events including reduced birth weight (Segal, 1989). Similarly, a stochastically acting "biasing" gene could also account for the reported incidence of left-handed offspring of left-handed parents. Since both parents have atypical asymmetry, then they and their offspring would all lack the hypothetical "biasing" gene. Traditional genetic concepts would predict that 100% of the offspring would be left-handed, but if a stochastically acting allele is involved, the handedness of the offspring should be determined stochastically. The predicted 1:1 ratio of right-handed to left-handed offspring agrees well with the observed proportions (77:50) in the 12 studies summarized by McManus (1985b), especially when the varied criteria for classification in these 12 studies are considered.

Two hypotheses for handedness and language dominance (Annett, 1985a; McManus, 1985b) have exploited the concept of a stochastically-acting biasing gene, and these hypotheses have generated a healthy debate in the literature (Annett, 1985b; Corballis and Morgan, 1978; McManus, 1985a,b) with regard to which hypothesis best accounts for the available data. Both hypotheses posit the existence of a single locus with two alleles. In the Annett model, the hypothetical dominant allele "shifts" a continuous distribution of skills to the right (in the case of handedness) from an otherwise neutral position. The theoretical shift occurs whether one or two copies of the dominant allele are present. However, for subjects that are homozygous for the recessive allele no such right shift occurs.

The McManus model also has a dominant allele that when homozygous produces right-handedness, and a recessive allele that when homozygous results in an equal chance of becoming right- or left-handed. However, the McManus hypothetical dominant allele is actually codominant, and, hence, the phenotypes of the heterozygote are intermediate between the two homozygotes (i.e., 75% chance of becoming right-handed). Both the Annett and the McManus models (Annett, 1985a; McManus, 1985c) deal with the relative prevalence of asymmetries at the population level by calculating the proportional representation in the population of their respective hypothetical alleles. Because of this both models do considerably better at fitting the monozygotic twin data and the offspring of two left-handed parents data than does the simplistic analogy to the *iv* locus described above. Additionally, the McManus model (1985c) can manipulate the strength of the hypothetical heterozygote effect, although he points out that if a zero value is chosen for the heterozygote effect and the population frequency of the recessive allele is adjusted, his model becomes identical to the Annett model and yields identical predictions. Interestingly, these two models, despite their similarity, are based on quite different phenotypes—the significance of the difference is not clear (cf. Annett, 1985a; McManus, 1985a,b).

The main point, however, is that both the Annett and the McManus models are *plausible*. They each present a simple genetic model that fits well to the available data. This indicates that a "biasing" gene that acts like *Iv* in the mouse viscera could participate in the production of asymmetries in the developing nervous system. Discriminating between these two models by using either genetic means or by further evaluation of the phenotype may indeed help to refine them, but it is difficult to believe that they can be truly tested without applying entirely different evaluation approaches and without establishing an independent way to evaluate the relevant phenotype. For example, both models involve only two alleles at a single locus and, moreover, that one locus controls more than one asymmetry (e.g., handedness and language dominance). However, there are no a priori reasons for these assumptions, and they were made in order to be able to suggest the simplest model consistent with the most data. For example, the existence of a third allele that shifts the distribution to the right a little less or produces a less than 100% incidence of right-handedness is also plausible. Presumably, discrimination between these two models or other more complicated suggestions (Levy and Nagylaki, 1972) awaits examination of a large pedigree with molecular techniques. An additional clue as to possible complexity of this issue comes from the observation that humans with Kartagener's syndrome, who have reversed symmetry of the viscera produced by an autosomal recessive gene that acts stochastically (Afzelius, 1976), do not have a higher incidence of left-handedness than the general population (Torgersen, 1950). Thus, if a hypothetical genetic locus that affects the symmetry of the CNS exists, it is not identical to at least one locus that affects the symmetry of the viscera. Thus, it is conceivable that a family of stochastically acting "biasing" genes, each of which contributes to a different subset of CNS and/or non-CNS, asymmetries could exist.

How Can Asymmetries Develop?

For a gene to produce an asymmetry in the CNS it must clearly act on some developmental process in order to have its effect in the adult. In particular, there must be plausible molecular and cell-biological mechanisms that can be exploited. Some insight into a possible mechanism that would result in asymmetrical development of the CNS can be obtained by considering the development of a much simpler organism—the nematode *Caenorhabditis elegans*. The development of *C. elegans* has been extensively studied. Indeed, since the entire organism has an invariant anatomy and a short developmental period, the lineage of each cell in the adult organism is known (Kimble and Hirsh, 1979; Sulston, 1976; Sulston and Horvitz, 1977; Sulston et al., 1980; Sulston et al., 1983). These observations and experimental manipulations using laser microsurgery (e.g., Sulston and White, 1980) have established that during nematode development the fate of most cells is specified by cell–cell interactions. Moreover, for the most part, the fate of each cell in the organism is specified and stereotypic. An exception to the general principle of stereotypical determination of fate is provided by two cells (Z1.ppp and Z4.aaa) whose fate seems to be determined stochastically. This pair of cells differentiates into a mature pair consisting of one anchor cell (AC) and one ventral uterine cell (VU). Each cell has the potential to become either the AC or the VU and follows each of the two possible fates with equal probability; i.e., in half of the embryos the AC develops from cell Z1.ppp, and in the other half from cell Z4.aaa (Kimble, 1981). A recent series of papers (Seydoux and Greenwald, 1989; Sternberg and Horvitz, 1989) has established the series of cell–cell interactions that results in this stochastic determination of fates; a schematic summary of this sequence of events, generalized to apply to any cell system, is shown in Fig. 2.

Initially, both immature cells have the capacity to synthesize a signal and a receptor (lin-12 in the nematode system) capable of recognizing that signal. An asymmetric relationship between the two cells is initiated when one cell becomes the first to synthesize (or perhaps release) the signal. The signal binds to the (lin-12) receptor on the other cell and thereby inhibits the synthesis of that signal by the second cell. The continued production of the signal by the cell that happened to have been the first to produce it establishes the basis for further asymmetric development of the two initially equipotential cells.

Conceptually, cell–cell interactions similar to those acting in nematode development could be envisioned as sufficient to produce an asymmetry in the CNS. Observed correlations among handedness, hemispheric functional asymmetry (for review see Bryden, 1982; Hécaen et al., 1981), gross morphological features of the brain (for review see Witelson, 1980; Witelson and Kigar, 1988), and area of the corpus callosum (Witelson, 1985, 1989; Witelson and Goldsmith, 1990) has formed the basis for a hypothesis (Witelson and Nowakowski, 1991) that elimination of callosal axons during early development is a determinant of functional asymmetries and further that variation in both the direction and magnitude of functional asymmetries is the result (at least in part) of variation in the extent of commissural axon elimination. The basis of this hypothesis is that during development axons of the corpus callosum originally are present in greater numbers than they are in the adult (Killackey, 1990). Then, as development proceeds, they are lost.

The proposal (Witelson and Nowakowski, 1991) is that a difference in the loss of axons from one side (or the other) could produce or accentuate an asymmetry. To extend that proposal one might postulate that the cell–

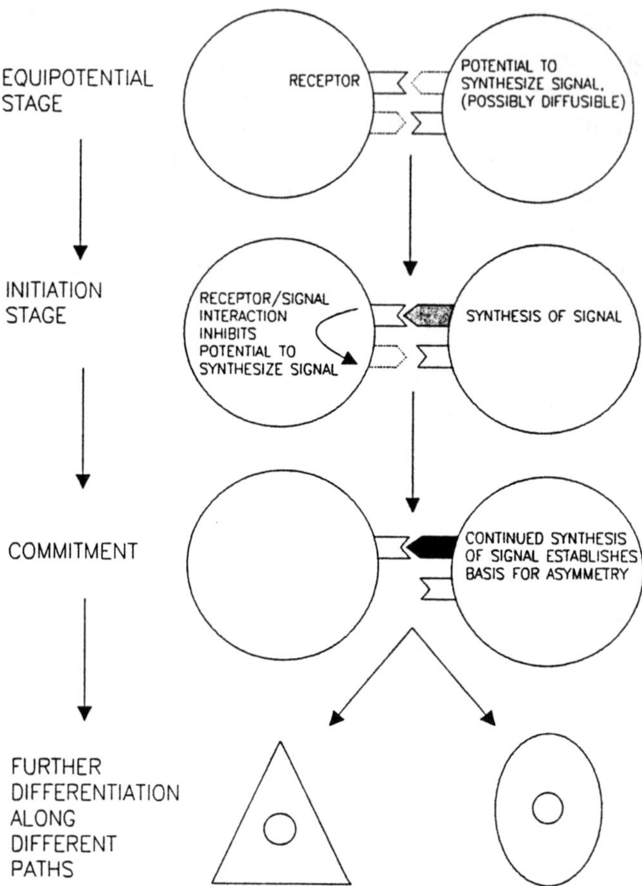

Figure 2. Cell–cell interactions during development can be stochastic and can produce asymmetries. This schematic diagram is a generalized view of the interaction of a signal with the lin-12 receptor during the development of *C. elegans* (Seydoux and Greenwald, 1989). The fate of two cells which are initially equipotent can be stochastic if the synthesis of a signal by one cell happens to begin (or reach some critical threshold) first. The synthesis of the signal initiates an asymmetry in the system by interacting with a receptor into the system by interacting with a receptor on the other cell. The receptor/signal interaction then inhibits the potential to synthesize signal in the now target cell, which eventually leads to the commitment of the originally identical cells to different fates.

cell contacts established by the axon extending across the midline provide the means for interaction between neurons whose "fate" is in some way coupled, and the commitment for further differentiation could be made similarly to the way that the AC and VU commitments are made in the nematode.

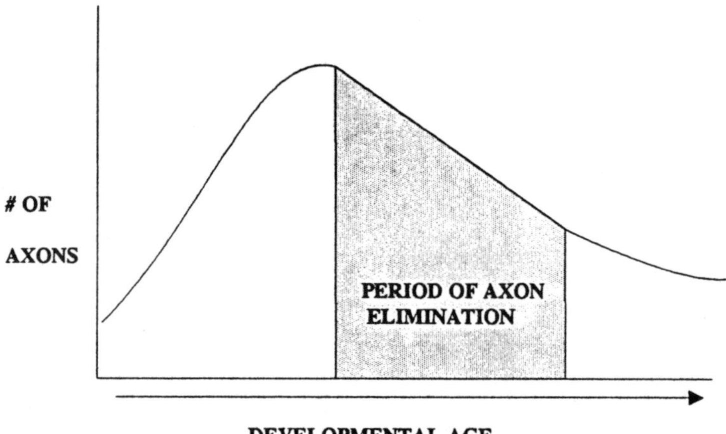

Figure 3. During the development of the CNS the number of axons in many white-matter tracts reaches a peak level some time during development. Following this peak the numbers of axons in reduced to the adult level. This reduction in axon number is referred to as axon elimination.

A Unifying Hypothesis?

Although the case is far from complete, there appear to be sufficient genetic, cell biological, and neuroembryological precedents to suggest that multiple alleles at a single genetic locus acting during development could be a sufficient basis for the presence of systematic directionally biased asymmetries in CNS organization.

A plausible sequence of events that is consistent with a great deal of the available data would consist of: (1) inheritance of dominant (or recessive) alleles at a "biasing" locus, (2) cell–cell interactions exploiting a signal-receptor mechanism similar to the lin-12 system in *C. elegans*, (3) the establishment of asymmetric (or symmetric) fates, (4) modification in the amount of axon elimination and in (5) the size of the corpus callosum, and (6) the establishment of a functional or structural asymmetry. Each of the hypothesized steps in this cascade of events is underpinned by a large number of additional hypotheses and speculations, but it is hoped that this basic outline of a plausible path from gene through cell–cell interactions and development to structure and behavior can serve as a useful target for further research.

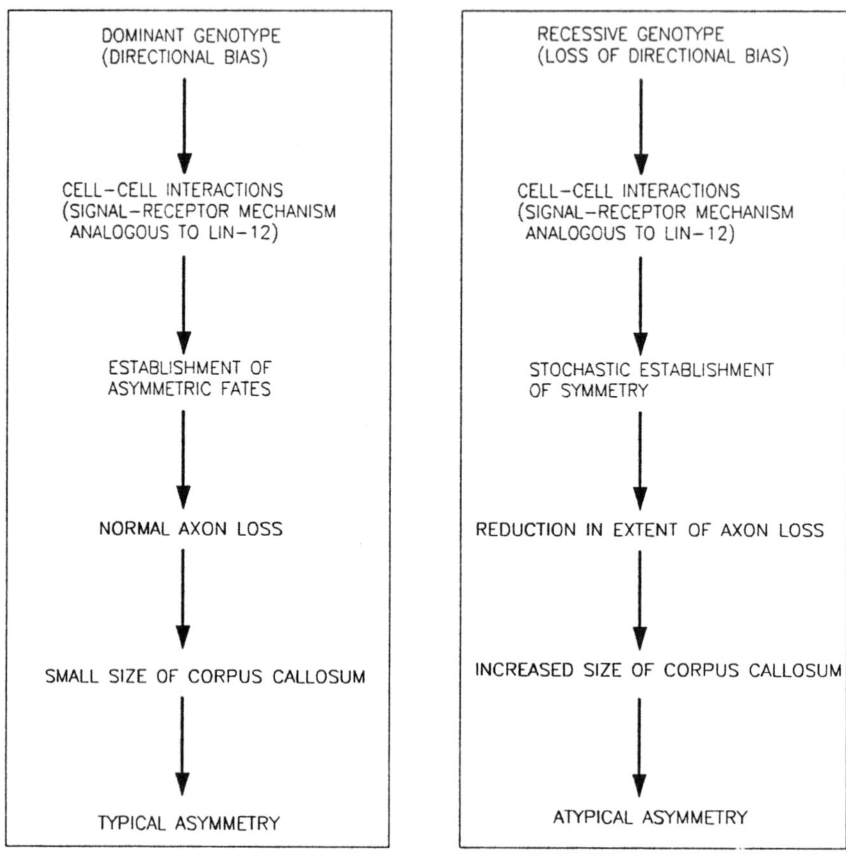

Figure 4. A speculative sequence of events from gene through cell–cell interactions in CNS development to the establishment of typical and atypical asymmetries in the mature CNS. See the text for details.

References

Afzelius BA (1976): A human syndrome caused by immotile cilia. *Science* 193:317–319

Annett M (1985a): *Left–Right, Hand and Brain: The Right Shift Theory.* New Jersey: Erlbaum

Annett M (1985b): Which theory fails? A reply to McManus. *Brit J Psychol* 76:17–29

Bianki VL, Kaidanov LZ, Novikov SN (1979): Genetic analysis of right- and left-handedness in the house mouse (*Mus musculus L.*) *Zhurnal Vysshei Nervnoi Deyatel'nosti* 29:1239–1247 (in Russian)

Brueckner M, D'Eustachio P, Horwich AL (1989): Linkage mapping of a mouse gene, iv, that controls left-right asymmetry of the heart and viscera. *Proc Natl Acad Sci* USA 86:5035–5038

Bryden MP (1982): *Laterality: Functional Asymmetry in the Intact Brain.* New York: Academic Press

Chi Je G, Dooling EC, Gilles FH (1977a): Gyral development of the human brain. *Ann Neurol* 1:86–93

Chi Je G, Dooling EC, Gilles FH (1977b): Left-right asymmetries of the temporal speech areas of the human fetus. *Arch Neurol* 34:346–348

Collins RL (1977): Toward an admissible genetic model for the inheritance of the degree and direction of asymmetry. In: *Lateralization in the Nervous System*, New York: Academic Press, pp. 137–150

Corballis MC, MJ Morgan (1978): On the biological basis of human laterality: I. Evidence for a maturational left-right gradient. *Behav and Brain Sci* 2:261–336

Green E (1981): *Genetics and Probability in Animal Breeding Experiments.* London: Macmillan Publishers Ltd

Hécaen H, De Agostini M, Monzon-Montes A (1981): Cerebral organization in left-handers. *Brain Lang* 12:261–284

Killackey H (May, 1990). *Development of callosal pathways in experimental primates.* Paper presented at the Third International Brain Workshop: Neurodevelopment, Aging, and Cognition; Dubrovnik, Yugoslavia

Kimble J (1981): Alterations in cell lineage following laser ablation of cells in the somatic gonad of *Caenorhabditis elegans.* PhD Thesis, University of Colorado, Boulder

Kimble J, Hirsh D (1979): Postembryonic cell lineages of the hermaphrodite and male gonads in *Caenorhabditis elegans. Dev Biol* 70:396–417

Lander ES, Bottstein D (1989): Mapping Mendelian factors underlying quantitative traits using RFLP linkage maps. *Genetics* 121:185–199

Layton WM Jr (1976): Random determination of a developmental process. *J Hered* 67:336–338

LeMay M, Culebras A (1972): Human brain. Morphologic differences in the hemispheres demonstrable by carotid arteriography. *New Eng J Med* 287: 168–170

Levy J (1977): The origins of lateral asymmetry. In: *Lateralization in the Nervous System*, Harnad S, ed., New York: Academic Press, pp. 195–212

Levy J, Nagylaki T (1972): A model for the genetics of handedness. *Genetics* 72:117–128

McManus IC (1985a): Right- and left-hand skill: Failure of the right shift model. *Brit J Psychol* 76:1–16

McManus IC (1985b): On testing the right shift theory: A reply to Annett. *Brit J Psychol* 76:31–34

McManus IC (1985c): Handedness, language dominance and aphasia: a genetic model. *Psychol Med Monograph Suppl* 8:1–40

Milner B (1974): Hemispheric specialization: scope and limits. In: *The Neurosciences: Third Study Program*, Schmitt FO, Worden FG, eds. Cambridge, MA: MIT Press, pp. 75–89

Morgan M (1977): Embryology and inheritance of asymmetry. In: *Lateralization in the Nervous System*, Harnad S, ed., New York: Academic Press, pp. 173–194

Nowakowski RS (1987): Basic concepts of CNS development. *Child Development* 58:568–595

Paterson AH, Lander ES, Hewitt JD, Peterson S, Lincoln SE, Tanksley SD (1988): Resolution of quantitative traits into Mendelian factors by using a complete linkage map of restriction fragment length polymorphisms. *Nature* 335:721–726

Petersen SE, Fox PT, Posner MI, Mintun M and Raichle ME (1988): Positron emission tomographic studies of the cortical anatomy of single-word processing. *Nature* 331:585–589

Plomin R (1990): The role of inheritance in behavior. *Science* 248:183–188

Segal N (1989): Origins and implications of handedness and relative birth weight for IQ in monozygotic twin pairs. *Neuropsychologia* 27:549–561

Seydoux G, Greenwald I (1989): Cell autonomy of lin-12 function in a cell fate decision in *C. elegans*. *Cell* 57:1237–1245

Sternberg PW, Horvitz HR (1989): The combined action of two intercellular signaling pathways specifies three cell fates during vulval induction in *C. elegans*. *Cell* 58:679–693

Sulston JE (1976): Post-embryonic development in the ventral cord of *Caenorhabditis elegans*. *Phil Trans R Soc Lond (Biol)* 275:287–297

Sulston JE, Horvitz HR (1977): Postembryonic cell lineages of the nematode *Caenorhabditis elegans*. *Dev Biol* 56:110–156

Sulston JE, White JG (1980): Regulation and cell autonomy during postembryonic development in *Caenorhabditis elegans*. *Dev Biol* 78:577–597

Sulston JE, Albertson DG, Thompson JN (1980): The *Caenorhabditis elegans* male: Postembryonic development of nongonadal structures. *Dev Biol* 78:542–576

Sulston JE, Schierenberg E, White JG, Thompson JN (1983): The embryonic cell lineage of the nematode *Caenorhabditis elegans*. *Dev Biol* 100:64–119

Torgerson J (1950): Situs inversus, asymmetry, and twinning. *Am J Human Genetics* 2:361–370

Witelson SF (1980): Neuroanatomical asymmetry in left handers: A review and implications for functional asymmetry. In: *Neuropsychology of Left-handedness*, Herron J, ed. New York: Academic Press, pp. 79–113

Witelson SF (1985): The brain connection: The corpus callosum is larger in left handers. *Science* 229:665–668

Witelson SF (1989): Hand and sex differences in the isthmus and genu of the human corpus callosum: A postmortem morphological study. *Brain* 112:799–835

Witelson SF, Goldsmith CH (1990): The relationship of hand preference to anatomy of the corpus callosum in men. *Brain Research*, in press.

Witelson SF, Kigar DL (1988): Asymmetry in brain function follows asymmetry in anatomical form: Gross, microscopic, postmortem and imaging studies. In: *Handbook of Neuropsychology*, Vol. 1, Boller F, Grafman J, eds. The Netherlands: Elsevier Science, pp. 111–142

Witelson SF, Nowakowski RS (1991): Left out axons make men right: an hypothesis for the origin of handedness and functional symmetry. *Neuropsychologia* 29:327–333

Neuroanatomical Bases of Hemispheric Functional Specialization in the Human Brain: Developmental Factors

Sandra F. Witelson

Functional Asymmetry: Nature and Ontogeny

Hemispheric functional specialization or functional asymmetry is a well-established characteristic of functional organization in the human brain. In essence, the right and left hemispheres have different roles in mediating various behaviors and higher mental processes. Tasks involving speech production, phonemic discrimination, comprehension of oral and written language, ability to write, performance of voluntary finger, limb, and oral movements, and perception of sequences of stimuli are more dependent on left- than right-hemisphere functioning in most people. In contrast, tasks involving the perception of two- and three-dimensional visual or tactual shapes, spatial position and orientation of stimuli, perception of faces and colors, mental rotation of three-dimensional shapes, the ability to direct attention to both lateral sensory fields, perception of musical chords and melodies, perception of emotional stimuli and prosodic aspects of speech, the ability to dress oneself and to construct block models are more dependent on the right hemisphere. In the past decade, numerous books have summarized these findings based on the study of brain-damaged people with unilateral lesions, people with commissurotomy, and neurologically intact people who were tested with various behavioral and perceptual tests involving right- and left-sided input or output (e.g., Beaton, 1985; Boller and Grafman, 1988–90; Bradshaw and Nettleton, 1983; Bryden, 1982; Corballis, 1983; Damasio and Damasio, 1990; Geschwind and Galaburda, 1984; Hannay, 1986; Heilman and

Valenstein, 1985; Hellige, 1983; Kolb and Whishaw, 1990; Molfese and Segalowitz, 1988; Ottoson, 1987).

However, several key issues related to hemispheric specialization remain unresolved. One, continually expanding lists of tests are being generated that are more dependent on either one or the other hemisphere. It is not demonstrated, however, what specific cognitive functions are common to each set of skills and are indicative of the type or types of processing that each hemisphere is specialized for. The numerous dimensions postulated for hemispheric differences, such as "verbal" versus "nonverbal" or "analytic" versus "synthetic," are merely inferred hypotheses from the numerous tasks studied of what are the essential functional characteristics of the hemispheres. Moreover, it has been shown that it is not only the nature of the task stimuli, but also the strategy or process the subject uses to perform the task that determines which hemisphere is dominant in processing that task. Currently, a widely used working hypothesis is that the left hemisphere is specialized for analytic, sequential, time-dependent functions and the right hemisphere for synthetic, spatial perceptual functions (e.g., Bradshaw and Nettleton, 1983).

A second issue concerns whether the specialization is based on a greater contribution or a better, more efficient contribution: that is, whether the functional difference is quantitative or qualitative. A third issue is whether the functional differences between the hemispheres are absolute or relative. Finally, recent work has underlined the existence of multiple functional asymmetries, with some degree of dissociation even among cognitive functions usually related and lateralized to the same hemisphere. Although most right-handers have speech production and manual motoric control represented in the left hemisphere, the majority of left-handers, which indicates at least some right-hemisphere involvement, have speech production represented mainly in the left hemisphere (Rasmussen and Milner, 1977). A small proportion of right-handers may have speech production or other typically left-sided functions mediated by the right hemisphere (e.g., Fischer et al., 1991).

The issue of dissociation among subsets of functional asymmetries leads to consideration of individual differences in hemispheric specialization. This relates to two dimensions, direction and degree. There is a typical pattern of direction of functional asymmetry. The large majority of people, about 80% (approximately 90% of right-handers who constitute about 90% of the population), have the pattern described previously. The pattern of "analytic" and "synthetic" functions may be reversed for at least some components in a small minority of the population [e.g., 15% of left-handers who have speech represented mainly in the right hemisphere (Rasmussen and Milner, 1977)]. Whether such left-handers have "synthetic" functions represented mainly in the left hemisphere is not known. Statistical analysis of verbal and spatial deficits following unilateral brain damage in large groups of right- and left-handers suggests that there is considerable independence in the lateralization of verbal and spatial skills; that is, lateralization of one set of skills does not necessarily predict lateralization of the other set of skills (Bryden et al., 1983).

In addition to these directional differences, there is evidence that the degree of functional asymmetry between the two hemispheres may be less in some people than in others. For example, as a group, left-handers show less marked functional asymmetry (or greater bihemispheric representation of functions). This is evidenced in studies of brain-damaged patients by a greater prevalence of aphasic deficits following right-sided lesions in left compared to right-handers (Hecaen et al., 1981), or in neurologically intact individuals by less right–left difference in perceptual tests of auditory or visual perception (Bryden, 1982).

Differences in lateralization between the sexes have also been noted. In studies of groups of neurologically intact children (e.g., Witelson, 1976) and adults (e.g., Levy and Reid, 1978), performance on tests of right–left perception indicate less marked asymmetry for both verbal and spatial skills. Studies with brain-damaged people have indicated sex differences in the prevalence and possibly duration of aphasia and other deficits associated with right- and left-sided damage (e.g., McGlone, 1980, 1986). Other studies have indicated that some sex differences in functional organization are not in degree nor direction, but in intrahemispheric organization (e.g., Hecaen et al., 1981; Kimura and Harshman, 1984). For example, Kimura and Harshman (1984) have suggested that women have a more focussed representation of speech and praxis (motor programming) functions in anterior left regions, compared to men who have representation in both anterior and posterior regions of the left hemisphere.

Some clinical groups show variation in patterns of functional asymmetry. For example, children with developmental dyslexia show less asymmetry on tasks of spatial perception (e.g., Hynd and Cohen, 1983; Witelson, 1977a); individuals with chromosomal abnormalities, such as Turner and Klinefelter syndromes, show different patterns of asymmetry on tasks of verbal and nonverbal perception compared to matched control groups (Netley and Rovet, 1988); and homosexual men and women show a greater prevalence of non-right-handedness (Lindesay, 1987; McCormick et al., 1990) and less asymmetry on verbal perceptual tasks (McCormick, 1990).

The ontogenetic course of hemispheric specialization is a controversial issue. Is it present in children or in neonates and, if so, does functional asymmetry change in magnitude with age? Early in the history of developmental neuropsychology, it was generally accepted that hemispheric specialization developed during childhood (e.g., Lenneberg, 1967). This position was consistent with the little data available at the time. The phenomenon of functional plasticity, that is, recovery of function lost subsequent to brain damage, was documented particularly following brain insult early in development. This was taken to mean that hemispheric specialization was not yet established. Moreover, the generally accepted conception that hemispheric specialization was based on speech and language functions was considered as further evidence that hemispheric specialization could not exist in preverbal children. The empirical results of studies with brain-damaged and neurologically intact children done in the past few decades indicate that hemispheric specialization does indeed exist within the first few months of life. These results have been reviewed elsewhere (e.g., Hiscock, 1988; Molfese and Segalowitz, 1988; Witelson, 1977b, 1985b, 1987; Witelson and Swallow, 1988). Moreover, the demonstration of functional asymmetry in nonhuman primates supports the idea that it is not verbal or handedness functions per se, but processes such as analytic perception that are the essence of left-hemisphere functioning (e.g., Hamilton and Vermeire, 1988) and that functional asymmetry may well exist from the start.

In the case of children, the clearest results are from studies with neurologically intact groups because such results indicate what the organization of the brain is at the time of testing. These results demonstrate asymmetry not only in children and infants, but in neonates. For example, four-day-old neonates showed evidence of functional asymmetry on tests of dichotic (simultaneous right and left auditory stimulation) discrimination as evidenced by response involving a sucking procedure (Bertoncini et al., 1989).

Studies of performance of brain-damaged children provide less clear evidence related to functional asymmetry because such results are frequently confounded with the effects of neural plasticity. Psychological testing is often assessed months after the initial insult and, consequently, observed performance reflects both hemispheric specialization and the functional consequences of neural plasticity. These are two aspects of brain functioning that coexist (Witelson, 1987). These two aspects of brain functioning are further complicated by the situation that not all mental functions follow the long-held "Kennard principle" that lesions sustained early in life lead to less deficit than do later lesions (Kennard,

1942). This may hold for specific functions, such as sensory or motor skills studied experimentally in animals or for speech production in children, but it does not hold for more general intellectual functions such as concept formation or problem solving (Witelson, 1987). Figure 1 presents schematically the different courses of recovery (or functional plasticity) depending on the nature of the cognitive skills in question, with recovery decreasing for specific skills as age of insult increases (Fig. 1c) and recovery increasing for general skills as age of insult increases (Fig. 1d). This operates on a brain that is already functionally asymmetric, a pattern that remains invariant over development (Fig. 1a), but in which the extent of cognition that becomes lateralized increases over time (Fig. 1b).

There is no evidence of greater bihemispheric representation of cognitive functions early compared to later in development, which suggests that magnitude of asymmetry does not change. However, this is difficult to assess unequivocally as baseline performance changes with age and measurement issues become complex (see Hiscock, 1988). It is also possible that the intrahemispheric organization for the representation of cognitive functions changes with age (see, e.g., Brown and Jaffe, 1975), although the extent of hemispheric specialization does not.

Two facts concerning functional asymmetry suggest that there is a genetic contribution to some neural substrate for the maturation of this aspect of organization of the human brain. (1) The same directional pattern of functional asymmetry exists in the majority of people across cultures and generations. (2) This same pattern is present in neonatal life. Elucidation of the origins and mechanisms of functional asymmetry may be aided by recent neuroanatomical studies that are beginning to reveal what the neuroanatomical substrates of functional asymmetry may be.

Neuroanatomical Asymmetry

Neuroanatomical asymmetry in the human brain has been noted and studied since the mid-19th century. Little attention was paid to anatomical asymmetry as a possible basis of functional asymmetry until a few decades ago. Since then, numerous studies have revealed reliable right–left morphological differences and some of these have been found to correlate with neuropsychological measures of functional asymmetry. These are described in detail in a recent review (Witelson and Kigar, 1988a). For example, numerous measures of gross anatomical asymmetry are reported for weights, volumes, and lengths of the hemispheres and for torque of the occipital lobes. These are of small magnitude and have

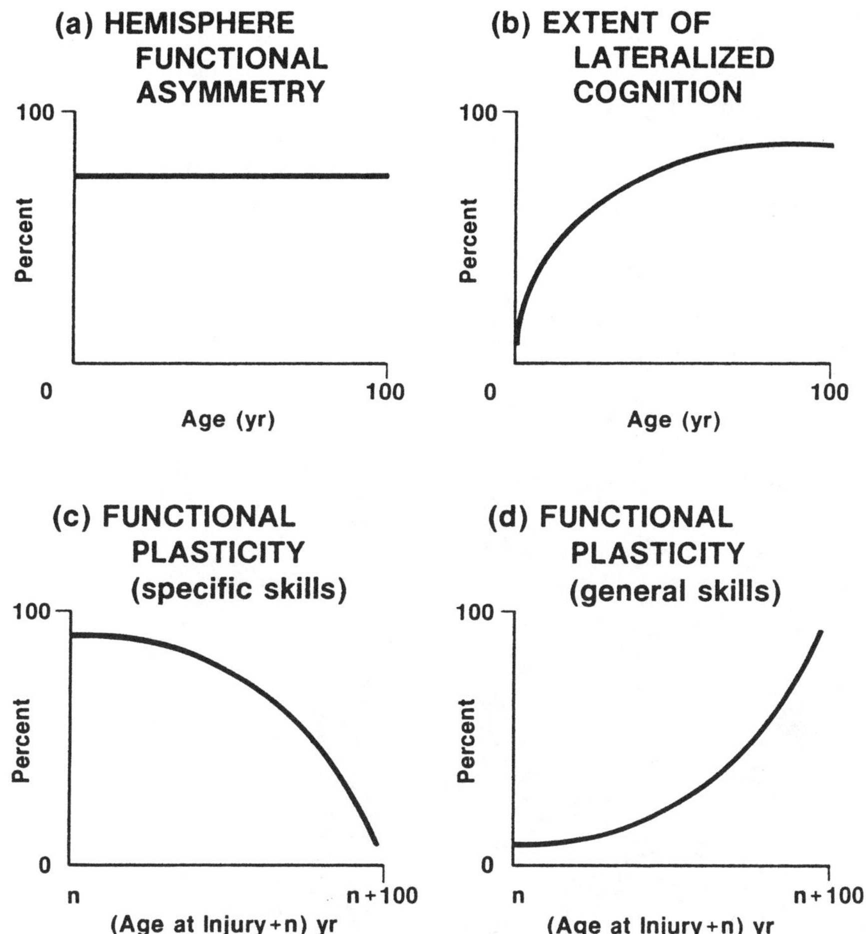

Figure 1. Theoretical scheme of hemispheric specialization and plasticity over time. (a) Degree of hemispheric specialization over time in which the slope of the curve is zero. (b) Extent of cognitive skill having lateralized brain representation over time: an epiphenomenon of cognitive development in a brain having invariant hemisphere specialization. The decreasing positive slope represents the more rapid rate of cognitive development earlier in life. (c) Degree of functional plasticity for *specific* skills *n* years after brain damage; the hypothesized increasing negative slope represents the slower rate of decrease in plasticity early in life. (d) Degree of functional plasticity for *general* skills *n* years after damage; the hypothesized increasing positive slope represents the relatively greater impact of early brain damage. (Adapted from Witelson, 1987.)

some unreliability. In contrast, marked and very reliable differences have been found, particularly in the parieto-temporal region as measured by the morphology and length of the Sylvian fissure and the extent of the parietal operculum. The posterior region of the superior surface of the superior temporal gyrus, termed the planum temporale, has been studied most extensively. Figure 2 presents the right and left hemispheres of an adult brain showing the longer posterior portion of the Sylvian fissure on the left side compared to the more anterior and sharply angulated upturn of the posterior ramus on the right side, and the larger posterior parietal operculum and planum temporale in the left rather than right hemisphere. The cortical areas in the parieto-temporal region are well documented to be involved in mediating cognitive functions that are lateralized in the brain (Damasio and Damasio, 1990).

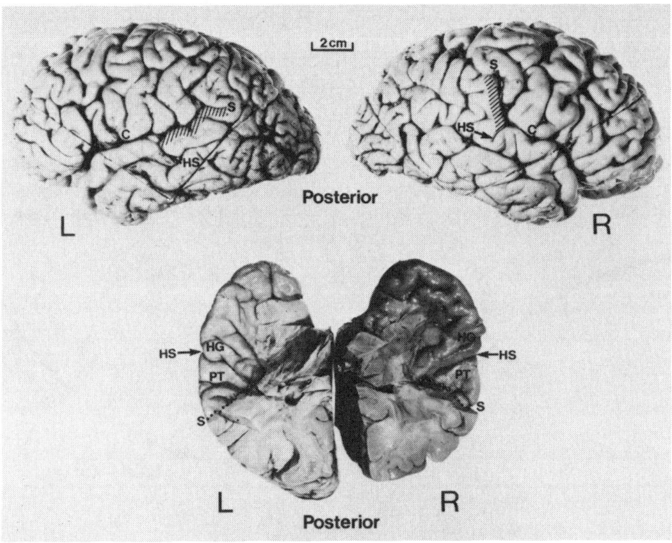

Figure 2. Views of the lateral surface of the two hemispheres and of the exposed superior surface of each temporal lobe of a human adult brain showing the typical asymmetries of a more anterior upswing of the Sylvian fissure on the right side; larger planum temporale on the left side; and a larger posterior parietal operculum on the left side (the region of the parietal lobe on the bank opposite the planum temporale). Note that the anterior end of the right temporal block is positioned with an upward tilt so that the planum is photographed in a horizontal plane. S, posterior end of the Sylvian fissure; C, inferior tip of the central sulcus; HS, Heschl sulcus; HG, Heschl gyrus; PT, planum temporale; dotted lines indicate the posterior boundary of the planum; hatched lines indicate the lateral edge of the planum. (Adapted from Witelson and Kigar, 1988a.)

Several studies have assessed neuroanatomical asymmetries in frontal opercular regions (pars opercularis and pars triangularis) involved in the lateralization of speech production. In general, there is a tendency for cortical extent in the left region to be greater than in the right, but without statistical significance. This lack of a reliable gross asymmetry in the general area of Broca's language region underlines the extent and reliability of the asymmetry in the parieto-temporal region (see Witelson and Kigar 1988a, for a review of the empirical reports).

In the few studies that addressed asymmetry in the histology of cortical regions, some indication of differences in *extent* of a particular cytoarchitectonic area has been noted. Little work has addressed the issue of quantitative microscopic right–left differences within a cytoarchitectonic area (Witelson and Kigar, 1988a).

The prominent parieto-temporal asymmetries are present early in life. For example, the planum temporale is reliably larger on the left side in neonates (e.g., Wada et al., 1975; Witelson and Pallie, 1973). Asymmetry in the development of Heschl's sulcus (the transverse sulcus, see Fig. 2), which delineates the planum temporale from Heschl's gyrus, occurs by about 31 weeks, when Heschl's sulcus appears on the right side. This is about two weeks earlier than its appearance on the left side in the majority of cases (Chi et al., 1977). Other asymmetries have been described as occurring even earlier. Fontes (1944) indicated that the course of the Sylvian fissure is more vertical (in relation to the anterior-posterior axis) on the right than left side in many cases in a series of 39 human fetal brains ranging in age from 16 to 40 months. LeMay and Culebras (1972) observed that the posterior end of the Sylvian fissure was higher on the right than left side in each of 10 fetal brains studied. Figure 3 presents this asymmetry in a 16-week-old fetus.

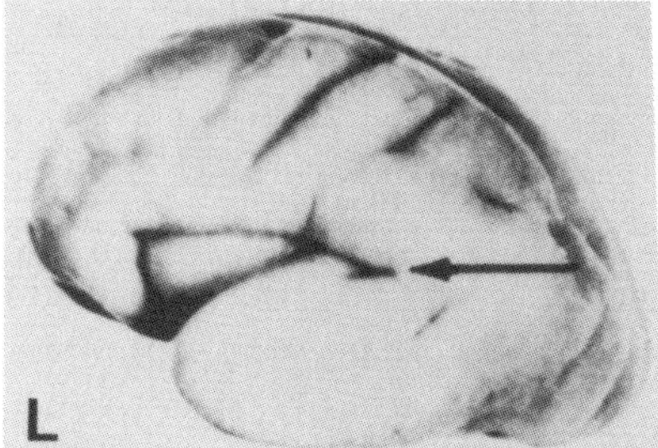

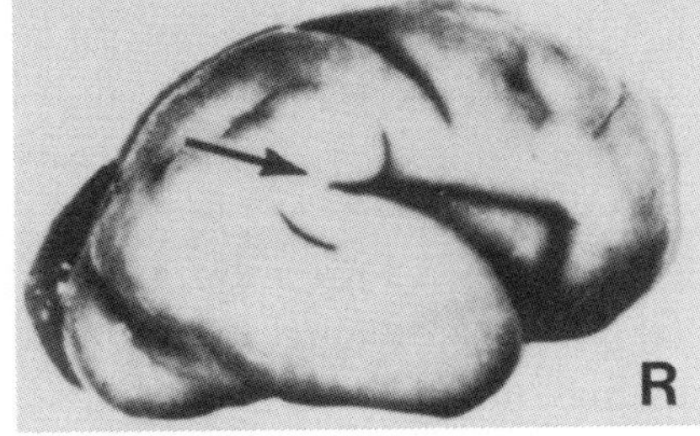

Figure 3. Views of the lateral surface of the left and right cerebral hemispheres of a 16-week-old human fetus, showing that the posterior end of the Sylvian fissure, indicated by arrow, is higher in the right hemisphere. (Adapted from LeMay and Culebras, 1972.)

Functional Asymmetry in Relation to Neuroanatomical Asymmetry

The relationship between functional and anatomical asymmetries is not delineated. Handedness was found to be associated with asymmetry in the inferred size of the parietal operculum (as determined by arteriograms) (Hochberg and LeMay, 1975; LeMay and Culebras, 1972); speech lateralization was found to be associated with asymmetry in the parietal operculum (Ratcliff et al., 1980) and in Sylvian fissure morphology (Strauss et al., 1985); but an equal number of studies failed to find similar associations (see Witelson and Kigar, 1988a). All these findings were based on measures from various *in vivo* imaging techniques. Correlational study of structure and function based on direct study of brain specimens is difficult to obtain.

Variation in Callosal Anatomy and Functional Asymmetry

Over the past ten years, work in my laboratory has been directed at studying relationships between variation in several gross neuroanatomical and histological measures derived from study of postmortem brains, and variation in neuropsychological measures of functional asymmetry in cognitively normal people. One focus has been on the corpus callosum. This structure is the main interhemispheric commissure connecting the neocortex of the right and left hemispheres. It was hypothesized that the gross anatomy of the corpus callosum might differ between right- and left-handers because left-handers as a group differ from right-handers in the pattern and magnitude of functional asymmetry.

Although the majority of studies with brain-damaged people have defined handedness predominantly by the hand used for writing, the model of handedness proposed by Annett (1985) was used in the studies in my laboratory. Annett's model suggests that people with consistent-right-hand (CRH) preference comprise a genetically different group from those showing mixed- or left-hand preference, that is, non-CRH preference. Specifically, the working hypothesis was that the less marked functional asymmetry or greater bihemispheric representation of cognitive functions in non-CRH compared to CRH may be associated with a larger corpus callosum, thus allowing for more interhemispheric communication.

Anatomical measurements were made from postmortem study of brain specimens. Many current MR imaging systems do not have sufficient resolution capabilities to measure the size of some of the regions that vary among people. The approach of the research involves prospectively testing seriously ill cancer patients. Although the patients have metastases, they are tested when they feel well and are ambulatory. The patients agree to participate in this research involving neuropsychological testing and an autopsy in the event of death. The details of our method are presented elsewhere (Witelson and McCulloch, 1991).

A larger total area of the midsagittal section of the corpus callosum was found in non-CRH than in CRH, based on a sample of 42 cases (Witelson, 1985a). This was found for the anterior and posterior halves of the corpus callosum, but not for the region of the splenium. It was suggested that the association of callosal area with handedness indicates that variation in callosal size is an anatomical substrate of variation in functional asymmetry.

To test this hypothesis further, it was suggested that particularly those regions of the corpus callosum that house axons from cortical regions predominantly involved in asymmetrical representation of cognitive functions should be particularly different between right- and non-right-handers. The posterior region of the body of the callosum, the isthmus, includes axons from parieto-temporal regions that are involved in the asymmetrical representation of many verbal, spatial, emotional, and attentional skills (see Fig. 4, region 6).

In an expanded group of 50 cases (Witelson, 1989), it was found that the posterior half of the body of the callosum was greater in non-CRH than CRH, particularly in the isthmus. The splenium and genu were not larger in non-CRH. Also noteworthy is the finding that the region of the body of the corpus callosum that houses the motor and premotor interhemispheric fibers was not different between CRH and non-CRH. This result suggests that it is not the motoric aspect of handedness per se that is related to callosal anatomy, but rather handedness as it reflects variation in asymmetrical representation of praxic or linguistic skills. With this larger group and further regional division of the corpus callosum, sex differences in callosal anatomy and in structure–function relationships were also noted. These are discussed in a later section.

Subsequent to the group of 50 cases, the brain specimens of seven more men were studied. Using discriminant functions derived from analysis of the previous group of 15 men in the sample of 50 cases (Witelson, 1989), hand-preference category was predicted correctly using isthmal area in six of the seven new male cases. In addition, the correlation between isthmal area and a hand-preference score (min/max = –12 to +12) was $r = -0.67$ ($n = 21$, $p = <.01$), indicating a strong relationship between anatomy of the isthmus of the corpus callosum and handedness in men

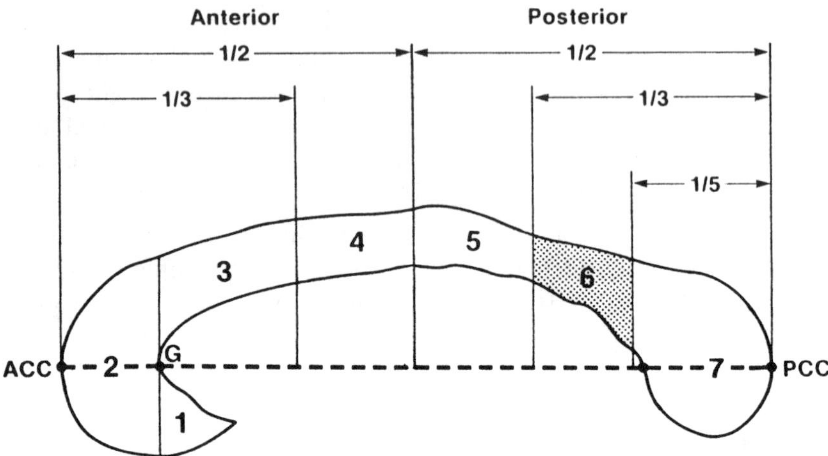

Figure 4. Diagram of the midsagittal view of the corpus callosum of the human adult, divided into seven subdivisions, numbered 1 to 7. ACC–PCC, defined by the anteriormost and posteriormost points of the callosum, was used as the linear axis to subdivide the callosum into anterior and posterior halves; anterior, middle, and posterior thirds; and the posterior one-fifth region (region 7), which is roughly congruent with the splenium. The line perpendicular to the axis at point G, the anteriormost point on the inner convexity of the anterior callosum, was used to define the anteriormost divisions of the callosum, roughly congruent with the genu (region 2) and the rostrum (region 1). Regions 3 and 4 constitute the anterior body of the callosum; regions 5 and 6, the posterior body. The region of interest here, the isthmus (region 6, stippled area), was defined as the posterior one-third of the corpus callosum minus the posterior one-fifth. The isthmus is 2/15 of the total area of the callosum, situated just anterior to the more posterior splenial region. (Adapted from Witelson, 1989.)

(Witelson and Goldsmith, 1991). The one MR study using relatively thin slices (5 mm) and a CRH versus non-CRH dichotomy similarly found larger callosal areas in non-CRH (Habib et al., 1991).

Axon Loss, Callosal Anatomy, and Functional Asymmetry

The corpus callosum varies with a behavioral index of hemispheric specialization at least in men. A larger corpus callosum could indicate a greater total number of axons. A larger callosal area could, of course, reflect other histological features, such as thicker axons, thicker myelin, or more myelinated relative to nonmyelinated fibers. There is extensive evidence that axon loss occurs as part of the regressive phase of brain development in prenatal and early postnatal periods (Cowan et al., 1984). In this neurobiological context, it was suggested that the larger corpus callosum of non-CRH is due, at least in part, to less axon elimination (Witelson, 1985a, 1989). There is unfortunately very little information on the histology of the human corpus callosum, even without regard to any possible association with behavior. The few reports available used light microscopy (Luttenberg, 1965; Tomasch, 1954), which could miss the smaller nonmyelinated fibers that have been demonstrated to exist in the monkey corpus callosum (LaMantia and Rakic, 1990a). Luttenberg (1965) did report smaller densities of axons in the corpus callosum at maturity than in early life, but he did not report total numbers of axons.

Considerable recent experimental work on nonhuman species has studied the developmental course of change in callosal axon number or in callosal connectivity patterns. Both types of studies indicate that few, if any, axons cross the midline after birth, and that there is extensive early loss of axons. For example, with quantitative analysis, there is evidence in cats (Koppel and Innocenti, 1983) and in monkeys (LaMantia and Rakic, 1990b) that the total number of fibers is greater soon after birth than at maturity, indicating that, as in other fiber tracts, there is a loss of exuberant axons, which develop during embryonic stages of brain development.

Loss of early developed axons is also indicated by the demonstration that specific connectivity patterns of callosal axons develop early in life through the loss of connections that were present at an earlier stage. The timing of the emergence of the adult patterns varies among different sensory modalities and different species. For example, in lower mammals such as rats and cats, the mature pattern of discontinuous distribution of callosal connections is finalized during early postnatal months (e.g., Innocenti and Caminiti, 1980; Ivy and Killackey, 1981). In contrast, in monkeys, the adult patterns of callosal connectivity emerge before birth. For example, they are present in the primary and secondary somatosensory cortex one month before birth (Killackey and Chalupa, 1986), and in the secondary visual cortex (area 18) within the last prenatal month (Dehay et al., 1988).

However, these findings do not indicate that environmental factors have no influence on anatomical development. With environmental conditions typical for the species, a particular pattern develops; but with altered sensory input, callosal connectivity is affected. For example, with reduction of peripheral stimulation by prenatal removal of the eyes in the monkey on embryonic (E) day 77, area 18 increases in size relative to area 17, and also develops with more neurons having callosal projections than in normal animals (Dehay et al., 1989).

These experimental results support the hypothesis that to the extent that variation in human callosal anatomy reflects differences in axon number, the variation may be due to a difference in the extent of axon loss, and not to a difference in the number of axons produced postnatally. These results also suggest that axon loss in the human brain may occur completely or mainly in the prenatal period. Accordingly, right-handers may undergo a greater loss of callosal axons during prenatal development compared to non-right-handers. To summarize, it was hypothesized that the majority of people have a relatively smaller corpus callosum related to relatively greater axon loss, and eventually manifest with CRH preference the typical pattern of functional asymmetry and with the pattern of cognitive skills Witelson associated with right-handedness. The minority of people have a relatively larger corpus callosum related to less axon loss and manifest as non-CRH with atypical patterns of functional asymmetry (Witelson, 1985a, 1989). Figure 5 presents this hypothesis schematically.

Timing of Axon Loss

The corpus callosum is visibly present as early as 10 weeks in the human brain (e.g., Rakic and Yakovlev, 1968). The size of the midsagittal section of the corpus callosum increases most rapidly in prenatal development. By approximately the end of the mid-trimester (26 E weeks), the corpus callosum has more than doubled in size. It then doubles by birth, and again by about age 2 years, when it reaches adult minimal values. Figure 6 presents a summary of the results of studies charting the growth of the corpus callosum in the human brain from conception to maturity (Witelson and Kigar, 1988b).

Axons may be lost due to death of the parent neuron or by loss of one collateral of a neuron (Purves and Lichtman, 1985). Neuron death may be associated with the development of synapses and the competition for connections. The first synapses in neocortex in the human brain appear by 23 E weeks (Molliver et al., 1973). Thus some callosal axons could die consequent to neuron death as early as 23 E weeks. On the basis of the timing of the emergence of adult patterns of callosal connectivity in the monkey, callosal axon loss may be essentially complete by 30–35 E weeks in human development. These results suggest a window from about 25 to 35 E weeks when most axon loss may occur in humans. Consistent with this proposed time frame of axon loss is Luttenberg's (1965) report that the density of axons in the human corpus callosum starts to decrease beginning at 30 E weeks.

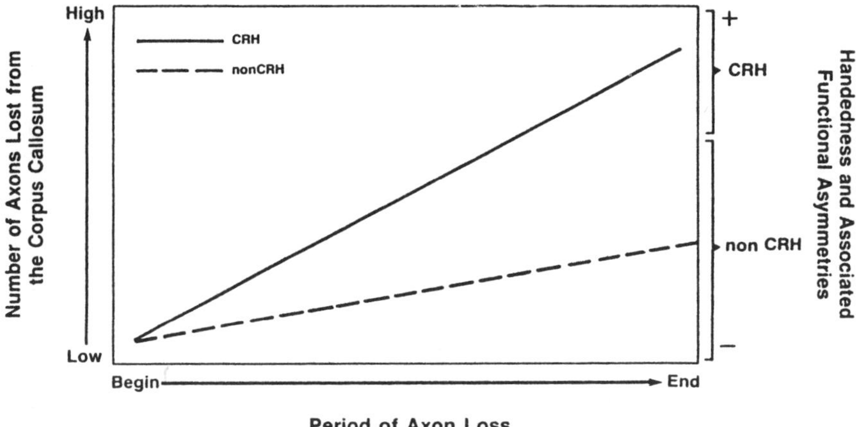

Figure 5. A schematic diagram of the proposed relationship between extent of axon loss in the corpus callosum and the pattern of handedness and associated functional asymmetries as a function of early developmental age in males. The abscissa represents the prenatal and early postnatal periods during which axons are lost. The left ordinate indicates the number of callosal axons lost. The greater the axon loss, the smaller the corpus callosum in the adult. The right ordinate represents the direction and magnitude of functional asymmetries (e.g., preference for use of the right hand or representation of linguistic functions in the left hemisphere), with "+" representing typical, and "–" atypical, asymmetry. The solid line shows the hypothetical developmental course of axon loss in consistent right-handers (CRH). In the case of hand preference, up to a certain point of axon loss, people manifest as non-CRH, and beyond, as CRH, as indicated on the right ordinate. The point above which CRH is manifested was chosen such that it is at 2/3 the range of the left ordinate, to represent the greater range of callosal size in non-CRH than CRH (see Fig. 3, Witelson and Goldsmith, 1991). The dashed line represents the developmental course for non-CRH. (Adapted from Witelson and Nowakowski, 1991.)

Prematurity and Axon Loss

Many retrospective and prospective studies of the cognitive consequences of prematurity indicate that among the many behavioral sequelae is a raised prevalence of left-handedness. A recent report is relevant to the present hypothesis of an association between loss of callosal axons and functional asymmetry. The effects of prematurity were studied as a function of birth weight and indirectly of gestational age (O'Callaghan et al., 1987). A group of very premature infants (birth weight less than 1000 g and gestational age at birth of about 26 weeks) had a 50% prevalence of left-handedness. This was significantly higher than for prematurely born infants with birth weight greater than 1000 g and than for full-term

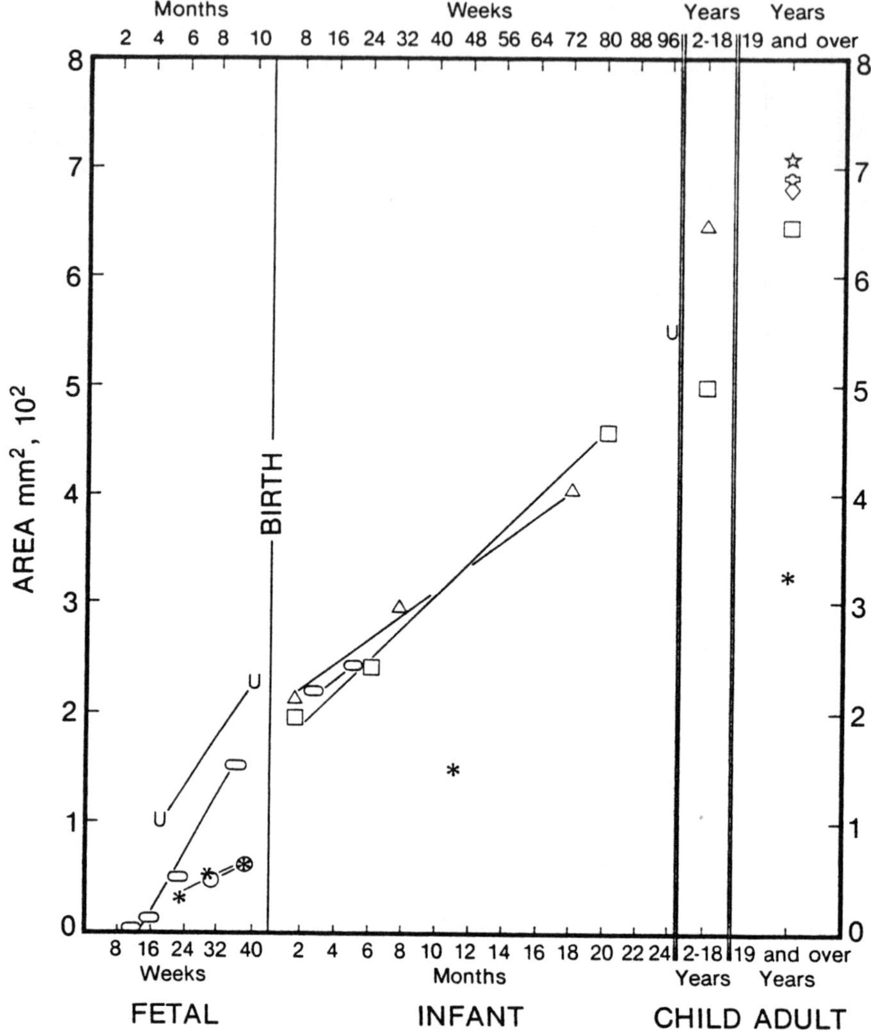

Figure 6. Summary of studies measuring total midsagittal area of the human corpus callosum to show its growth over development. Most data points are mean values; a few are scores for individual cases. The location of scale marks on the X-axis for fetal and infant age intervals corresponds to exactly the same time interval (8 weeks or 2 months). Data points of different studies are indicated by different symbols. Note that callosal area increases rapidly in prenatal life, doubling with the last trimester, and doubling again by postnatal age 2 years, by which time it is quite close to adult minimal values. (Adapted from Witelson and Kigar, 1988b.) Symbols: □ Bean (1906); △ Bell and Variend (1985); ∪ Clarke et al. (1986); ⊙ De Lacoste et al. (1986); ☆ De Lacoste-Utamsing and Holloway (1982); ◇ Holloway and de Lacoste (1986); ⊂⊃ Luttenberg (1965); * Rakic and Yakovlev (1968); ⌓ Witelson (1985a)

infants. If a definition of non-CRH were to be used, the proportion of left-handers to right-handers would likely be even greater in the 26-E-week group of infants. The intensive and varied stimulation that neonates necessarily receive compared to stimulation in the *in utero* condition could result in a different progressive course of synaptogenesis and regressive course of neuron death and axon loss. Thus, birth at 26 E weeks could interfere with the typical course of axon loss, resulting in the observed left-handedness and a larger corpus callosum. This issue has been elaborated elsewhere (Witelson and Nowakowski, 1991). A recent report of thumb sucking *in utero* observed via ultrasound investigation reported a preference for sucking the thumb of the right hand. Inspection of the results table indicates a tendency for a decreasing proportion of left thumb sucking in the sample as gestational age advanced from 15 weeks to term (Hepper et al., 1990). These results are consistent with the hypothesis presented here.

Clearly, it is not yet known whether the corpus callosum is, in fact, larger in very prematurely born infants, but this is a testable hypothesis. An association between prematurity and atypical neuroanatomical development was demonstrated in monkeys who showed atypical synaptogenesis subsequent to birth three weeks before term (approximately day E 146) (Bourgeois et al., 1989). Callosal anatomy was not considered in this report, but the monkeys in this study were likely not sufficiently premature for callosal anatomy to be affected. In summary, it is suggested that in human development under normal conditions there is variation in the rate or extent of axon loss, and individuals with less axon loss manifest themselves as non-right-handers and with cognitive characteristics associated with left-handedness. Under atypical conditions there may be an altered course of the typical progression, and this also results in less axon loss, subsequent non-right-handedness, and possible cognitive correlates.

Callosal Inhibition or Excitation in Hemispheric Specialization

The concept of interhemispheric inhibition via callosal activity was a widely held notion to account for the apparent increase in hemispheric specialization with chronological development (e.g., Moscovitch, 1977), which is more likely a manifestation of the decrease in functional plasticity (see Fig. 1). More recently, callosal inhibition has been proposed as a mechanism for the active suppression of secondary systems in the minor hemisphere; for example, of a secondary language system in the right hemisphere (Levy, 1985). This model is different from the one proposed (Witelson, 1987) in which functional asymmetry is present and constant from birth and recovery of function depends on the recruitment or realization of alternate systems given conditions of sufficient impetus such as brain damage, and not on the release of inhibited secondary systems present from the start.

In a model involving callosal inhibition of secondary systems, right-handers, having greater lateralization of functions, would be expected to have more callosal interconnections (although inhibitory), and a larger corpus callosum. This prediction is not consistent with the present anatomical results.

An alternate hypothesis is that the atypical pattern of relatively less or differently organized functional asymmetry in left-handers is associated with greater callosal excitation for greater interhemispheric communication. This model is consistent with the empirical result of a larger corpus callosum in non-right-handers. Two aspects of hemispheric specialization support the excitation hypothesis. Greater bihemispheric representation could be of two patterns: one, a "twinned" situation, in which each hemisphere has independent representation of a particular function; two, a "shared" situation in which regions of the two hemispheres form a combined substrate for a function. The data support the latter pattern. Results of speech representation on the basis of intracarotid amytal testing reveal that bilateral representation of speech is associated with some disruption of speech with injections of either side (Milner, 1974), indicative of neither side being sufficient alone. A "twinned" situation would be supported by the observation of a lack of speech disruption with injections to either hemisphere, which does not occur. Similarly, the higher incidence of transient aphasia in left- than in right-handers (Hecaen et al., 1981) is compatible with a "shared" type of representation. It would seem that a "shared" pattern of lateralization of related functions would require greater facilitation between hemispheres.

The second aspect supporting the callosal excitation model concerns the possibly greater dissociation of lateralization of related functions among left- than right-handers. In most right-handers, there is a strong association between the hemispheric representation of language and praxic functions. In contrast, in some left-handers, it appears that the right hemisphere plays the major role in motor learning and praxis, even when the left hemisphere is the major hemisphere for speech and language functions (Heilman et al., 1973). The fact that the majority of left-handers have predominant language representation in the left hemisphere but are still left-handed indicates that at least some part of the network of motor programming is represented in the non-language hemisphere. There is little study of apraxia following right- and left-sided lesions in large groups of left-handers, which would help clarify this issue. Greater interhemispheric communication may be needed for the execution of written language with such dissociated neural organization for language perception and praxis. The question remains what advantage(s) led to the development and continuation of such an apparently inefficient neural network in which information must cross the midline at some level for manual output of linguistic information represented in the other hemisphere.

Some right-handers also show a dissociation between hemispheric representation of language and praxis. Some right handers with left-hemisphere damage do not become apraxic (Heilman and Rothi, 1985). The proportion of right-handers who do not show aphasia with left-sided lesions is not known. Numerous "exceptional" cases have been documented of "crossed" aphasia, that is right-handers who suffer aphasia following right hemisphere damage (e.g., Fischer et al., 1991). The existence of such cases is consistent with the report of a minority (about 5 to 10%) of right-handers who show right-hemisphere speech with amytal testing (Rasmussen and Milner, 1977). Such cases of right-handers would also seem to require greater interhemispheric excitation or facilitation. The prediction follows that such right-handers may have a larger corpus callosum than those with typical lateralization. Such "crossed dextrals" may be the right handers who are not genotypically CRH, but in whom random or environmental factors led to a right-hand bias (Annett, 1985).

Callosal Anatomy and Neuroanatomical Asymmetry

Some association exists between the various hemispheric neuroanatomical asymmetries and measures of functional asymmetry. Some of these results were summarized in a previous section and are reviewed in detail elsewhere (Witelson and Kigar, 1988a). It follows that callosal anatomy and neuroanatomical asymmetry may show some relationship. Such studies remain to be done.

The question may be raised whether either anatomical variation (callosal or hemispheric asymmetry) drives the other. In this context it is noted that some anatomical asymmetries have been investigated early in ontogenetic development. In infants, asymmetry has been demonstrated in the size of the planum temporale on the right and left sides (e.g., Wada et al., 1975; Witelson and Pallie, 1973). Asymmetry in the planum temporale has been demonstrated as early as 31 weeks gestation (Chi et al., 1977). The earliest gross anatomical asymmetry noted is in Sylvian fissure morphology. By 16 weeks, the posterior end of the right Sylvian fissure appears higher than the left (see Fig. 3), the same pattern of asymmetry noted in adults (Witelson and Kigar, 1988a). This asymmetry is likely genetically determined and could result from any one of several factors, such as asymmetric cell proliferation, cell migration, cell differentiation, or some other histological factor. It is likely that regressive events in neural development have not yet started. This suggests that the early rapid phase of development of callosal fibers and the variation in axon loss postulated to be associated with functional asymmetries both occur in an already asymmetric brain. If in future studies, a correlation is found between measures of anatomic asymmetry and callosal anatomy, it would suggest that there may be some common factor leading to both, or that the pruning of callosal fibers is determined to some degree by asymmetry in hemispheric anatomy.

A Sex Factor in Structure–Function Relationships

The relationship of callosal anatomy to hand preference is different in the sexes. In men, hand preference can be predicted by callosal size (Witelson and Goldsmith, 1991). In women, no relationship was observed. This is so even though the variation in callosal anatomy is as great in women as in men. Callosal anatomy may prove to be correlated with other behavioral or cognitive measures in women, which may or may not also be found in men. But the finding remains that the anatomical basis of handedness is different between the sexes. Similar results were found for Sylvian fissure anatomy. In men, the morphology of the Sylvian fissure was different between CRH and non-CRH groups. In women, CRH and non-CRH did not show a difference in Sylvian fissure morphology (Witelson and Kigar, 1990).

Such sex differences in structure–function relationships are different than differences in absolute size per se, such as the findings of a larger isthmus or possibly larger splenium in women compared to men (see reviews in Clarke et al., 1989; Witelson, 1989).

What appears to be the most marked difference between the sexes is not differences in anatomy or anatomic asymmetry per se, but the relationship of anatomy to functional asymmetry. Such results require some factor to account for the sex difference. *Some biological factor related to sex must differentially and specifically influence neuroanatomical variation in relation to functional asymmetry in males versus females.* The most direct evidence for the operation of such a factor is the observation of right–left differences in the level of androgen receptors in the frontal and temporal association neocortex in the male fetal monkey brain, but of no consistent pattern in the distribution of receptors in females (Sholl and Kim, 1990). These results are strikingly analogous to the sex differences in human neuroanatomy observed in my laboratory. The sex difference in distribution of androgen receptors could conceivably allow the presence of early sex hormones to differentially affect brain development in the sexes: to affect asymmetry in males, and not in females. The key here is the asymmetric localization of receptors, at least in males. It is difficult to conceptualize how variation only in the level of sex hormones throughout a system could lead to anatomic asymmetry.

Summary of a Two-Factor Model

There may be two factors leading to brain anatomy as a substrate of functional asymmetry. One is genetic, leading to typical asymmetry in the human brain, whether male or female. This appears to be operative as the same basic anatomic asymmetry is present in both sexes. Therefore, it is unlikely that any factor that is different in the sexes, such as early levels of sex steroids, produces the basic pattern of anatomic asymmetry. A second factor is sex-related, either a genetic factor acting relatively directly, or a sex hormonal factor, acting on the neuroanatomically asymmetric brain and differentially influencing the relation of anatomy to function in the sexes. There is some evidence that the pattern or degree of anatomical asymmetry per se is minimally different between the sexes (e.g., Wada et al., 1975; Witelson and Kigar, 1990). This could be due to the operation of the sex-related factor. Such a two-factor model would allow for a typical neuroanatomical substrate in both sexes but a different relationship of the anatomy to at least some aspects of functional asymmetry. It would also be consistent with sex differences in the pattern of functional asymmetries (e.g., Beaton, 1985) and the sex differences in patterns of cognitive skills that are asymmetrically represented in the hemispheres (e.g., Hyde and Linn, 1988; Linn and Petersen, 1985).

The observation of neuroanatomical asymmetry in higher apes (e.g., LeMay et al., 1982) suggests that functional asymmetry in these species may also exist. The required neuropsychological studies are difficult to do. The observation of a consistent pattern of functional asymmetry within species of lower apes, even if less marked (e.g., for auditory perception, Heffner and Heffner, 1986; for visual perception, Hamilton and Vermeire, 1988), suggests some neuroanatomical asymmetry may also be present. These results could be related to the operation of a genetic factor. The operation of an additional sex-related factor might be operative in nonhuman animals too, but it would be less visible if operative on a much less asymmetric substrate to begin with.

Other long-standing difficult questions remain. Why neuroanatomical asymmetry? Why neuroanatomical asymmetry in a consistent direction? Why neuroanatomical asymmetry in the particular direction it is? What advantages accrue from the typical right–left organization of the human brain, and what, if any, advantages may be associated with atypical brain organization?

Acknowledgments. This research was supported by contract NS62344 and Grant NS18954 from the US National Institutes of Health, and Grant MA-10610 from the Medical Research Council of Canada.

References

Annett M (1985): *Left. Right. Hand and Brain: The Right Shift Theory.* New Jersey: Erlbaum

Bean RB (1906): Some racial peculiarities of the Negro brain. *American Journal of Anatomy* 5:353–432

Beaton A (1985): *Left Side. Right Side: A Review of Laterality Research.* London: Batsford

Bell AD, Variend S (1985): Failure to demonstrate sexual dimorphism of the corpus callosum in childhood. *Journal of Anatomy* 143:143–147

Bertoncini J, Morais J, Bijeljac-Babic R, McAdams S, Peretz I, Mehler J (1989): Dichotic perception and laterality in neonates. *Brain and Language* 37:591–605

Boller F, Grafman J, eds. (1988–1990): *Handbook of Neuropsychology*, Vols. 1–4, New York: Elsevier Science

Bourgeois J-P, Jastreboff PJ, Rakic P (1989): Synaptogenesis in visual cortex of normal and preterm monkeys: Evidence for intrinsic regulation of synaptic over production. *Proceedings of the National Academy of Sciences* 86:4297–4301

Bradshaw JL, Nettleton NC (1983): *Human Cerebral Asymmetry.* New Jersey: Prentice-Hall

Brown W, Jaffe J (1975): Hypothesis on cerebral dominance. *Neuropsychologia* 13:107–110

Bryden MP (1982): *Laterality. Functional Asymmetry in the Intact Brain*. Toronto: Academic Press

Bryden MP, Hécaen H, de Agostini M (1983): Patterns of cerebral organization. *Brain and Language* 20:249–262

Chi JeG, Dooling EC, Gilles FH (1977): Gyral development of the human brain. *Annals of Neurology* 1:86–93

Clarke S, Kraftsik R, Innocenti G, van der Loos H (1986): Sexual dimorphism and development of the human corpus callosum. *Neuroscience Letters* 26:S 299 [Abstract]

Clarke S, Kraftsik R, Van der Loos H, Innocenti G (1989): Forms and measures of adult and developing human corpus callosum: Is there sexual dimorphism? *Journal of Comparative Neurology* 280:213–230

Corballis MC (1983): *Human Laterality*. New York: Academic Press

Cowan WM, Fawcett JW, O'Leary DDM, Stanfield BB (1984): Regressive events in neurogenesis. *Science* 225, 1258–1265

Damasio H, Damasio A (1990): *Lesion Analysis in Neuropsychology* Oxford: Oxford University Press

Dehay C, Horsburgh G, Berland M, Killackey H, Kennedy M (1989): Maturation and connectivity of the visual cortex in monkey is altered by prenatal removal of retinal input. *Nature* 337:265–267

Dehay C, Kennedy H, Bullier J, Berland M (1988): Absence of interhemispheric connections of area 17 during development in the monkey. *Nature* 331:348–350

De Lacoste MC, Holloway RL, Woodward DJ (1986): Sex differences in the fetal human corpus callosum. *Human Neurobiology* 5:93–96

De Lacoste-Utamsing C, Holloway RL (1982): Sexual dimorphism in the human corpus callosum. *Science* 216:1431–1432

Fischer RS, Alexander MP, Gabriel C, Gould E, Milione J (1991): Reversed lateralization of cognitive functions in right-handers. *Brain* 114:245–261

Fontes V (1944): *Morfologia do Cortex Cerebral*. Lisbon

Geschwind N, Galaburda AM, eds. (1984): *Cerebral Dominance the Biological Foundations*. Cambridge: Harvard University Press

Habib M, Gayraud D, Oliva A, Regis J, Salamon G, Khalil R (1991): Effects of handedness and sex on the morphology of the corpus callosum: A study with brain magnetic resonance imaging. *Brain and Cognition* 16:41–61

Hamilton CR, Vermeire BA (1988): Complementary hemispheric specialization in monkeys. *Science* 242:1691–1694

Hannay HJ, ed. (1986): *Experimental Techniques in Human Neuropsychology*. New York: Oxford University Press

Hécaen H, de Agostini M, Monzon-Montes A (1981): Cerebral organization in left-handers. *Brain and Language* 12:261–284

Heffner HE, Heffner RS (1986): Effect of unilateral and bilateral auditory cortex lesions on the discrimination of vocalizations by Japanese macaques. *Journal of Neurophysiology* 56:683–701

Heilman KM, Coyle JM, Gonyea EF, Geschwind N (1973): Apraxia and agraphia in a left-hander. *Brain* 96:21–28

Heilman KM, Rothi LJG (1985): Apraxia. In: *Clinical Neuropsychology*, 2nd ed., Heilman KM, Valenstein E, eds. New York: Oxford University Press, pp. 131–150

Heilman KM, Valenstein E, eds. (1985): *Clinical Neuropsychology*, 2nd ed. New York: Oxford University Press

Hellige JB, ed. (1983): *Cerebral Hemisphere Asymmetry. Method, Theory, and Application*. New York: Praeger

Hepper PG, Shahidullah S, White R (1990): Origins of fetal handedness.*Nature* 347:431

Hiscock M (1988): Behavioral asymmetries in normal children. In: *Brain Lateralization in Children*, Molfese DL, Segalowitz SJ, eds. New York: Guilford Press pp. 85–169

Hochberg FH, LeMay M (1975): Arteriographic correlates of handedness. *Neurology* 25:218–222

Holloway RL, de Lacoste MC (1986): Sexual dimorphism in the human corpus callosum: An extension and replication study. *Human Neurobiology* 5:87–91

Hyde JS, Linn MC (1988): Gender differences in verbal ability: A meta-analysis. *Psychological Bulletin* 102:53–69

Hynd GW, Cohen M (1983): *Dyslexia: Neuropsychological Theory Research and Clinical Differentiation*. Toronto: Grune and Stratton

Innocenti GM, Caminiti R (1980): Postnatal shaping of callosal connections from sensory areas. *Experimental Brain Research* 38:381–394

Ivy GO, Killackey HP (1981): The ontogeny of the distribution of callosal projection neurons in the rat parietal cortex. *Journal of Comparative Neurology* 195:389–389

Kennard MA (1942): Cortical reorganization of motor function: Studies on series of monkeys of various ages from infancy to maturity. *Archives of Neurology and Psychiatry* 48:227–240

Killackey HP, Chalupa LM (1986): Ontogenetic change in the distribution of callosal projection neurons in the postcentral gyrus of the fetal rhesus monkey. *Journal of Comparative Neurology* 244:331–348

Kimura D, Harshman RA (1984): Sex differences in brain organization for verbal and non-verbal functions. *Progress in Brain Research* 61:423–441

Kolb B, Whishaw IQ (1990): *Fundamentals of Human Neuropsychology*. 3rd ed. New York: WH Freeman

Koppel H, Innocenti GM (1983): Is there a genuine exuberancy of callosal projections in development? A quantitative electron microscopic study in the cat. *Neuroscience Letters* 41:33–40

LaMantia A-S, Rakic P (1990a): Cytological and quantitative characteristics of four cerebral commissures in the Rhesus monkey. *Journal of Comparative Neurology* 291:520–537

LaMantia A-S, Rakic P (1990b): Axon overproduction and elimination in the corpus callosum of the developing rhesus monkey. *Journal of Neuroscience* 10:2156–2175

LeMay M, Billig MS, Geschwind N (1982): Asymmetries of the brains and skulls of nonhuman primates. In: *Primate Brain Evolution*, Armstrong E, Falk D, eds. New York: Plenum Press, pp.263–277

LeMay M, Culebras A (1972): Human brain - morphologic differences in the hemispheres demonstrable by carotid arteriography. *New England Journal of Medicine* 287:168–170

Lenneberg EH (1967): *Biological Foundations of Language*. New York: Wiley

Levy J (1985): Interhemispheric collaboration: Single-mindedness in the asymmetric brain. In: *Hemispheric Function and Collaboration in the Child*, Best C, ed. New York: Academic Press, pp.11–31

Levy J, Reid M (1978): Variations in cerebral organization as a function of handedness, hand posture in writing, and sex. *Journal of Experimental Psychology: General* 107:119–144

Lindesay J (1987): Laterality shift in homosexual men. *Neuropsychologia* 25:965–969

Linn MC, Petersen AC (1985): Emergence and characterization of sex differences in spatial ability: A meta-analysis. *Child Development* 56:1479–1498

Luttenberg J (1965): Contribution to the fetal ontogenesis of the corpus callosum in man. II. *Folia Morphologica* 13:136–144

McCormick CM (1990): A neuropsychological study of sexual orientation: Neurobiological implications. PhD thesis, McMaster University

McCormick CM, Witelson SF, Kingstone E (1990): Left-handedness in homosexual men and women: Neuroendocrine implications. *Psychoneuroendocrinology* 15:69–76

McGlone J (1980): Sex differences in human brain asymmetry: a critical survey. *The Behavioral and Brain Sciences* 3:215–263

McGlone J (1986): The neuropsychology of sex differences in human brain organization. In: *Advances in Clinical Neuropsychology*, Vol. 3, Goldstein G, Tarter RE, eds. New York: Plenum Press

Milner B (1974): Hemispheric specialization: Scope and limits. In: *The Neurosciences: Third Study Program*,Schmitt FO, Worden FG, eds. Cambridge: MIT Press, pp. 75–89

Molfese DL, Segalowitz SJ, eds. (1988): *Brain Lateralization in Children. Developmental Implications*. New York: Guilford Press

Molliver ME, Kostović I, Van der Loos H (1973): The development of synapses in cerebral cortex of the human fetus. *Brain Research* 50:402–407

Moscovitch M (1977): The developmental lateralization of language functions and its relation to cognitive and linguistic development: A review and some theoretical speculations. In: *Language Development and Neurological Theory*, Segalowitz SJ, Gruber FA, eds. New York: Academic Press, pp. 193–211

Netley C, Rovet J (1988): The development of cognition and personality in X aneuploids and other subject groups. In: *Brain Lateralization in Children. Developmental Implications*, Molfese DL, Segalowitz SJ, eds. New York: Guilford Press, pp. 401–416

O'Callaghan MJ, Tudehope DI, Dugdale AE, Mohay H, Burns Y, Cook F (1987): Handedness in children with birth weights below 1000 g. *Lancet* 1:1155

Ottoson D, ed. (1987): *Duality and Unity of the Brain*. England: MacMillan Press

Purves D, Lichtman JW (1985): *Principles of Neural Development*. Massachusetts: Sinauer Assoc

Rakic P, Yakovlev PI (1968): Development of the corpus callosum and cavum septi in man. *Journal of Comparative Neurology* 132:45–72

Rasmussen T, Milner B (1977): The role of early left-brain injury in determining lateralization of cerebral speech functions. *Annals of the New York Academy of Sciences* 299:328–354

Ratcliff G, Dila C, Taylor L, Milner B (1980): The morphological asymmetry of the hemispheres and cerebral dominance for speech: A possible relationship. *Brain and Language* 11:87–98

Sholl SA, Kim KL (1990): Androgen receptors are differentially distributed between right and left cerebral hemispheres of the fetal male rhesus monkey. *Brain Research* 516:122–126

Strauss E, LaPointe JS, Wada JA, Gaddes W, Kosaka B (1985): Language dominance: Correlation of radiological and functional data. *Neuropsychologia* 23:415–420

Tomasch J (1954): Size distribution, and number of fibres in the human corpus callosum. *Anat Record* 119:119–135

Wada JA, Clarke R, Hamm A (1975): Cerebral hemispheric asymmetry in humans. Cortical speech zones in 100 adults and 100 infant brains. *Archives of Neurology* 32:239–246

Witelson SF (1976): Sex and the single hemisphere: Right hemisphere specialization for spatial processing. *Science* 193:425–427

Witelson SF (1977a): Developmental dyslexia: Two right hemispheres and none left. *Science* 195:309–311

Witelson SF (1977b): Early hemisphere specialization and interhemisphere plasticity: An empirical and theoretical review. In: *Language Development and Neurological Theory*, Segalowitz SJ, Gruber FA, eds. New York: Academic Press, pp. 213–287

Witelson SF (1985a): The brain connection: The corpus callosum is larger in left-handers. *Science* 229:665–668

Witelson SF (1985b): On hemisphere specialization and cerebral plasticity from birth. Mark II. In: *Hemispheric Function and Collaboration in the Child*, Best C, ed. New York: Academic Press, pp. 33–85

Witelson SF (1987): Neurobiological aspects of language in children. *Child Development* 58:653–688

Witelson SF (1989): Hand and sex differences in the isthmus and genu of the human corpus callosum: A postmortem morphological study. *Brain* 112:799–835

Witelson SF, Goldsmith CH (1991): The relationship of hand preference to anatomy of the corpus callosum in men. *Brain Research* 545:175–182

Witelson SF, Kigar DL (1988a): Asymmetry in brain function follows asymmetry in anatomical form: Gross microscopic, postmortem and imaging studies. In: *Handbook of Neuropsychology*, Vol. 1, Boller F, Grafman J, eds. The Netherlands: Elsevier Science

Witelson SF, Kigar DL (1988b): Anatomical development of the human corpus callosum: Implications for individual differences and cognition. In: *Developmental Implications of Brain Lateralization*, Molfese DL, Segalowitz SJ, eds. New York: Guilford Press, p. 35–57

Witelson SF, Kigar DL (1990): Hand and sex differences in Sylvian fissure morphology. *Society for Neuroscience Abstracts* 6:381.13

Witelson SF, McCulloch PB (1991): Method for pre- and post-mortem measurement to study structure–function relations: A normal human brain collection. *Schizophrenia Bulletin* 583–591; commentary, 593–596

Witelson SF, Nowakowski RS (1991): Left-out axons make men right: A hypothesis for the origin of handedness and functional symmetry. *Neuropsychologia* 29:327–333

Witelson SF, Pallie W (1973): Left hemisphere specialization for language in the newborn: Anatomical evidence of asymmetry. *Brain* 96:641–646

Witelson SF, Swallow JA (1988): Neuropsychological study of the development of spatial cognition. In: *Spatial Cognition: Brain Bases and Development*, Stiles-Davis J, Kritchevsky M, Bellugi U, eds. Hillsdale, New Jersey: Lawrence Erlbaum, pp. 373–409

PART II
Neuroscience

Development and Aging in Down's Syndrome

K. E. Wisniewski and M. Dambska

The delayed development and impaired maturation of the central nervous system (CNS) in Down's syndrome (DS) is presented from the prenatal period until premature aging. The parameters examined by many authors, such as head measurements, brain weight and shape, and light and electron microscopic examination of selected cortical areas are discussed. The data suggest that the first differences between DS and non-DS brains appear during the second half of fetal life when several minor abnormalities in the number of cortical neurons and synaptic structure are seen. Postnatal development in DS looks evidently disturbed and retarded. The decreased neuronal density, reduced number of synapses, and abnormal features of these synapses are most accentuated during and after midinfancy. The distribution of neuronal changes reveals a particular deficit of granular neurons, leading to a deep reorganization of cortical structure. The heterogeneity of anomalies occurring during brain development that never reach a normal level overlap with the process of aging, which occurs in the third decade of life and even before this age. Neuropathological changes of the Alzheimer type, such as fibrillary tangles and senile plaques, develop 20 to 30 years earlier in the DS than the non-DS population, and in DS dementia occurs later than morphological features of aging. We hope that the review of the complex events occurring during delayed and even more abnormal development in DS cases overlapping with premature aging will be useful for further genetic study of the pathomechanism of this syndrome.

The incidence of children born with dysmorphic features that are characteristic enough for diagnosis of Down's syndrome is 1–1.4 per 1000 live births and includes various types and degrees of developmental disabilities. They often have congenital heart diseases and skeletal, gastrointestinal, brain, and other minor and major malformations. Impaired gonadal structure resulting in infertility (eventually sub-fertility), impaired immune response often leading to various infections, and finely increased incidence of leukemia also occurs in DS individuals (Benda, 1969). It is known that this syndrome is associated predominantly with trisomy of chromosome 21, and only a small percent of cases are associated with a translocation of a part of this chromosome or a mosaic of normal and trisomic cells (Kemper, 1988).

Developmental disabilities in DS are characterized by a various degrees of cognitive dysfunction and impairment of fine and gross motor coordination. In DS children, a delay of developmental milestones, particularly of language, is observed, and it never reaches a normal level. This level, although not satisfactory in a significant part of the DS population, may start very early to diminish. This diminishment may progress to dementia in middle-aged DS subjects. DS patients often also die earlier than the general population. Until now, this unfavorable clinical course, leading to a fatal end, had no proper pathogenetic explanation. This is above all expected from geneticists because of the chromosomal abnormality generally known as a cause of DS. They have to explain why development in DS subjects with trisomy 21 is disturbed, producing minor and major anomalies and impairment of central nervous system (CNS) maturation occurring at various stages. It is supposed that not a "consensus segment" of chromosome, but more probably an abnormal gene dosage in this aneuploidy syndrome, is responsible for this long-lasting process (Optiz and Gilbert-Barnes, 1990).

We will discuss the neuropathological changes described occurring in the CNS of DS individuals during their whole life from their prenatal development until aging with the view that such a review might be useful for further studies on this well-known but not definitively understood syndrome.

The brains of DS patients have been examined and described by many authors from the beginning of this century (Apert, 1914). The studies were continued (Gans, 1925), increasing in number particularly in recent years (Benda, 1969; Courchesne, 1988; Crome et al., 1966; Jacob, 1956; Kemper, 1988; Ross et al., 1984; Scott et al., 1983; Wisniewski,1990; Wisniewski et al., 1984, 1986; Wisniewski and Schmidt-Sidor, 1989). Morphological features of the CNS—gross, microscopic and ultrastructural—were investigated. Brain shape; brain weight; cortical organization, including number of neurons, synaptic density, and their particular features; myelination of brain pathways; occurrence of degenerative changes—all were analyzed at various ages.

Prenatal Development

Observations concerning prenatal development of the CNS in DS are rare. Sylvester (1983) found the brain smaller and the hippocampus less mature in one DS case than in a control case at 18 weeks of developmental age. A more exhaustive morphometric study performed by Schmidt-Sidor et al. (1990) in 15 DS and 8 non-DS age-matched fetuses from 15 to 22 weeks of gestational age demonstrated that during this span of time no evident differences in brain growth and maturation occurred. In both groups, the heads of fetuses were brachycephalic, with similar circumferences and configurations of frontal and temporal lobes and almost the same size cerebellum and brainstem. The development of the cerebral and cerebellar cortex, as far as myelination of the spinal cord and brain stem tracts, occurred at the same stage. Biochemical studies of the composition of gangliosides, neuronal cell adhesion molecules, and the activity of acetylcholinesterase also did not find differences in DS and non-DS fetal brains at this stage of development (Brooksbank et al., 1989). The first difference in brain weight between the DS and non-DS groups was observed by Wisniewski et al. (1986) and Schmidt-Sidor et al. (1990) at birth. They estimated the weights of DS newborn brains to be lower than those of non-DS controls, although they could be considered in the lower normal weight range. Benda (1971) considered the weight of DS newborn brains to be nearly normal.

Microscopic examination has permitted the discovery of some differences between DS and non-DS cerebral cortices in newborns. Neuronal density in the occipital and frontal areas was demonstrated as lower in DS than non-DS subjects at birth (Wisniewski et al., 1984, 1986). The morphology of neurons in these areas and their dendrites and synaptic spines estimated in the Golgi-rapid method by Takashima et al. (1981) looked until birth to be similar in both groups. In contrast, the same features estimated in the DS cases at the ultrastructural level revealed at the first day of life a decreased synaptic density, presynaptic length, and average surface area per synaptic contact (Wisniewski et al., 1986). Petit et al. (1984), also found anomalies in synaptic morphology but only a slight reduction in their density. The disturbed electrical membrane properties of the fetal brain were observed very early, since 16 weeks of gestational age (Scott et al., 1982, 1983).

The results cited above revealed the first observed differences between DS and non-DS newborn brains and permitted the estimate that the differences start during the second half of fetal development. The data related by various authors, which were in some points controversial, depending on the population examined and the methods used for examination, prove that the degree of observed changes was moderate. Nevertheless, particular features found in the neuronal population of the cerebral cortex in DS newborns testify that the changes occurring so early during brain development are profound and may be important for further development.

Postnatal Development from Birth to Adulthood

After birth, the changes occurring in developing DS brains become much more evident than those observed during fetal life. They are particularly accentuated during early childhood. This is the time when general development of DS children is slow, and their differences from the normal population increase with age. Cognitive function in DS children declines in comparison with that in controls. The body length and body weight drops from the 0.5 standard deviation (SD) at birth to 1.5 and 2.0 SD at 7 years of age (Cronk, 1978; Cronk and Pueschl, 1983). Head circumference became even 3.0 SD lower than expected during the first two years of age (Roche, 1966) and in 20% it remains in the lower normal range (Wisniewski, 1990). The decrease in expected brain weight in DS infants is parallel to the increasingly evident small head. The differences between DS and non-DS brain weight observed by some authors increase after birth. This is particularly visible between two and four years of age, and it was concluded that the "most evident brain weight retardation is seen after midinfancy" (Schmidt-Sidor et al., 1990; Wisniewski et al., 1986, 1990). The degree of brain weight reduction was evaluated as 24% (Crome et al., 1966) or 30% (Wisniewski et al., 1986) in DS infants and children as compared to control age-matched subjects. Therefore, the DS cases can often be diagnosed as microcranial and microencephalic.

Figures 1 and 2 present data from our recent observations concerning the reduction in brain weight of DS infants and children, increasing after 5–8 months and continuing up to 5 years. As a result of examining in both DS and non-DS brains, Schmidt-Sidor et al. (1990) have verified that lower brain weight is associated with gastrointestinal malformations and with congenital heart diseases so far in both DS and non-DS children. The shape of DS brains observed in the majority of cases presents some particular features. The first one observed by many authors, shortened dimension of antero-posterior diameter, was attributed to some reduction of the frontal lobes (Benda, 1971; Crome and Stern, 1972; Schmidt-Sidor et al., 1990; Wisniewski et al., 1986; Wisniewski, 1990), which was

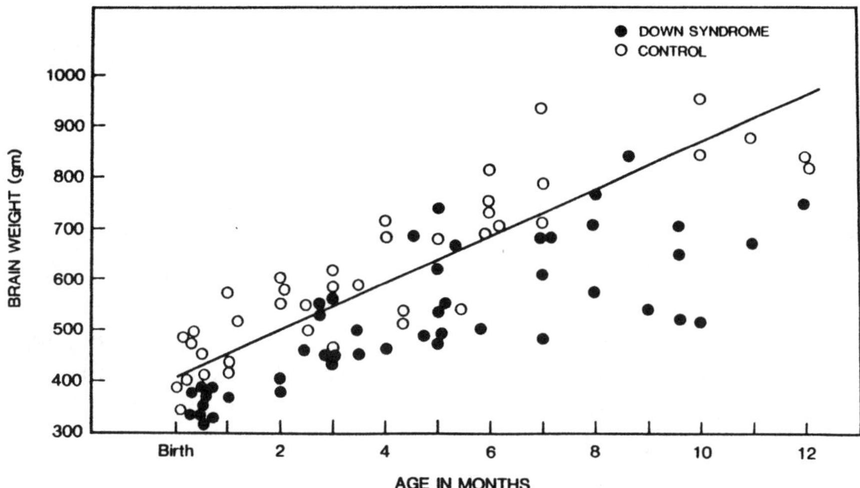

Figure 1. Brain weight in 47 cases of Down's syndrome and 40 control cases from birth to 1 year.

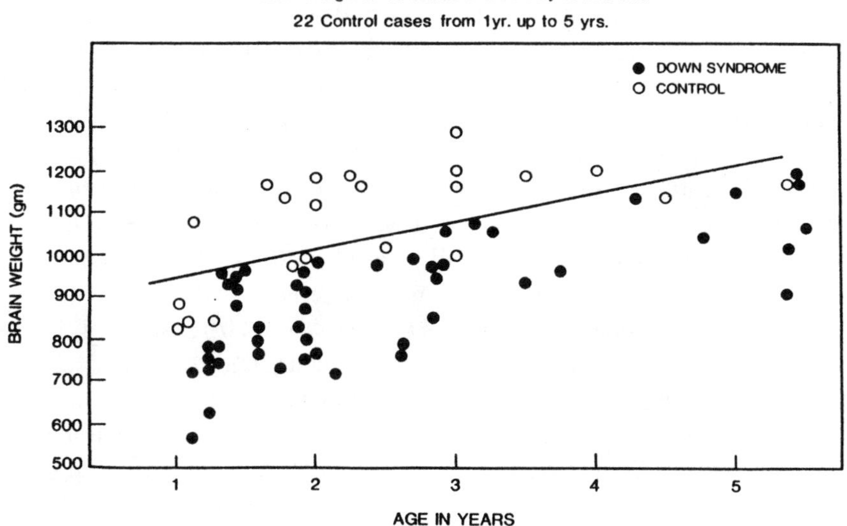

Figure 2. Brain weight in 49 cases of Down's syndrome and 22 control cases from 1 year up to 5 years.

not evident in other studies (Kemper, 1988). Superior temporal gyri in one or both hemispheres were often estimated as evidently narrow. This phenomenon was found in 33% of cases in Schmidt-Sidor material and in 50% of brains examined by others (Friede, 1989; Zellweger, 1977). The small size of the cerebellum and brainstem in DS as compared with non-

DS individuals has been reported many times (Crome et al., 1966; Friede, 1989; Gulotta and Rehder, 1974; Wisniewski, 1990). Even the ratio of cerebellum and brainstem weight to cerebrum weight was estimated by Crome et al. (1966) who found it to be 1:9 in DS and 1:7 in non-DS subjects.

The developmental disability in DS children encouraged studies concerning neuronal density in various brain structures, particularly in the cortex, as was previously cited in newborns. The reduced neuronal density in DS was reported very early (Apert, 1914; Davidoff, 1928). Sylvester (1983), described a decreased total number of neurons in the hippocampus. Zellweger (1977) noted the difficulty in recognizing cortical layers because of cytoarchitectonic changes. The morphometric study of neuronal density in three areas of the cerebral cortex from birth to 14 years of age demonstrated its statistically significant decrease in the occipital cortex in all groups, in the frontal cortex in the youngest group, and in the hippocampus after 11 years of age in DS children compared with control cases (Wisniewski et al., 1984, 1986). The neuronal distribution in examined cortical areas showed that the neuronal density in the visual and hippocampal cortex occurred particularly in the II and IV layers. In the frontal cortex, which normally contains few granular cells, the decrease in number of neurons was not as advanced as compared with controls. This type of changes was in agreement with results of Ross et al. (1984), who found the deficit of small neurons in the cortex in two DS brains. The cortical neurons in DS are not only reduced in number, but they present also the reduction of synaptic spines. Such observations were done using the Golgi-rapid method by Takashima et al. (1981), who found the dendritic spines after the fourth postnatal month not only reduced in number but also very thin. The results of Becker et al. (1986), performed on the visual cortex in DS children from four months to five years of age, were similar. Suetsugu and Mehraein (1980) observed the diminished number of synaptic spines in hippocampal and cingulated gyri in DS subjects from 3 to 25 years of age. The changes not only persist to adulthood, but they also overlap with the aging process, as was found by Takashima et al. (1989) in DS cases older than 30 years of age. Ultrastructural observations of synaptic morphology in the visual cortex from the first day of life until 18 years of age, evaluated with morphometric methods, were made by Wisniewski et al. (1986). They confirmed that synaptic length and surface area per synaptic contact were lower in DS cases than in non-DS controls. Myelination of cerebral pathways, another important parameter of brain maturation, was examined by Wisniewski and Schmidt-Sidor (1989). They found myelination delay in

22% of examined cases. The myelination was most evidently retarded during the first year of life, the time of the most intensive progress of myelination within cerebral hemispheres, but in some cases was seen even later. The topography of tracts retarded in the myelination process concerned, (Wisniewski and Schmidt-Sidor, 1989) "tracts with late beginning and protracted cycle of myelination." This phenomenon was observed in cortico-pontine and cortico-spinal tracts, corona radiata, corpus callosum, and frontal and temporal lobes. Less advanced myelination was also found in the U-fibers of three of seven cases.

Patients with DS were previously known rarely to survive to adulthood. Improvements in clinical treatment of congenital heart disease which often occurs in DS, comprising treatment of infections, for which DS patients are particularly susceptible, resulted in longer survival of DS patients. DS individuals reaching the fifth, sixth, and even seventh decades of life are no longer the exception they used to be, although 41% suffer from congenital heart disease or other serious diseases (McGrother and Marshall, 1990; Thase, 1988). The problem of aging in this group of mentally retarded people became actual, although the "precipitated senility" has been observed in DS since this syndrome was characterized (Evenhuis, 1990).

The clinical diagnosis of dementia in mentally retarded DS individuals brings many difficulties. Premature regression in memory, temporal orientation, and other capacities can be evaluated only in the group of patients with relatively high IQ values (Lai and Williams, 1989). In more profoundly mentally retarded patients a decline in everyday activities, loss of motivation, decrease of social interaction, and apathy were cited as signs of progressing dementia, but seizures were found as its most distinctive feature (Lott, 1986; Rabe et al., 1990). The computerized tomography, if performed, revealed progressive brain atrophy accompanying dementia in older DS patients (Lott, 1986; Shapiro et al., 1989).

Before the clinical observations concerning dementia in DS individuals were made of representative groups of patients, the premature appearance of age-related changes in their brains was reported (Jervis, 1948). Bertrand and Koffas (1946) found in DS brains senile plaques (SP) and also perivascular mineralization in the globus pallidus. In a more recent study, the basal ganglia calcifications that are accepted to be symptoms of premature aging were described in all of 30 examined DS brains, even in the first decade of life (Wisniewski et al., 1982). Brain atrophy, documented by narrow pallial gyri and dilated lateral ventricles, was also seen (Haberland, 1969; Kemper, 1988; Wisniewski et al., 1982). This observation is in agreement with the age-related decrease in brain weight in DS verified by Solitare (1990). The most important observations were those concerning morphological changes of the Alzheimer type in DS brains.

Alzheimer's disease could be diagnosed in cases with progressive presenile dementia (starting before the age of 65 years). The older cases had senile dementia, which often is of the Alzheimer type. This diagnosis is based on the similarity of the changes in the CNS, consisting of the appearance of two characteristic features: neurofibrillary tangles (NFT) and senile plaques (SP) (Khachatur-iants, 1985). NFT could be seen in lower numbers in old but not demented people and in individuals with various neurological conditions, e.g., metabolic and same viral infections. Senile plaques composed of amyloid fibers which occur in aging, are very numerous in Alzheimer's disease and were found in some experimental unconventional viral diseases (Wisniewski and Merz, 1984). The excessive number of senile plaques was recently recognized as being caused by overproduction of amyloid β-protein, which seems to be the main etiological factor in AD (Rabe et al., 1990).

The appearance of NFT and a great number of SP in young adult DS individuals has been described (Kemper, 1988). The studies performed to correlate the clinico-pathological data became the point of principal interest. This type of prospective longitudinal study performed on seven DS patients (Wisniewski et al., 1985a) demonstrated that they all developed dementia. Postmortem examination of their brains, including a morphometric study, revealed a great number of SP and NFT (more than 20/1.5 x 106 nm^2 in their prefrontal and hippocampal cortices). Plaques and tangles were also found in other regions of the brains. This finding led to the diagnosis of Alzheimer's disease in developmentally retarded DS patients, who started experiencing dementia between 43 and 57 years of age. The earlier studies were reviewed (Dalton and Wisniewski, 1990; Wisniewski and Rabe, 1986) and a new one, a retrospective clinico-pathological study of 49 cases was performed concerning the incidence of dementia (Wisniewski et al., 1985a). It was found that more aging DS subjects became demented than did individuals in the general population. In Lai and Williams's study (1989) material dementia was diagnosed in 51% of 96 persons with DS examined; in other groups of cases, it was clinically detected even more frequently (Evenhuis, 1990; Wisniewski et al., 1985b). In the same material, more than 50% of cases, some of them even younger than 30 years of age, demonstrated plaques and tangles; they were visible sporadically in DS subjects younger than 15 years of age. It was concluded that "plaques and tangles developed 20–30 years earlier... than is known in the non-DS population" (Wisniewski et al., 1985b). The large number of plaques is consistent with pathological production of amyloid β-protein for which the particular gene was detected on chromosome 21 (Rabe et al., 1990). The reduction in the number of dendritic spines in the hippocampal cortex in DS patients with associated AD changes was found to be greater than in age-matched controls without dementia (Ferrer and Gulotta, 1990). The correlation of AD changes with the clinical course demonstrated that dementia in DS occurs later than would be expected according to the age in which morphological changes were detected. A review of previous studies reveals that the great majority of DS patients develop a large number of plaques and tangles in the second half of the fourth decade of life, and in about 50% of them dementia could be observed 10 to 20 years later (Rabe et al., 1990).

This discrepancy between the clinical and morphological features of AD prompted comparisons of other eventual abnormalities in AD and DS. The deficiency in the brain neurotransmitters known in AD was studied in DS individuals. It was found that in adult and aging DS subjects the deficiency is similar to those observed in AD patients from the general population. This deficiency concerned cortical deficit in choline acetyltransferase and serotonin whose level was lower than normal, and norepinephrine and somatostatine, which also were present in lower than normal levels. The number of neurons was reduced in the regions in which the level of neurotransmitters was lower than in controls, as mentioned above. Only the dopamine levels looked normal. Some investigators suggest that deficits in neurotransmitters could be congenital in DS (Godridge et al., 1987; Pierotti et al., 1986; Rabe et al., 1990; Shortridge et al., 1985; Yates et al., 1980, 1983). This idea has to be verified in relation to the previously cited study (Brooksbank et al., 1989), which suggests that neuronal maturation looks normal from the biochemical point of view until midgestation stage.

Other recent studies try to estimate various aspects of brain metabolism. Interesting results have been found regarding the levels of Vitamin E, which seem to be lower in demented than nondemented DS subjects. All correlations between DS aging individuals and AD dementia occurring in the normal population could not change the evidence that only some older DS patients develop a progressive dementia despite the nearly general appearance in DS subjects of SP and NFT earlier than in the normal population (Dalton and Wisniewski, 1990). The topography of those changes also requires further study and interpretation particularly as far as the whole problem of relations between DS and AD.

Discussion

The course of CNS development and maturation in DS as presented above appears to be similar to normal until midgestation, unless a malformation appears during organogenesis. Morphological observations, particularly those concerning the prenatal development and maturation of the CNS in DS subjects, are, after that age, not normal. We are constrained to compose a mosaic of well-established fragments and to try to find those features that offer an approach for reconstructing the mechanism of events.

Such parameters as head measurements, brain weight and shape, microscopic pictures of the cerebral and cerebellar cortex, and myelination of brainstem and spinal cord indicate apparently normal CNS development until midgestation age (Schmidt-Sidor et al., 1990). The first presage of functional deviation observed earlier than morphologic anomalies were abnormal electrophysiological membrane properties in the fetal brain observed by Scott et al. (1982, 1983).

The second stage of development, examined in many different ways, was brain development and maturity at birth. Many reported data were controversial, but to a moderate degree, depending on the material examined. Nevertheless, it was possible to note that at this stage of development the brain weight starts to be lower in the DS than in the normal population, and some differences in the cortical structure in both groups are present. Lower neuronal and synaptic density (Wisniewski et al., 1984, 1986), and reduced synaptic length and average surface area per synaptic contact were noted (Petit et al., 1984; Wisniewski et al., 1986). Although only moderate, the above-mentioned differences indicate that structural changes in the DS brains start during the second half of fetal life.

Postnatal development in DS is much more evidently disturbed and retarded. Brain weight and shape indicate that the intensive changes start early after birth and are most accentuated after midinfancy (Wisniewski et al., 1986; Wisniewski, 1990). It is not rare that DS children are becoming microencephalic. The essential changes are observed on the CNS cortical level. A detailed analysis of other parts of the CNS grey matter is until now only fragmentary or does not exist at all. Cortical maturation comprises growth of the neuronal body, axons and dendrites, formation of synapses, final arrangement of neuronal contacts, and myelination of nerve tracts preceded by myelination gliosis. While all these processes occur, the abnormal features of DS cortex come to light. Decreased neuronal density has been reported by many authors (Sylvester, 1983;

Kemper, 1988; Wisniewski et al., 1984, 1986). Reduced numbers of synapses and their abnormal features have also been seen (Becker et al., 1986; Takashima et al., 1981; Wisniewski et al., 1986). It was found that these parameters follow the trends appropriate for normal development but are reduced at each stage as compared with controls. The disturbed synaptic morphology, as noted above, persists until adulthood (Suetsugu and Mehraein, 1980; Wisniewski et al., 1986) and even overlaps with the process of aging (Takashima et al., 1989).

The distribution of neuronal changes in the cortex of DS brains appears as a particular deficit of small neurons. The decrease in neuronal density in so-called granular layers II and IV is more important than in other layers, as compared with age matched controls (Ross et al., 1984; Wisniewski et al., 1984, 1986). From the morphological point of view, we observed a deep reorganization of cortical structure, which is best visible in areas of granular cortex, where are also observed the changes in synaptic density. These observations can be made in both the DS and normal cases with parallel regularity: an increase in number of synapses and then a reduction, reflecting physiological elimination of neurons and neuronal contacts. At each stage, all those parameters are lower in DS cases. Many authors emphasize that this abnormality involving cortical interneurons may play an important role in any impairment of cognitive functions and behavioral abnormalities.

The process of impaired development documented in selected areas of cerebral cortex could be the result of diminished production of neurons and/or their elimination greater than in the general population. The first eventuality cannot be excluded. This hypothesis is possible because of the evidently diminished neuronal density in DS newborns in the frontal cortex, which is the "youngest" from the histogenetic point of view and is where neuronal migration is prolonged. Therefore, it is possible that the last division occurs relatively late in fetal life.

On the other hand, the normal appearance of cerebral cortex in DS fetuses after half gestation, when the majority of cortical neurons are postmitotic, allows us to question an additional cause of neuronal deficit. Until now, the described changes have concentrated for the most part on a definite population of nerve cells, and they are particularly pronounced in the visual cortex, not according to division and migration timing, but rather morphological and physiological cell type. The granular cells from cortical layers II and IV, supposed to be phylogenetically of rather recent origin, develop their important intracortical connections relatively late. The other types of neurons seem to be less involved in changes occurring in the DS cortex during definitive cortical organization.

This is the time when elimination of supernumerary neurons, their axons, dendrites, and synapses and rearrangement of connections occur (Huttenlocher, 1984). It is very probable that these granular neurons are more likely to be excessively eliminated in DS individuals than in the general population. The cortex that is poor in neurons and of a worse quality could cause a delayed and less intensive myelination, which was found in the cortico-fugal, association, and intercortical fibers of the temporal and frontal lobes, which were particularly delayed in the myelination process (Wisniewski and Schmidt-Sidor, 1989). This was also observed in some cases in U-fibers. The time of the most advanced developmental deficit of the brain fits well with the time of the normally most intensive neuronal growth and myelination of hemispheric pathways.

In living subjects with DS, a more or less advanced degree of developmental disability is present from the time of infancy because their nervous system is not able to develop to a normal level. Moreover, they manifest early the tendency to degenerative change. The Alzheimer type of pathology appears in DS brains 20 to 30 years earlier than in the general population; NFT tangles and SP plaques become numerous (Wisniewski et al., 1985a,b). The eventual positive correlations between DS and Alzheimer's disease are of particular interest because both these entities require further studies to clarify the process of aging. The tendency to consider the changes in the CNS of DS individuals as a natural model of AD, a cerebral form of amyloidosis, is currently accepted (Rabe et al., 1990). On the other hand, the genetic studies concerning both entities have until now lacked enough data that could contribute to a better understanding of real relationships between Down's syndrome and Alzheimer's disease.

Acknowledgments. This work was supported by NIH grant PO1HD22634. The authors wish to express their appreciation to Maureen Stoddard Marlow for editorial assistance, Madeline Tinney for excellent secretarial assistance, and Lawrence Black for bibliographical assistance.

References

Apert E (1914): Mongolism. *Le monde medicale* 24:201

Becker LE, Armstrong DL, Chan F (1986): Dendritic atrophy in children with Down's syndrome. *Ann Neurol* 20:520–526

Benda CE (1969): *Down's Anomaly*, 2nd Edition, New York: Grune Stratton

Benda CE (1971): Mongolism. In: *Pathology of the Nervous System*, Vol. 2. Mencken J, ed. McGraw-Hill Inc., pp. 1361–1371

Bertrand T, Koffas D (1946): Cas d' idiotie mongolienne adulte avec nombreuses plaques senileg et concretions calcarine pallidales. *Rev Neurol* 78:338–345

Brooksbank BWL, Walker D, Balazs R, Jorgensen OS (1989): Neuronal maturation in the fetal brain in Down's syndrome. *Early Hum Dev* 18:237–246

Courchesne E (1988): Physioanatomical considerations in Down's syndrome. In: *The Psychobiology of Down Syndrome*, Nadel L, ed. Cambridge, MA: MIT Press, pp. 291–313

Crome L, Covie V, Slatter E (1966): A statistical note on cerebellar and brain stem weight in mongolism. *J Ment Defic Res* 10:69–72

Crome L, Stern J (1972): *Pathology of Mental Retardation.* Baltimore: William and Wilkins

Cronk CE (1978): Growth of children with Down syndrome: birth to age 3. *Pediatrics* 61:564–568

Cronk E, Pueschel S (1983): Autopometric studies. In: *A Study of the Young Child with Down Syndrome.* Pueschel SSM, ed). New York: Hum Sci Press, pp. 105–141

Dalton AJ, Wisniewski HM (1990): Down's syndrome and the dementia of Alzheimer's disease. *Int Rev Psychiatr* 2:41–50

Davidoff ML (1928): The brain in mongolian idiocy. *Arch Neurol Psvchiatr* 20:1229–1257

Evenhuis HM (1990): The natural history of dementia in Down's syndrome. *Arch Neurol* 47:237–263

Ferrer J, Gulotta F (1990): Down's syndrome and Alzheimer's disease: dendritic spines in the hippocampus. *Acta Neuropath* 79:680–685

Friede RL (1989): *Developmental Neuropathology.* Berlin: Springer-Verlag

Gans A (1925): Anatomische Beobachtungen bei der mongoloiden Idiotic. *Nederl Tijdschr Geneesk* 69:922–925

Godridge H, Reymonds GP, Czudek C, et al. (1987): Alzheimer-like neurotransmitter deficit in adult Down's syndrome brain tissue. *J Neurol Neurosurg Psychriatr* 50:775–778

Gulotta F, Rehder H (1974): Chromosomal anomalies and central nervous system. *Birth Pathol* 152:74–80

Haberland C (1969): Alzheimer's disease in Down's syndrome: clinical-neuropathological observations. *Acta Neurol Bela* 69:369–380

Huttenlocher PR (1984): Synapse elimination and plasticity in developing human cerebral cortex. *Am Ment Def* 88:488–496

Jacob H (1956): Mongolism. In: *Hanbuch der speziellen pathologischen Anat. u. Histologie.* Vierter Teil, Lubarsch O, Henke F, Rossle E, eds. Berlin: Springer-Verlag, pp. 82-98

Jervis GA (1948): Early senile dementia in mongoloid idiocy. *J Psychiatr* 145: 102–106

Kemper TL (1988): Neuropathology of Down syndrome. In: *The Psychobiology of Down's Syndrome.* Nadel L, ed. Cambridge, MA: The MIT Press, pp. 269–289

Khachatur-iants LS (1985): Diagnosis of Alzheimer disease. *Ann Neurol* 42:1097–1105

Lai F, Williams R (1989): Alzheimer's dementia in Down's syndrome. *Arch Neurol* 46:849–853

Lott JT (1986): The neurology of Down syndrome. In: *The Neurobiology of Down Syndrome.* Epstein CJ, ed. New York: Raven Press, pp. 17–25

McGrother CW, Marshall B (1990): Recent trends in incidence, morbidity and survival in Down's syndrome. *J Neurol Def Research* 34:49–57

Optiz JM, Gilbert-Barnes EF (1990): Reflections on the pathogenesis of Down syndrome. *Am Med Genet* (in press).

Petit TL, LeBoutillier JC, Alfano DP, Becker LE (1984): Synaptic development in the human fetus. A morphometric analysis of normal and Down' s syndrome neocortex. *Exp Neurol* 83:13–23

Pierotti AR, Harmor AJ, Simpson J, et al. (1986): High-molecular-weight forms of somatostatin are reduced in Alzheimer's disease and Down's syndrome. *Neurosci Lett* 63:141–146

Rabe A, Wisniewski KE, Schupf N, Wisniewski HM (1990): Relationship of Down's syndrome to Alzheimer's disease. In: *Application of Basic Neuroscience to Child Psychiatry*, Deutsch S, Weizman A, eds. R. Weizman Plum Publ Corp, pp. 325–340

Roche AF (1966): The cranium in mongolism. *Acta Neurol Scand* 42:62–78

Ross MH, Galaburda AM, Kemper TL (1984): Down's syndrome: Is there a decreased population of neurons? *Neurology* 34:909–916

Scott BS, Petit TL, Becker LE, Edwards BAV (1982): Abnormal electric membrane properties of Down's syndrome DRG neurons in cell culture. *Dev Brain Res* 2:257–270

Scott BS, Beeker LE, Petit TL (1983): Neurobiology of Down's syndrome. *Progress in Neurobiology* 21:199–237

Schmidt-Sidor B, Wisniewski KE, Shepard TH, Sersen EA (1990): Brain growth in Down syndrome subjects 15 to 22 weeks of gestational age and birth to 60 months. *Clin Neuropath* (in press).

Shapiro MB, Luxemburg JS, Kaye JA, Haxby JV, Friedland RP, Rapaport SJ (1989): Serial quantitative CT analysis of brain morphometrics in adult Down's syndrome of different ages. *Neurology* 39:1349–1353

Shortridge BA, Vogel FS, Burger PC (1985): Topographic relationship between neurofibrillary change and acetylocholinesterase-rich neurons in the upper brain stem of patients with senile dementia of the Alzheimer's type and Down's syndrome. *Clin Neuropath* 4:227–237

Solitare JB (1990): Aging and brain weight in Down's syndrome. *Neurology* 40:729

Suetsugu H, Mehraein P (1980): Spine distribution along the apical dendrites of the pyramidal neurons in Down's syndrome. *Acta Neuropathol* 50:207–210

Sylvester PE (1983): The hippocampus in Down's syndrome. *J Ment Defic Res* 27:227–236

Takashima S, Becker LE, Armstrong DL, Chan F (1981): Abnormal neuronal development in the visual cortex of the human fetus and infant with Down's syndrome. A quantitative and qualitative Golgi study. *Brain Res* 225:1–21

Takashima S, Jeshima A, Nakamura H, Becker LE (1989): Dendrites, dementia and Down syndrome. *Brain Dev* 11:131–133

Thase ME (1988): The relationship between Down's syndrome and Alzheimer's disease. In: *The Psychobiology of Down Syndrome.* Nadel L, ed. Cambridge, MA: The MIT Press, pp. 345–368

Wisniewski KE, French JH, Rosen JF, Kozlowski PB, Tenner M, Wisniewski HM (1982): Basal ganglia calcification (BGC) in Down's syndrome (DS)—Another manifestation of premature aging. *Ann NY Acad Sci* 396:179–189

Wisniewski HM, Merz GS (1984): Aging, Alzheimer's disease and developmental disabilities. In: *Aging and Developmental Disabilities.* Janieki MP, Wisniewski HM, eds. Baltimore: Brookes, pp. 177–184

Wisniewski HM, Rabe A (1986): Discrepancy between Alzheimer type neuropathology and dementia in persons with Down's syndrome. *Ann NY Acad Sci* 447:247–259

Wisniewski KE, Laure-Kamionowska, Wisniewski HM (1984): Evidence of arrest of neurogenesis and synaptogenesis in brains of patients with Down's syndrome. *N Engl J Med* 18:1187–1188

Wisniewski KE, Dalton AJ, McLachlan C, Wen GY, Wisniewski HM (1985a): Alzheimer's disease in Down's syndrome. *Neurology* 35:957–961

Wisniewski KE, Wisniewski HM, Wen GY (1985b): Occurrence of neuropathological changes and dementia of Alzheimer's disease in Down's syndrome. *Ann Neurol* 17:278–282

Wisniewski KE, Laure-Kamionowska M, Connell F, Wen GY (1986): Neuronal density and synaptogenesis in the postnatal stage of brain maturation in Down syndrome. In: *Neurobiology in Down Syndrome*, Epstein CJ, ed. New York: Raven Press, pp. 29–44

Wisniewski KE, Schmidt-Sidor B (1989): Postnatal delay of myelin formation in brains from Down syndrome infants and children. *Clinical Neuropathology* 2:55–62

Wisniewski KE (1990): Down syndrome children often have brain with maturation delay, retardation of growth and cortical dysgenesis. *Am J Med Genet* 7:274–281

Yates CML, Simpson J, Maloney AFY, et al. (1980): Alzheimer-like cholinergic deficiency in Down's syndrome. *Lancet* 11:979

Yates CM, Simpson J, Gordon A, et al. (1983): Catecholamines and cholinergic enzymes in pre-senile and senile Alzheimer-type dementia and Down's syndrome. *Brain Res* 280:119–126

Zellweger H (1977): Down syndrome. In: *Handbook of Clinical Neurology.* Vinken P, Bruyn GW, eds. Vol. 31 Pt II, Amsterdam, NY; Oxford: North Holland Publ Co., pp. 367–469

Alzheimer's Disease and the Cerebral Amyloidoses

Thomas M. Wisniewski and Henryk M. Wisniewski

Alzheimer's disease (AD) is the most common illness characterized by the accumulation of amyloid. It was first described by Alois Alzheimer in 1907, when it was regarded as a rare and inconsequential neuropathological entity. However, since that time there has been a major reappraisal of the pathological basis of common senile dementia. AD is now known to cause well over half of the senile dementias and has become a medical, social, and economic problem of enormous dimensions. Nearly 5% of the population over the age of 65 is severely demented and approximately another 10% moderately so. Furthermore, in most Western countries it is the population over the age of 75 that is growing at the fastest rate. With advancing age, a significant proportion of the population returns to a kind of second childhood, handing over responsibility for their own care to the rest of society. Indeed, some researchers view AD as a reversal of development. There are many analogies between childhood and dementia, such as the dependency and the infantile position of advanced AD patients (Reisberg et al., 1990). Studies of aged animals and humans reveal that with the passage of time the cells, organs, and organism accumulate changes that are not seen during their peak function. One such substance that accumulates is amyloid.

Amyloid and the Systemic Amyloidoses

Amyloid has diverse biochemical origins, as well as occurring in a diverse group of diseases. Virchow first applied the term amyloid or "starch-like" to tissue deposits that had the staining properties resembling starch and cellulose. Amyloids, regardless of their biochemical composition, share the properties of a pleated sheet secondary structure, green birefringence under polarized light with Congo red staining, and a fibrillar quaternary structure. The systemic amyloidoses have been worked on by general pathologists for a long time and their pathogenesis is relatively well understood. Hence, they provide insights into the possible pathogenesis of AD and the other cerebral amyloidoses. The amyloid of primary amyloidosis, AL, was the first to be biochemically characterized. AL was found to be composed of the amino terminal fragment of monoclonal immunoglobulin light chains with a molecular weight ranging from 5 to 23 kdaltons (Glenner et al., 1971a; Sletten et al., 1974). It has been shown that some of the light chains are highly amylogenic and others not at all (Glenner et al., 1971b). The precursor protein for secondary or reactive amyloidosis (AA) was found to be serum precursor protein (SAA) (Benditt et al., 1971; Levin et al., 1972). SAA is an acute phase reactant, whose physiological role is unknown. The serum levels of SAA are tremendously increased by inflammatory processes, hence this form of amyloidosis occurs as a complication of common chronic inflammatory conditions such as tuberculous, osteomyelitis, or rheumatoid arthritis. In both these amyloidoses abnormal proteolytic cleavage of the serum precursors by the reticuloendothelial (RE) system are thought to lead to the accumulation of amyloid (Glenner, 1980). In addition, it has been shown that the monocytes of patients with secondary amyloidosis are unable to completely degrade SAA, leading to the release of protein fragments with the same molecular weight of AA (Lavie et al., 1978). Another large group of systemic amyloidoses are the familial amyloidotic polyneuropathies (FAP). These all share the same precursor for the amyloid, transthyretin (also called pre-albumen). Each of the various FAPs has a different point mutation in the transthyretin. It is thought that these abnormalities of transthyretin are prone to abnormal aggregation or proteolytic cleavage, which in turn leads to the amyloid formation and deposition in the systemic vessels and peripheral nerves (Pras et al., 1981). The common thread in all these systemic amyloidoses is the existence of a circulating soluble precursor protein, a portion of which has an amylogenic primary structure. The incomplete degradation of the precursor, leading to amyloid formation, occurs either as a consequence of a mutation in the primary structure of the precursor, rendering it amyloidogenic or because of abnormal processing of the precursor. This abnormal processing of the precursor may only be a problem at times of overproduction. Bacterial infections induce a 1000-fold increase in SAA mRNA (Morrow et al., 1981) and with a corresponding increase in circulating levels of SAA protein (Benson and Kleiner, 1980). The processor cell in both instances is thought to be part of the RE system. It is likely that similar considerations are important in the cerebral amyloidoses.

Cerebral Amyloidoses

Amyloid fiber deposition is noted in a number of different conditions of the central nervous system. A partial list is given in the following table:

Disease	Amyloid-Related Protein
Alzheimer's	β-peptide
Down's Syndrome	β-peptide
Creutzfeldt-Jacob	PrP
Kuru	PrP
Gerstmann-Straussler	PrP
Scrapie	PrP
Hereditary cerebral hemorrhage with amyloidosis (Dutch)	β-peptide
Hereditary cerebral hemorrhage with amyloidosis (Icelandic)	Cystatin C
Sporadic Cerebral amyloid angiopathy	β-peptide

The first of the cerebral amyloidoses to have its amyloid protein biochemically characterized was hereditary cerebral hemorrhage with amyloidosis of Icelandic type (HCHWA-I). HCHWA-I is an autosomal-dominant disease identified in 128 affected individuals within eight families from one geographical area in Iceland. The neuropathological findings are amyloid deposition within the walls of small arteries and arterioles of the cerebral cortex and leptomeninges. This deposition weakens the vessel walls leading to massive hemorrhage and death usually before the age of 40 (Gudmundsson et al., 1972). The amyloid protein isolated from the leptomeninges of this condition was found to start at position 11 of the normal urinary cystatin C and has an amino acid substitution glutamine for leucine at position 68 (Cohen et al., 1983; Ghiso et al., 1986; Grubb and Lofberg, 1984). Hence, this disease appears to be similar to the familial amyloid polyneuropathies in that a point mutation leads to the production of an amyloidogenic protein.

Hereditary cerebral hemorrhage with amyloidosis of the Dutch type (HCHWA-D) has several similarities to the Icelandic type. It is also autosomal dominant with deposition of amyloid in the walls of small arteries and arterioles of the cerebral cortex and leptomeninges leading to hemorrhage and death in the fifth and sixth decades of life. A significant additional feature is parenchymal amyloid deposits, which resemble diffuse senile plaques of the Alzheimer's type and the sporadic cerebral amyloid angiopathy (Luyendijk et al., 1986; Van Duinen et al., 1987; Wattendosff et al., 1982). Characteristically, neurofibrillary tangles are not found in this disease. In association with the senile plaques some of the patients have been noted to be demented. Levy et al. (1990) have succeeded in cloning and sequencing the two exons of the APP gene that encode the β-peptide revealing a point mutation that causes a single amino acid substitution of glutamine instead of glutamic acid at position 22 of the amyloid protein. They obtained high molecular weight genomic DNA from the brains of two female HCHWA-D patients from different families and performed polymerase chain reaction (PCR) experiments with oligonucleotide primers designed to amplify exons 14 and 15. The resultant sequences demonstrated that both patients had a single point mutation, cytosine instead of guanine, as well as having one normal allele each. Such a point mutation could always be the result of polymorphism in the gene encoding the amyloid α-protein and not be the cause of the specific amyloid deposition in HCHWA-D. However, polymorphism in the gene encoding β-peptide has not been observed, and more importantly Van Broeckhaven et al. (1990) have carried out genetic linkage analysis in two of the Dutch families and found that the APP gene is tightly linked to the disease. Hence, these two cerebral amyloidoses are clearly the result of point mutations within different amyloid precursor proteins.

Alzheimer's Disease

The dominant neuropathological lesions of AD are the intracellular neurofibrillary tangles, which contain paired helical filaments (PHF) and the β-amyloid plaques. The β-amyloid plaques may be diffuse, neuritic, or classical in type. Diffuse plaques consist of wisps of extracellular amyloid in the absence of neuritic changes at the light microscopic level. In neuritic plaques, more amyloid as well as dystrophic neurites, often containing neurofibrillary tangles (NFTs), microglia, macrophages, and astrocytes are evident. Classical plaques in addition contain a core, also called a central star, of amyloid. Another change frequently noted in AD is the deposition of cerebrovascular amyloid in small arteries and arterioles (Alzheimer, 1907; Glenner et al., 1981). Other lesions are also seen in AD, but do not seem related to amyloid deposition and do not correlate with dementia. These include the granulovacuolar degeneration of neurons, which occurs mainly within the pyramidal cell layer of the hippocampus (Hooper and Vogel, 1976; Simchowicz, 1911) and Hirano bodies, which also are found at the latter site. Hirano bodies are intracytoplasmic, eosinophilic rod-shaped inclusions containing actin and actin-associated proteins (Galloway et al., 1988; Hirano et al., 1968).

Much literature has been devoted to whether the primary change in AD is the neuronal pathology, particularly in the formation of NFT, or the deposition of amyloid, which then leads to neuritic plaques. There is much reason to think that amyloid deposition occurs first. In Down's syndrome (DS) patients AD changes are noted very early in life. When young DS brains are examined it is the neuritic plaque, which is the first change. The NFT is identified in diverse neuropathological conditions and appears to be part of the neuron's limited repertoire of responses to injury. NFTs are well known to occur in subacute sclerosising panencephalitis, post-encephalitic Parkinson's disease (Corsellis, 1962), Guam Parkinson-Dementia Complex (Hirano and Zimmerman, 1962), dementia pugalistica (Wisniewski et al., 1976), supranuclear palsy and hydrocephalus (Hirano et al., 1961), as well as occasionally in brain neoplasms and around other mass lesions such as arterio-venous malformations (Wisniewski, et al., 1979). Hence the NFT although important as a marker of neuronal injury appears rather non-specific. Further support for this line of thinking comes from various immunohistochemical studies using antibodies to native and synthetic β-peptide. These reveal a much more extensive distribution of β-peptide deposition in AD than was previously suggested by conventional stains (Tagliavini et al., 1988; Yamaguchi et al., 1988). The finding of diffuse β-peptide deposits, without any surrounding cellular changes and in clinically asymptomatic regions of the brain, such as the cerebellum, strongly suggests that β-peptide deposition is an early event and that the local response to its presence might vary from region to region (Wisniewski et al., 1989a).

Biochemical and Molecular Studies of Alzheimer's Disease

The amyloid associated with AD was first characterized from partially purified preparations of meningeal cerebrovascular amyloid. A peptide of 4 kdaltons was extracted and the sequence of its first 28 residues was determined (Glenner and Wong, 1984b; Wong et al., 1985). This peptide has been named β-peptide. The vascular and plaque fibrils have seen been found to differ slightly in the molecular composition; the vascular form of the fibril has a 39 amino acid length (Prelli et al., 1988), while the plaque form has 43 amino acids (Kang et al., 1987). Although it is known that the neuritic plaque contains several other components besides the β-peptide, there are many studies to indicate that it is the most important component. Numerous antisynthetic β-peptide antiserums have been shown to stain neuritic plaques and cerebrovascular amyloid more extensively than any conventional stain (Allsop et al., 1986; Joachim et al., 1988). In addition in vitro studies using synthetic β-peptide have shown that sequences of as little as 14 amino acids can spontaneously form amyloid fibrils, with the typical Congo staining and green birefringence, in water and saline solutions without any further additives (Castano and Frangione, 1988).

With the use of oligonucleotide probes four independent groups were able to isolate and characterize complementary DNA (cDNA) clones coding for the precursor protein to the β-peptide, known as the amyloid precursor protein (APP). Kang et al. (1987) used cerebral cortical tissue from a five-month-old aborted fetus to isolate messenger RNA (mRNA) to construct a cDNA library. With an oligonucleotide probe designed from the N-terminus of the β-protein they pulled out a clone encoding for a 695 amino acid residue protein. This protein had the properties of glycosylated cell surface receptor that spanned the membrane once, with a large, extra-membranous NH2-terminal portion and a short tail in the cytoplasm. Since this gene was localized to chromosome 21 they hypothesized that the deposition of the β-protein in AD and DS was the result of abnormal catabolism of the APP.

In a similar fashion Goldgaber et al. (1987) used an oligonucleotide probe to the first 20 amino acids of the protein. They isolated four clones from a human brain cDNA library. They put forward the possibilities that either abnormal post-translational modification of its product or increased gene dosage might lead to an accumulation of β-protein and hence AD.

With their oligonucleotide probe Tanzi et al. (1987) isolated two cDNA clones. They found that although gene expression of APP was most pronounced in the brain, the spleen, kidney, and heart it was also found in many other tissues. Because of this finding it was clear that the amyloid of the NP or cerebrovascular amyloid might originate from outside the brain. In an attempt to explain why plaques are deposited only in the brain they hypothesized that the specific amyloidogenic protein cleavage of APP might occur only in the brain.

Robakis et al. (1987) also used this strategy to pull out a clone coding for a 412 amino acid protein. A suggestion they put forward was that a point mutation in APP could alter the primary structure rendering it amylogenic, or that the enzymes involved in processing the protein could be defective in AD and DS patients.

In addition to the APP of 695 amino acid residues (APP-695) other researchers have isolated cDNAs that contain inserts of 168 base pairs (APP-751) (Ponti et al., 1988; Tanzi et al., 1988) or 225 base pairs, APP-770 (Kitaguchi et al., 1988). The insert on the APP-751 encodes an additional internal domain having the sequence characteristics of a Kunitz-type serine protease-inhibitor (KPI). A further computer search and in vitro experiments revealed that the large NH_2-terminal portion of the APP-751 is secreted by some cells in culture and is identical to the proteinase inhibitor, protease nexin II (Ponti et al., 1988). This APP-751 was found by Tanzi et al. (1987) in both brain and peripheral tissue of normal and AD patients. The APP-770, which was isolated from a cDNA of a human glioblastoma cell line, had an insert highly homologous to basic trypsin inhibitors. The finding of these other APPs with protease inhibitor inserts raises the possibility that these proteins can promote amyloidogenesis by inhibiting proteases that degrade β-peptide. In this regard it is noteworthy that the major component of the Icelandic variant of amyloid is a type of cysteine protease inhibitor. Palmert et al. (1988) has found that in AD, neurons of the nucleus basalis of Meynert and locus ceruleus contain twice as much APP mRNA containing the KPI insert, compared to controls. Further *in situ* hybridization studies demonstrate a two-fold increase in the ratio of APP-751 to APP-695 in mRNA within the hippocampal pyramidal neurons in AD versus control subjects (Johnson et al., 1990). However, the importance of these KPI inserts containing APPs remains unclear. It is also possible that they could oppose amylogenesis by inhibiting proteases that convert APP to β-peptide, so that the increased levels of APP-751 could occur as a reaction to amyloid deposition rather than as its cause. A further hypothesis is that the true APP for β-peptide has not been found and that the APPs belong to a large family of proteins with slightly different sequences, such as the cell adhesion molecule group (Cunningham et al., 1987).

Another question that must be answered is which cell generates the amyloid? The classical neuritic plaque is composed of the amyloid star, microglial cells, abnormal neuritic processes, and reactive astrocytes; these are all occasionally in close association with blood vessels. (Wisniewski and Terry, 1973; Wisniewski et al., 1981). Each of these cellular elements has researchers who support it as the producer of amyloid: the circulatory system (Allsop et al., 1988; Glenner and Wong, 1984a; Selkoe et al., 1986), the neuronal elements (Bahmanyar et al., 1987; Glenner, 1979; Masters et al., 1985; Powers, 1986; Powers and Sheen, 1988), the perivascular cells (Miyakawa et al., 1974), the circulating platelets (Van Nostrand et al., 1990), the degenerating neurites (Itagaki et al., 1989), and finally the microglial/macrophage cells (Wisniewski et al., 1989b; 1990). There is compelling evidence to suggest that the microglial/macrophage cell is the most important. Ultrastructural, three-dimensional reconstructions of cells surrounding the amyloid star in classical plaques in AD show that the fibers appear within the smooth endoplasmic reticulum (ER) and infolding of cytoplasmic membranes of the microglia/macrophage (Wisniewski et al., 1990, 1991). Histochemical studies of the localization of nucleoside diphosphatase (NDP-ase), an enzyme normally associated with the smooth ER and cell membranes, reveal that it is associated with the newly formed amyloid fibers within the distended cisternae of the ER. These studies also clearly demonstrate that the microglial cells had the greatest interface with the amyloid stars. The portion of the microglial cells adjacent to the amyloid possess a specific amyloid-making pole. The cell membrane at this pole is extended by apparent connection to the membranes of cytoplasmic channels filled with amyloid fibers. This ultrastructural appearance is similar to that noted in Kupffer cells, which also belong to the RE system, in murine amyloidosis. Amyloid fibers were found not only in the extracellular space but also very close to the cell surface or actually within the cytoplasm of the Kupffer cells (Ranlov and Wanstrup, 1967; Sorenson et al., 1964; Uchino et al., 1985). It has often previously been argued that the microglial cell is engaged only in phagocytic activity of the amyloid. However, these ultrastructural studies indicate that the microglial/macrophage is the origin of β-amyloid fibers, mirroring the importance of the RE system in the systemic amyloidoses.

A question that also has received some attention is how β-amyloid leads to the degeneration of neuronal processes and ultimately to neuronal death. The primary structure of the β-peptide provides some possible answers. The β-peptide corresponds to residues 597–636 of the Kang et al. (1987) sequence. It has 28 hydrophilic residues (597–624) in the extracellular domain and 12–14 hydrophobic amino acid residues in the transmembrane region. If the β-peptide results from proteolytic cleavage of a larger precursor, then one of the cuts must be in the membrane domain, as the entire membrane cannot be spanned by a 12-membrane amino acid helix. This arrangement gives the β-peptide interesting amphipathic, detergent-like properties, which may be important in the lysis of membranes. It may be objected that all membrane-associated proteins have amphipathic structures; however, these proteins do not appear to reach the concentrations of the β-peptide. In addition, the β-peptide C-terminal hydrophobic sequence shows a homology with the hydrophobic region of fusion peptides of many enveloped RNA viruses, in particular paramyxovirus (PMV) F1 protein sequences and the orthomyxovirus (OMV) HA2 sequence (Currie et al., 1990; Kingsbury, 1985; Morrison, 1988; Wiley, 1985; White et al., 1983). Like the β-peptide the F1 protein of PMV and the HA2 protein of OMV results from the proteolytic cleavage of longer precursor proteins. The cleavage event results in the generation and exposure of a hydrophobic N-terminus, which is the most highly conserved region among the PMV and OMV studied to date. It is not known whether this homology merely represents some common ancestor or rather a common function.

It seems plausible that with the slow accumulation of α-peptide there would be a modification of the structure and function of the membrane by possibly changing the membrane permeability or the enzyme or receptor activity. This sequence homology to viral proteins invites speculation as to whether AD results from the activation of an inserted, truncated viral sequence or from the production of a truncated AAP gene product due to cell transformation. Other ways by which amyloid can damage neurons and their processes include compression damage as is seen within liver and kidney in the systemic amyloidosis, and perhaps by the immunocompetent cells of the plaques.

What caused APP to be metabolized to β-peptide in AD is not known. Calculations based on the amounts of APP mRNA in the brain would suggest that most of it is degraded without the formation of amyloid (Miller et al., 1989). Northern blots indicate that the APP mRNA comprised about 0.1% to 0.2% of total brain mRNA in the temporal and parietal cortex. Using average rates of protein turnover and protein content (Dunlop, 1983) one is led to the conclusion that the normal brain could produce about 3 μg of β-peptide per day per gram of tissue or enough peptide to form 30000 plaque cores, assuming the β-peptide is not degraded. Clearly this does not happen and the vast majority of APP in normal as well as in AD brains must be degraded without β-peptide formation. This has been confirmed in recent studies by Esch et al. (1990). They used human cells transformed with either APP-695 or APP-751 cDNAs and purified the large NH2-terminal fragment from culture medium, showing it to end at position 15 within the β-peptide. The remaining 9 kdalton COOH terminal fragment was purified from the cell membranes and found to begin at residue 17 of the β-peptide. The one amino acid between these termini could have been removed from one or the other fragment by an exopeptidase after the original cleavage. This experiment reveals the presence of an enzyme (called APP secretase) that catabolizes APP without the formation of β-peptide. If this pathway is confirmed in normal human tissue, it will presumably be the major pathway for APP cleavage. The alternative pathway that leads to β-peptide is probably a very minor one. One reason for the alternative processing of APP is a mutation in the gene. Recently, such a mutation was found in four families with early onset of familial AD (Goate et al 1991; Narlise et al., 1991). In these patients, a cytosine for thiamine mutation at codon 642 of APP 695 causes an Ila for Val substitution, a substitution not located in the β-peptide sequence. However, only a few of the familial AD cases have been linked to chromosome 21 (St. George-Hyslop et al., 1987). In Down's syndrome an increase of transcription of APP because of elevated gene dosage may cause more APP to be diverted to the alternative β-peptide-producing pathway leading to the early amyloid deposition. Hence, β-peptide deposition can occur both in the presence of a mutation and without it. With a better understanding of the events that lead to AD, it is likely that multiple different mutations and different abnormal other processes can lead to β-peptide deposition.

The success of medical science in extending the average human life span has led to the emergence of the β-peptide related diseases as a major health care issue. Rapid advances in the molecular and cellular events surrounding senile plaque formation have increased our understanding of the pathogenetic events that lead to AD. Current research opens up the possibility of therapeutic modalities that will control the formation of amyloid deposits in the victims of AD. Therapy is already available for some of the systemic amyloidoses like Familial Mediterranean Fever. The development of further transgenic animals and cellular models with the cDNAs available for the various APPs will provide models for the amyloidogenesis of AD. These should greatly enhance our understanding of the etiology and pathogenesis of AD.

Acknowledgments. Supported in part by funds from the New York State Office of Mental Retardation and Developmental Disabilities and Grant No. 1-PO1-AGO-4220-06 from the National Institute on Aging, National Institutes of Health.

References

Allsop D, Landon M, Kidd M, Lowe JS, Reynolds GP, Gardner A (1986): Monoclonal antibodies raised against a subsequence of senile plaque core protein react with plaque core, plaque periphery and cerebrovascular amyloid in Alzheimer's disease. *Neurosci Lett* 68:252–256

Allsop D, Wong CW, Ikeda SK, Landon M, Kidd M, Glenner GG (1988): Immunohistological evidence for derivation of a peptide ligand from the amyloid a-protein precursor of Alzheimer's disease. *Proc Nat Acad Sci USA* 85:2790–2794

Alzheimer A (1907): Uberline eigenartige erkrangkung der hirnrinde. *Allg Z Psychiat* 64:146–148

Bahmanyar S, Higgins A, Goldgaber D, Lewis DA, Morrison JH, Wilson MC, Shanker SK, Gadjusek DC (1987): Localization of amyloid β-protein messenger RNA in brains from patients with Alzheimer's disease. *Science* 237:77–80

Benditt EP, Erikssen N, Hermodson NA, Ericsson LH (1971): The major proteins of human and monkey amyloid substance: common properties including unusual N-terminal amino acids sequences. *FEBS Lett* 19:169.

Benson MD, Kleiner E (1980): Synthesis and secretion of serum amyloid protein A (SAA) by hepatocytes in mice treated with casein. *J Immunol* 124:495–499

Castano EM, Frangione B (1988): Biology of Disease: Human amyloidosis, Alzheimer's disease and related disorders. *Letts Invest* 58:122–132

Cohen DH, Feiner H, Jensson D, Frangione B (1983): Amyloid fibril in hereditary cerebral hemorrhage with amyloidosis (HCHWA) is related to the gastroenteropancreatic neuroendocrine protein gamma trace. *J Exp Med* 158:623.

Corsellis Jan (1962): *Mental Illness and the Aging Brain*. Institute of Psychiatry, Maudsley Monographs, No. 9, Oxford University Press

Cunningham BA, Hemperly TJ, Murray BA, Prediger EA, Brackenburg R, Edelman GM (1987): Neuronal cell adhesion molecule: structure immunoglobulin-like domains, cell surface modulation and alternative RNA splicing. *Science* 236:799–806

Currie J, Wisniewski HM, Chauhan A, Brockerhoff H, Wegiel J (1990): Alzheimer amyloid β-protein: Cell of origin and possible fusogenic and membranolytic properties. In: *Regulation of Gene Expression in the Nervous System*, Giuffida A, de Vellis J, eds. Alan R. Liss, in press.

Dunlop DS (1983): Protein turn-over in brain: Synthesis and degradation. In: *Handbook of Neurochemistry*, Lajtha A, ed. New York: Plenum Press, pp. 25–65

Esch FS, Keim PS, Beattie EC, Blacher RW, Culwell AR, Oltersdork T, McClure D, Ward PJ (1990): Cleavage of amyloid β-peptide during constitutive processing of its precursor. *Science* 248:1122–1124

Galloway PG, Perry G, Gambetti P (1988): Hirano body filaments contain actin and actin associated proteins. *J Neuropathol Exp Neurol* 46:185–199

Ghiso G, Jensson O, Frangione B (1986): Amyloid fibrils in hereditary cerebral hemorrhage with amyloidosis of Icelandic Type is a variant of gamma-trace basic protein (Cystatin C). *Proc Natl Acad Sci USA* 83:2974.

Glenner GG, Terry W, Harada M, Isersky C, Paye C (1971a): Amyloid fibril proteins: proof of nomology with immunoglobulin light chains by sequence analysis. *Science* 172:1150

Glenner GG, Ein D, Eanes ED, Bladen HA, Terry W, Page D (1971b): Creation of amyloid fibrils from Bence Jones proteins in vitro. *Science* 174:712.

Glenner GG (1979): Congophilic microangiopathy in the pathogenesis of Alzheimer's syndrome (presenile dementia). *Med Hypotheses* 5:1231–1236

Glenner GG (1980): Amyloid deposits and amyloidosis: The α-fibrilloses. *N Engl J Med* 302:1283

Glenner GG, Henry JH, Fujihara S (1981): Congophilic angiopathy in the pathogenesis of Alzheimer's degeneration. *Ann Pathol* 1:120–129

Glenner GG, Wong CW (1984a): Alzheimer's disease and Down's syndrome: Sharing of a unique cerebrovascular amyloid protein. *Biochem Biophys Res Commun* 122:1131–1135

Glenner GG, Wong CW (1984b): Alzheimer's disease: Initial report of the purification and characterization of a novel cerebrovascular amyloid protein. *Biochem Biophys Res Commun* 120:885–890

Goate A, Chartier-Harlin MC, Brown J, Crawford F, Fidami L, Giuffra L, Haynes A, Irving N, James L, Mant R, Newton P, Rooke K, Roques P, Talbot C, Pericak-Vance M, Roses A, Williamson R, Rosser M, Owen M, Hardy J (1991): Segregation of a missense mutation in the amyloid precursor protein gene with familial Alzheimer's disease. *Nature* 349:704–706

Goldgaber D, Lerman MI, McBride OW, Saffiotti U, Gajdusek DC (1987): Characterization and chromosomal localization of a cDNA encoding brain amyloid of Alzheimer's disease. *Science* 235:877–880

Grubb A, Lofberg H (1984): Human trace, a basic microprotein: Amino acid sequences and presence in the adenohypophysis. *Proc Natl Acad Sci USA* 79:3024

Gudmundsson G, Hallgrimsson J, Jonasson TA, Bjarnason O (1972): Hereditary cerebral hemorrhage and amyloidosis. *Brain* 95:587

Hirano A, Dembitzer HM, Kurland LT, Zimmerman HM (1968): The fine structure of some intraganglionic alterations: Neurofibrillary tangles, granulovacuolar bodies and "rod-like" structures as seen in Guam amyotrophic lateral sclerosis and Parkinson dementia complex. *J Neuropathol Exp Neurol* 27:167–182

Hirano A, Malamud N, Kurland LT (1961): Parkinsonism-dementia complex, an endemic disease on the Island of Guam. *Brain* 84:662–679

Hirano A, Zimmerman HM (1962): Alzheimer's neurofibrillary changes. *Arch Neurol* 7:227–241

Hooper MW, Vogel FS (1976): The limbic system in Alzheimer's disease: A neuropathologic investigation. *Amer J Pathol* 85:1–20

Itagaki S, McGeer PL, Akiyama H, Zhu S, Selkoe D (1989): Relationship of microglia and astrocytes to amyloid deposits of Alzheimer disease. *J Neuroimmunol* 24:173–182

Joachim CL, Duffy LK, Morris JH, Selkoe DJ (1988): Protein chemical and immunocytochemical studies of meningovascular-vascular β-amyloid protein in Alzheimer's disease and normal aging. *Brain Res* 474:100–101

Johnson SA, McNeil T, Cordell B, Finch CE (1990): Relation of neuronal APP-751/APP-695 mRNA ratio and neuritic plaque density in Alzheimer's disease. *Science* 248:854–857

Kang J, Lemaire HG, Unterbeck A et al. (1987): The precursor of Alzheimer's disease amyloid A4 protein resembles a cell-surface receptor. *Nature* 325:733–736

Kingsbury DW (1985): Orthomyxo and paramyxoviruses and their replication. In: *Virology*, Fields et al., eds. New York: Raven Press, pp. 1151–1178

Kitaguchi N, Takahashi Y, Tokushima Y, Shiojiri S, Ito H (1988): Novel precursors of Alzheimer's disease protein shows protease inhibitory activity. *Nature* 331:530–532

Lavie G, Zucker-Franklin D, Franklin EC (1978): Degradation of serum amyloid A protein by surface-associated enzymes of human blood monocytes. *J Exp Med* 148:1020–1023

Levin M, Franklin EC, Frangione B, Pras M (1972): The amino acid sequence of a major non-immunoglobulin component of some amyloid fibrils. *J Clin Invest* 51:2773–2776

Levy E, Carman MD, Fernandez-Madrid IJ, Power MD, Lieberburg I, Van Duinen SG, Bots GThAM, Luyendijk W, Frangione B (1990): Mutation of the Alzheimer's disease amyloid gene in hereditary cerebral hemorrhage, Dutch type. *Science* 248:1124–1126

Luyendijk W, Bots GThAM, Vegter-Vander Vlis M, Went LN (1986): Familiaire hersenblocdingen als gevolg van cerebral amyloid angiopathie. *Ned Tijdschr Geneesk* 130:1935–1940

Masters CL, Multraup G, Simms G, Pottgiesser J, Martius RW, Beyreuther K (1985): Neuronal origin of a cerebral amyloid: Neurofibrillary tangles of Alzheimer's disease contain the same protein as the amyloid of plaque cores and blood vessels. *EMBO J* 4:2747–2763

Miller DL, Currie JR, Iqbal K, Potempska A, Styles J (1989): Relationship among the cerebral amyloid peptides and their precursors. *Ann Med* 21:83–87

Miyakawa T, Sumiyoshi S, Murayama E, Deshimaru M (1974): Ultrastructure of capillary plaque-like degeneration in senile dementia. Mechanism of amyloid production. *Acta Neuropath (Berl)* 29:229–236

Morrison TG (1988): Structure, function and intracellular processing of paramyxovirus membrane proteins. *Virus Res* 10:113–136

Morrow JF, Stearman RS, Peltzman, CO, Potter DA (1981): Induction of hepatic synthesis of serum amyloid A protein and actin. *Proc Natl Acad Sci USA* 78:4718–4722

Narlise S, Igarashi S, Aoki K, Kaneko K, Ihara K, Miyatake T, Kobayashi H, Inuzuka T, Schimizu T, Kojima T, Tsuji S (1991): Missense mutation Val → Ile in exon 17 of amyloid precursor protein gene in Japanese familial Alzheimer's. *Lancet* 337:978–979

Palmert MR, Golde TE, Cohen ML et al. (1988): Amyloid protein precursor messenger RNAs: Differential expression in Alzheimer's disease. *Science* 241:1080–1084

Ponti P, Gonzalez-De Whitt P, Schilling J, Miller J, Hsu D, Greenberg B, Davis K, Wallace W, Leiderberg I, Fuller F, Cordell B (1988): A new A4 amyloid mRNA contains a domain homologous to a serine proteinase inhibitor. *Nature* 331:525–527

Powers JM (1986): Senile cerebral amyloid-evidence for a neuronal origin of the fibril protein. In: *Amyloidosis*, Glenner GG, Osserman EF, Benditt EP, Calkins E, Cohen AS, Zucker-Franklin, D, eds. Plenum Press, pp.743–749

Powers JM, Sheen JT (1988): Ultrastructural heterogeneity in cerebral amyloid of Alzheimer's disease. *Acta Neuropath* 76:613- 623.

Pras M, Franklin EC, Prelli F, Frangione B (1981): A variant of prealbumen from amyloid fibrils in familial polyneuropathy of Jewish type. *J Exp Med* 154:989

Prelli F, Castano E, Glenner GG, Frangione B (1988): Differences between vascular and plaque core amyloid in Alzheimer's disease. *J Neurochem* 15:648–651

Ranlov P, Wanstrup J (1967): Ultrastructural investigations on the cellular morphogenesis of experimental mouse amyloidosis. *Acta Pathol Microbiol Scand* 71:575–591

Reisberg B, Pattschull-Furlan A, Franssen E, Sclan S, Kluger A, Dingcong L, Ferris SH (1990). Cognition related functional, praxis and feeding changes in CNS aging and Alzheimer's disease and their developmental analogies. In: *Molecular Mechanisms in Aging,* K. Beyreuther ed. Berlin: Springer-Verlag, pp. 18–40

Robakis NK, Ramakrishna N, Wolfe G, Wisniewski HM (1987): Molecular cloning and characterization of a cDNA encoding the neuritic plaque amyloid peptides. *Proc Nat Acad Sci USA* 84:4190–4194

Selkoe DJ, Abraham CR, Podlismy MB, Duffly LK (1986): Isolation of low-molecular weight proteins from amyloid plaques fibers in Alzheimer's disease. *J Neurochem* 46:1820–1834

Simchowicz T (1911): Histologische studien uber die senile demenz. In: *Histologische und Histopathologische Arbeiten uber die grosshirnrinde*. Nissl F, Alzheimer A, eds. Gustav Fisher, 4:267–444

Sletten K, Husby G, Natvig JB (1974): N-terminal amino acid sequence of amyloid fibril protein AR, prototype of a new lambda-variable subgroup Vr1. *Scand J Immunol* 3:831–836

Sorenson GD, Hoefner WA, Kiewpatrick JP (1964): Experimental amyloidosis II. Light and electron microscopic observations of liver. *Am J Pathol* 44:629–644

St. George-Hyslop PH, Tanzi RE, Polinsky J et al. (1987): The genetic defect causing familial Alzheimer's disease maps on chromosome 21. *Science* 235:885

Tagliavini F, Giaccone G, Frangione B, Bugiani O (1988): Pre-amyloid deposits in the cerebral cortex of patients with Alzheimer's disease and non-demented individuals. *Neuroscience Lett* 93:191–196

Tanzi, RE, Gusella JF, Watkins PC et al. (1987): Amyloid β-protein gene: cDNA, mRNA distribution and genetic linkage near the Alzheimer locus. *Science* 235:880–884

Tanzi RE, McClatchey AI, Lamperti ED et al. (1988): Protease inhibitor domain encoded by an amyloid protein precursor mRNA associated with Alzheimer's disease. *Nature* 331:528–530

Uchino F, Takahashi M, Yokota T, Ishihara T (1985): Experimental amyloidosis. Role of hepatocyte and Kuffer cells in amyloid formation. *Appl Pathol* 3:78–87

Van Broeckhaven C, Haan J, Bakker E, Hardy JA, Van Hul W, Wehnert A, Vegter-Vander Vlis M, Roos RAC (1990): Amyloid β-protein precursor gene and hereditary cerebral hemorrhage with amyloidosis (Dutch). *Science* 248:1120–1122

Van Duinen SG, Castano EM, Prelli F, Bots GThAB, Luyendijk W, Frangione B (1987): Hereditary cerebral hemorrhage with amyloidosis in patients of Dutch origin is related to Alzheimer's disease. *Proc Natl Acad Sci USA* 84:5991–5994

Van Nostrand WE, Schmaier Farrow JS, Cunningham DD (1990): Protease Nexin II (Amyloid β-protein precursor): A platelet a-granule protein. *Science* 248: 745–748

Wattendosff AR, Bots GThAM, Went LN, Endtz LJ (1982): Familial cerebral amyloid angiopathy presenting as recurrent cerebral hemorrhage. *J Neurol Sci* 55:121

Wiley DC (1985): Viral membranes. In: *Virology*, Fields et al., eds. New York: Raven Press, pp. 45–67

White J, Kielian M, Helenius A (1983): Membrane fusion proteins of enveloped animal viruses. *Quart Rev Biophys* 16:151–195

Wisniewski HM, Terry RD (1973): The examination of the pathogenesis of senile plaques. In: *Progress in Neuropathology*, Vol 2. Zimmer HM, ed. New York: Grune and Stratton, pp. 1–26

Wisniewski HM, Narang HK, Corsellis JAN, Terry RD (1976): Ultrastructural studies of the neuropil and neurofibrillary tangles in Alzheimer's disease and post-traumatic dementia. *J Neuropathol Exp Neurol* Abst. 35:367

Wisniewski HM, Moretz RC, Lossinsky AS (1981): Evidence for induction of localized amyloid deposits and neuritic plaques by an infectious agent. *Ann Neurol* 10:517–522

Wisniewski HM, Bancher C, Barcikowska M, Wen GY, Currie J (1989a): Spectrum of morphological appearance of amyloid deposits in Alzheimer's disease. *Acta Neuropathol* 78:337–347

Wisniewski HM, Wegiel J, Wang KC, Kujawa M, Lach B (1989b): Ultrastructural studies of the cells forming amyloid fibers in classical plaques. *Can J Neurol Sci* 16:535–542

Wisniewski HM, Vorbrodt AW, Wegiel J, Morys J, Lossinsky AS (1990): Ultrastructure of the cells forming amyloid fibers in Alzheimer disease and scrapie. *Am J Med Genetics*, Suppl 7:287–297

Wisniewski HM, Barcikowska M, Kida E (1991): Phagocytois of B/A4 amyloid fibrils of the neuritic neocortical plaques, *Acta Neuropathol* 81:588–590

Wisniewski K, Jervis GA, Moretz RC, Wisniewski HM (1979): Alzheimer neurofibrillary tangles in diseases and other than senile and presenile dementia. *Ann Neurol* 5:228–294

Wong CW, Quaranta V, Glenner GG (1985): Neuritic plaques and cerebrovascular amyloid in Alzheimer's disease are antigenically related. *Proc Natl Acad Sci USA* 82:8729–8732

Yamaguchi H, Hirai S, Morimatsu M, Shoji M, Harigaya Y (1988): *Acta Neuropathol (Berl)* 71:113

Evolution of Cholinergic Cortical Innervation after nbM-Lesioning (An Experimental Alzheimer Model)

Adolfo Toledano

Introduction

The social and economic transcendence of Alzheimer's disease (AD) leads to an increasing interest in basic and clinical research on this neurodegenerative disorder. Scientific research programs on AD are continuously expanding and the number of AD-related publications and scientific meetings are multiplying year by year.

At present, research on AD involves many scientific fields. The multidisciplinary approach is a common feature of much research (Ordy et al., 1988). Scientific workers in many laboratories try to clarify the etiology as well as the pathogeny of AD. Practical aspects on AD, like early diagnosis and ethiopathogenic treatments, are also in the first line of the research interest.

Both postmortem human brain studies and clinical ones set the guidelines to experimental research on AD. Not every animal suffers AD spontaneously and no agent has been isolated to induce this disorder in every animal; however, many experimental AD models have been described to study selected aspects of the pathological process in the laboratory. In the first part of this review, the most common Alzheimer models will be introduced, before presenting the general morphofunctional implications of the lesions of the rat nucleus basalis magnocelularis (nbM), these being the most interesting of the AD models. The functional structure of the basal forebrain and of its nucleus basalis will be separately discussed. Nucleus basalis magnocellularis in rodents is the anatomofunctional equivalent of the Meynert nucleus described in man, and, in this sense, nbM—nucleus basalis of Meynert—can serve as a general denomination for different species.

In the second part, the evolution of the cortical cholinergic markers in an AD model produced by a single unilateral nbM lesion, which has been carefully studied by us in a morphofunctional comparative way, will be discussed.

Alzheimer Models

Different kinds of experimental Alzheimer models have been developed in research laboratories. Each of them bring, as support, one of the diverse theories that are been stated to explain the etiology and/or the pathogeny of the AD. In turn, these theories are based on the results obtained in human or animal studies.

The following models can be mentioned as the most important:

Normal aged animals. The first model we can consider is that of normal senile animals (Bertoni-Freddari et al., 1986; Briegon et al., 1984; Miquel et al., 1983; Pintor et al., 1988; Toledano et al., 1988), because a close relationship is observed between senility and sporadic AD. This topic could be contested, but all the authors agree on the need to differentiate normal and pathological aging. Moreover, in these two theoretical processes, identical involutive pathways are encountered (Bartus et al., 1982, 1987; Collerton, 1986; Chui et al., 1984; Mann et al., 1984b; Mash et al., 1985; Pintor et al., 1988; Springer et al., 1987; Toledano, 1988; Wisniewsky et al., 1981).

Experimental animals. The second group of Alzheimer models to consider are the animals with induced lesions or dysfunctions. A subgroup could link all the animals with cerebral beta fibrillosis together (Gajdusek, 1987). Both the viral and the toxic theories on AD led to the study of animal models in whose CNS amyloid or neuronal fibrils similar to tangles can be induced. For example, scrapie virus has been successfully inoculated into the mouse brain, promoting amyloid deposits, and aluminum preparations provoke β-fibrillosis in the rat. Close to these models one can place a few fibrillar involutive nervous diseases of several animals like the scrapie of sheep.

Cellular and subcellular models. Many other biological materials (cultured neurons, membranes, synaptosomes, blood cells, and so on) can be used to study cellular or molecular mechanisms that are interesting for the best understanding of AD. This last group will be of special interest for several purposes—early diagnosis, characterization of AD markers and mechanisms.

But the most useful models of AD are supported by the ethiopathogenic theories, which are proposed to explain the involutive changes observed in specific neuronal centers or in special neurotransmitter systems. Using several methods, the lesion or the dysfunction of selected neurons are provoked to study the changes promoted by the consecutive anatomical or functional deinnervation.

At present, neuronal lesioning is obtained by stereotactical injection of excitotoxins such as kainic, quisqualic, N-metil-D-aspartic, or ibotenic acids, and the section of neuronal pathways is obtained by microsurgery.

The cholinergic theory of AD (Bartus et al., 1982, 1985a; Coyle et al., 1983; Davies, 1979; Deutsch, 1971; Drachman and Leavitt, 1974) has outlined the main part of the Alzheimer experimental models. The best known examples of these cholinergic experimental models are:

(1) Animals with lesion or dysfunction in the forebrain areas in which cholinergic neurons are laid down (Dunnett et al., 1987; El Defrawy et al., 1985, 1986; Gower, 1986; Haroutunian et al., 1986a; Hepler et al., 1985; Smith, 1988; Sofroniew et al., 1987). The most commonly described cholinergic centers in the forebrain are the septum, which innervates the hippocampus and the amygdala, the diagonal band of Broca, which innervates the gyrus cingularis and other limbic structures, and the nucleus basalis of Meynert, which innervates the neocortex

(2) Animals with sections in the cholinergic pathways. The sectioned fimbria-fornix rat is the best-known model (Hefti et al., 1984).

(3) Animals with an impairment of the cholinergic neurotransmission system, after both a general or a local drug treatment. Animals under cholinergic drugs, such as scopolamine or arecoline (Ordy et al., 1988) or animals with stereotaxically or systemically injected hemicholinum or arizidinium salts (Chroback et al., 1988), are suitable examples of these models.

But other noncholinergic theories have created their own models, such as the adrenergic or somatostatinergic theories (Amstrong et al., 1989; Haroutunian et al., 1986b; Mann et al., 1984a).

Animals with lesions in adrenergic centers of the brainstem or with sections in the ascending adrenergic bundles are useful when studying the adrenergic involvement in AD; and animals under intraperitoneal

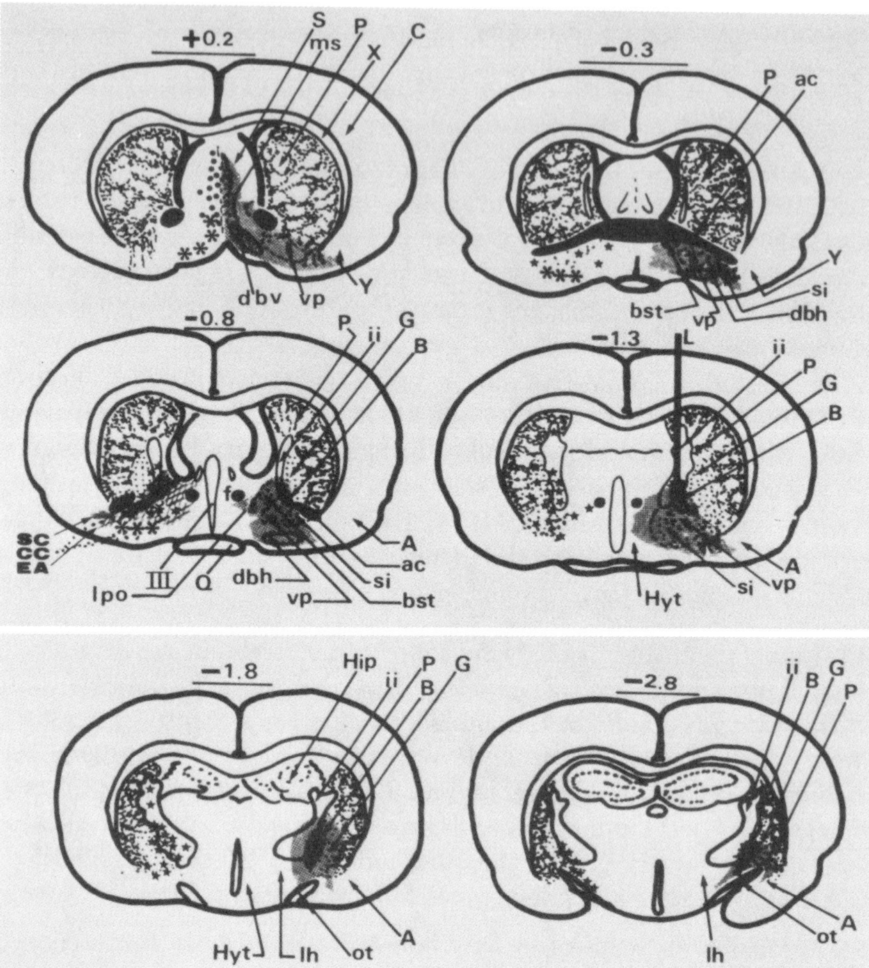

Figure 1. Schemes displaying the location of the basal forebrain (highlighted in black on the right side) and the cholinergic corticopetal neurons (stars, on the left side) at different levels of coronal sections of the rat brain (+0.2, –0.3, –0.8, –1.3, –1.8, and –2.8 mm planes, the 0.0 mm plane corresponding to the one passing by the bregma). The medial septal nuclei (ms), the vertical part of diagonal band of Broca (dbv), the horizontal part of diagonal band of Broca (dbh), and the nucleus basalis magnocellularis (B) are represented by different kinds of star-dots. Each dot represent a different number of cholinergic neurons (6–12 for ms, 4–8 for db, and 1–3 for B in a 40-micron cut). In the –0.8 level, the extended amygdala complex (EA), the striatopallidal complex (SC), and the corticopetal complex (stars–CC) of the basal forebrain (Alheid and Heimer, 1988) are represented. In the –1.3 level, the placement of the lesion referred in the text is indicated (L, arrow). Drawings are nonsymmetrical; each half was taken from a different case. In this way, the actual individual anatomical variations are pointed out.

Key to Figure 1: A = amygdala and related structures; ac = anterior commissure; B = nucleus basalis magnocellularis; bst = bed nucleus of stria terminalis; C = frontoparietal neocortex; db = Diagonal Band of Broca; G = globus pallidus; Hip = hippocampus; Hyt = hypothalamus; ii = internal capsule; Ih = lateral hypothalamus; lpo = lateral preoptic area; ms = medial septal nuclei; ot = optic tract; P = caudato putamen; Q = chiasma; S = septum; si = substantia innominata; vp = ventral pallidum; X = corpus callosum; Y = Pyriforme cortex and related structures.

cisteamine treatment, which suffer a depletion of the somatostatin storage, can serve to evaluate the role of somatostatin deficits.

The Nucleus Basalis of Meynert of the Basal Forebrain: Structure and Function

Looking for the best understanding of the AD cholinergic models, it is necessary to consider the special morphofunctional characteristics of the cholinergic centers in the basal forebrain.

In the forebrain, there exists a medial and basal region, from the septum to the midbrain, passing under the anterior commissure, that groups telencephalic and diencephalic structures (Fig. 1). This region, in the posterior part, is intermeshed with striatal and hypothalamic structures, and is laterally in close relationship with the amygdala and the medial part of the temporal lobe. This region is classically named the basal forebrain. In concrete areas of this region, the neuronal density is high enough to consider the existence of grey nuclei. The most outstanding of these are the neuronal groups in which large cholinergic neurons appear. Classically, from the anterior-medial part to the posterior-lateral part, the three above-mentioned groups are considered: the medial septal nuclei, the diagonal band of Broca nuclei (with a vertical and another horizontal set of neurons), and the nucleus basalis magnocellularis (or the Meynert nuclei, in man) (Bear et al., 1985; Bigil and Woolf, 1982; Butcher and Woolf, 1986; Johnston et al., 1981; Lamour et al., 1982; Wenk et al., 1980).

Shute and Lewis (1963, 1967) and Krnjević and Silver (1965) were the first to describe the cholinergic component in normal and experimental material.

Although the septum and diagonal band of Broca are relatively well defined, the nucleus basalis is extremely difficulty to define, especially in rodents. Cholinergic neurons presumably belonging to this nucleus are located in the internal capsule close to the globus pallidus, but other important clusters are laid down in areas ventral to the pallidum striatal complex (Mesulam et al., 1983a,b; Schwaber et al., 1987). These areas, which are referred to as the ventral pallidum and substantia innominata, are very poorly described and defined (Fig. 1).

With these morphological reasons and also by the demonstration, using new techniques, that clusters of neurons with neurotransmitters other than acetylcholine are intermeshed with the cholinergic ones (Bennett-Clarke, 1988; Crawley and Wenk, 1989; Fisher et al., 1988; Rye et al., 1984; Senut et al., 1989; Walker et al., 1989), several authors have revised the classical grouping in the last few years. Mesulam et al. (1983a,b) still consider these cholinergic groups but in a broader sense, while Heimer and co-workers (Alheid and Heimer, 1988) have changed the nuclear morphological division to a morphofunctional complex of three systems: the striopallidal system, the extended amygdala, and the magnocellular corticopetal complex (Fig. 1).

Briefly, the striatopallidal system is a rostroventral extension of both the caudate-putamen and globus pallidus structures, closely related to the cortex–basal ganglia organization. The ventral pallidum is epidermal growth factor immunoreactive, like the globus pallidus.

The centromedial amygdaloid complex and the bed nucleus of the stria terminalis are components of a continuum extending from the temporal lobe up to the stria terminalis having a stronger reaction to angiotensin II.

The magnocellular corticopetal system consist mainly of cells that use acetylcholine as their neurotransmitter but also include a varying percentage of neurons that are noncholinergic. Despite the current emphasis on the cholinergic neurons, increasing interest is attracted to these noncholinergic neurons.

Neurons that transmit by GABA, galanin (in cotransmission with acetylcholine or alone), somatostatin, neurotensin, enkefalins, and so on, have been demonstrated to lie intermeshed with cholinergic neurons in the magnocellular clusters.

The cholinergic corticopetal neurons have a precise correlation with the cortical areas, and probably so do the noncholinergic ones, such as the GABA set of the septum innervating the hippocampus or the neurotensin basalis group that innervate the frontoparietal cortex.

The endings of the cholinergic groups were very well described years ago, as was mentioned above. The septal group innervate the hippocampus, the diagonal band innervates the cyngulate cortex, and the basalis group innervates the fronto-parietal neocortex. Cholinergic axon terminals end in a diffuse manner on both intrinsic and extrinsic cortical neurons and blood vessels. But good topographic correlation is observed.

After this description of basal forebrain, the most important points to keep in mind in relation to AD and its experimental models are the following:

(1) Basal forebrain is an anatomical region in close relationship with striatal, amygdalar and hypothalamic systems in which a magnocellular corticopetal system exists.

(2) Of this corticopetal system three functional cholinergic groups, similar but more diffusely located than the classically described nucleus of the septum, the diagonal band and the nucleus basalis, are the most outstanding sets of neurons regulating cortical functions. Each group specifically ends in a part of the cortex.

(3) These cholinergic neurons have been involved in the ethiopathogeny of AD, because their number and their function is statistically diminished in AD patients (Coyle et al., 1983). Moreover, experimental Alzheimer models are obtained by lesioning areas in which these cholinergic neurons are placed.

(4) Intermeshed with these cholinergic neurons, other noncholinergic (with GABA, neurotensin, somatostatin, and so on, as chemical transmitters) neurons exist, as both interneurons or corticopetal system (Freund and Antal, 1988). In this last case agonist or antagonist effects of the cholinergic corticopetal innervation could be expected, both in normal young and senile animals and in experimental models.

This new point of view on the basal forebrain could be very useful for understanding the pathogeny of AD. On the one hand it can be used as an integrative clue for cholinergic and noncholinergic theories on AD: in AD important noncholinergic deficits are always present. Moreover, in many experimental studies it is pointed out that the consequences of the nbM-lesion or the septum-lesion have no complete explanation when considering only the cholinergic involution. Modifications in GABA, neurotensin, and somatostatin neurons are evoked as a keystone to understanding the alterations.

On the other hand, results obtained in experimental models or human Alzheimer brains could be explained more easily. All cholinergic neurons seem to be a functional unit (Hepler et al., 1985). Although AD has been described as an involution of the nbM, the cholinergic deinnervation appears in the neocortex as in the hippocampus, and involutive somata occur both in the nbM and the septum.

General Morphofunctional Implications of Lesions in the Nucleus Basalis

Septal lesions and hippocampal manipulations are very useful in studying the cortical changes induced by cholinergic deinnervation. But the models obtained after nucleus basalis lesion have attained much more attention because different kinds of lesions could be studied and a better follow-up of the cortical alterations can be done.

The most important features of the nbM-lesioned model are the following. A high deficit of presynaptic cholinergic markers is observed in the frontoparietal cortex. Acetylcholinesterase (AChE), cholinacetyltransferase (CAT), and the system of high affinity of choline uptake (HACU) can fall to 70% of the initial value in rats (Bartus et al., 1985b; Casamenti et al., 1988; Dunnett et al., 1987; El-Defrawy et al., 1985, 1986; Gardiner et al., 1987; Santos-Benito et al., 1988).

All of these cholinergic markers are presynaptic and only AChE could be located in postsynaptic neuronal places or blood vessels.

Muscarinic receptors diminish in lesser quantity (about 20%) (De Belleroche et al., 1985; McKinney and Coyle, 1982; Norman et al., 1986), the most important decrease being the M2 classes that were considered as presynaptic (Mash et al., 1985).

After lesioning, a disappearance of cholinergic neurons in nbM and surrounding forebrain and pallido striatal structures is observed. Moreover, other neurons disappear, and several authors declare these changes more important than the cholinergic ones. For example, Wenk et al. (1989) claim that the cortical neurotensin-fall is the cause of behavioral impairments. Other less intense lesions of the nbM promote fewer cortical deficits and memory impairments.

Factors that outline the characteristics displayed by each model are: (1) the unilateral or bilateral type of lesion (Casamenti et al., 1988; Orzi et al., 1988; Watson et al., 1985); (2) the location of the lesion in the nbM area and the extension of the lesion (a small lesion can be produced by a single injection; an extended lesion needs, in several cases, two points of toxin injection) from the anterior commissure to the postero medial pallidal area (El-Defrawy et al., 1985); and, (3) the kind of toxin employed and the concentration of the employed toxin (Dunnett et al., 1987; Robbins et al., 1989; Toledano, 1989).

Changes in these factors clearly promote cortical and/or basal forebrain differences in the models, and also different kinds of responses in mnemonic tests. In many cases, these differences have an easy explanation: the more extensive the lesion is in size and amplitude, the more extensive are both the morphofunctional cortical and behavioral deficits. However, paradoxical, contradictory, or unexpected results can be obtained. Cortical deficits and impairments of behavior are not always in correlation. In several cases, ibotenic acid promotes fewer cortical deficits and more behavioral impairment than quisqualic acid lesions (Robbins et al., 1989; Wenk et al., 1989).

Long-term morphological and functional modifications are also very variable in different studies. For example, Bartus et al. (1985b, 1986) described a persistent cholinergic impairment with a behavioral recovery, El-Defrawy et al. (1986) a lack of recovery of cholinergic function, and other authors a partial or total recovery (Casamenti et al., 1988; Gage et al., 1983; Gardiner et al., 1987; Pedata et al., 1982; Santos Benito et al., 1988; Vige and Briley, 1989; Wenk and Olton, 1984).

Evolution of the Cortical Cholinergic Marker in an Alzheimer Model Produced by Single Unilateral Lesion

To be able to study the factors involved in the variability of the results, we have tested several kinds of nbM-lesioned models, being the model with a unilateral single injection most useful for our purposes (Toledano, 1989). The model is obtained by stereotactical injection of 0.5 microliters of solutions with selected toxins in a portion of the nbM, into the internal capsule, that innervate the dorsal fronto parietal cortex (Table 1; Figs. 1, 2). We have employed ibotenic (7.5–40 nmols), quisqualic (7.5–100nmols) and N-methyl-D-aspartic acids (7.5–60 nmols), that are glutamatergic agonists acting on different classes of glutamate receptors that provoke a toxic entry of calcium (Zaczek and Coyle, 1982).

The cortical short-term alterations, between the 1st and the 14th days, depend on the toxin and its concentration. In Table 1 you can observe the decrease in cholinergic markers after different doses of quisqualic, N-methyl-D-aspartic, or ibotenic acids. Histochemically, three days after lesioning, a disappearance of cholinergic fibers is observed, and at EM level, necrotical axonal endings can be shown. M receptors have few changes, as is shown in Table 1, in comparison with other cholinergic markers.

In the nbM, dysplasic and necrotical cells can be observed in areas neighboring the injection point. No positivity of AChE or CAT is observed after 3 to 10 days, but few surviving neurons with slight activity could be seen at 14 days (Fig. 2). In surrounding places (globus pallidus,

Table 1

Lesion	AChE	CAT	M1	M2	HACU
15 nm Quis	12.3 ± 8.2	13.3 ± 7.8			
30 nm Quis	32.0 ± 14.2	34.2 ± 12.0	8.2 ± 1.2	12.8 ± 8.1	21.0 ± 7.3
60 nm Quis	47.5 ± 16.8	48.2 ± 9.8	8.3 ± 2.3	14.6 ± 6.0	42.0 ± 10.1
15 nm NMDA	28.1 ± 10.3	30.0 ± 9.8			
30 nm NMDA	41.3 ± 13.3	41.2 ± 9.0	8.2 ± 2.2	16.0 ± 4.1	
40 nm NMDA	52.8 ± 13.6	56.7 ± 9.7	8.2 ± 6.6	10.8 ± 6.8	
15 nm Ibo	21.1 ± 10.2	24.3 ± 7.0			
25 nm Ibo	42.5 ± 12.2	46.3 ± 13.6	8.2 ± 0.2	8.2 ± 0.2	41.4 ± 18.3
50 nm Ibo	51.6 ± 6.8	58.1 ± 14.2	8.3 ± 2.3	14.6 ± 6.0	49.8 ± 6.8

Variations of cortical cholinergic markers 7 days after nbM lesioning. Lesions were carried out by stereotactical injection (coordinates: AP = –1.3 mm; L = –2.5 mm; V = –6.7 mm; see text) of different quantities (7.5–60 nmols) of quisqualic (Quis), N-methyl-D-aspartic (NMDA) and ibotenic (Ibo) acids dissolved in 0.5 microliters, in the right nbM. Average values of ipsilateral lesioned side and standard deviation is given in per cent, 100% being in each case the absolute value obtained in the contralateral unlesioned side. Acetylcholinesterase (AChE) was determined by the Ellmans method (Ellmans et al., 1961), cholinacetyltransferase (CAT) by the Fonnum procedure (Fonnum, 1975), muscarinic receptors (M1, M2) were estimated by densitometry on autoradiograms obtained by the method of Palacios et al. (1987), and the high affinity choline uptake system (HACU) was determined as by Pedata et al. (1982). Average values were obtained in groups of 8–17 animals. The number of surviving animals after lesioning was higher than 92% in all cases except in 50 nmols Ibo and 40 nmols NMDA-lesioned animals.

substantia innominata, posterior part of the nbM close to optic tract, and so on), cholinergic neurons remain.

NGF receptor is increased in a short period, about four days after lesioning, in fibers and nbM cholinergic neurons, and also in cortical pyramidal neurons.

In a long-term period, from 14 to 180 days, the changes that we can observe are very variable, depending strongly on the kind of toxin and the concentrations employed (Table 2). Recovery of cholinergic markers is the most important characteristic in several cases: 7.5, 15, and 30 nMoles of quisqualic led to models that can recover the normal cortical values in 24, 30, and 45 days, respectively. But other cases do not show the recovery and N-methyl-D-aspartate shows a partial recovery at 180 days. In the quisqualic case, the recovery is variable, and the profiles displayed by individuals in histochemical and biochemical studies are quite different (Fig. 3). Both 15- and 30-nMoles-quisqualic-injected animals are useful

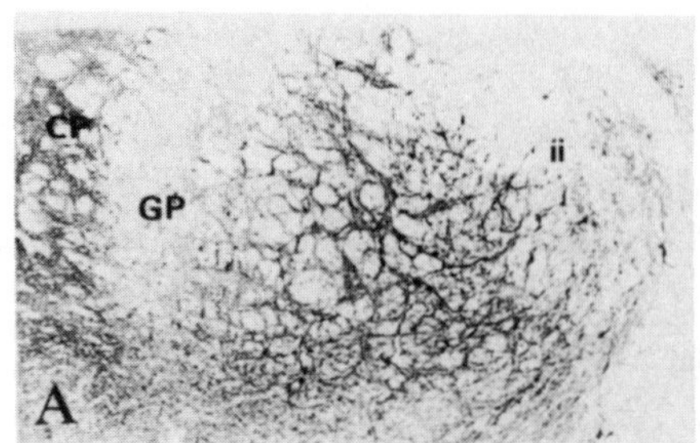

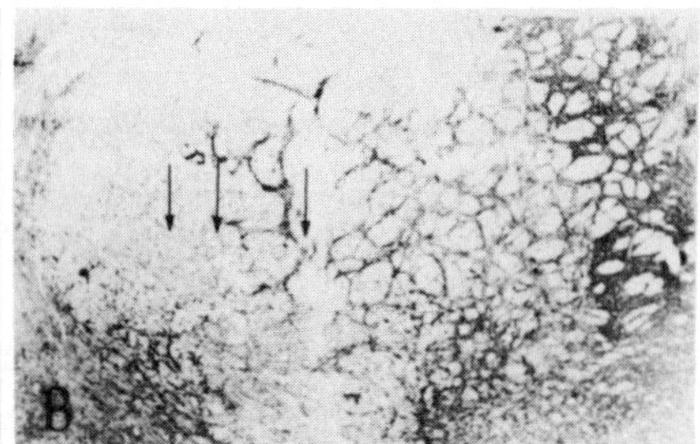

Figure 2. Loss of cholinergic neurons in the nbM 14 days after an injection of 30 nmols of quisqualic acid. A = nonlesioned left nbM displaying stellate AChE-positive neurons (the nbM) on the border of the globus pallidus (GP) and in the internal capsule (ii). B = lesioned right nbM showing a loss of cholinergic cells in the nbM areas. (Koelle and Friedenwald (1972) method; 20 micron slide of a 4% p-formaldehyde-perfused brain).

Table 2

Days	30 nm Quis	30 nm NMDA	25 nm Ibo
7	32.0 ± 10.2	41.3 ± 7.3	42.5 ± 12.2
14	33.4 ± 15.3		
28	22.3 ± 16.6	42.1 ± 8.8	44.5 ± 6.8
56	9.8 ± 17.2		
84	6.6 ± 12.8	22.0 ± 12.3	41.5 ± 10.8
180	6.8 ± 14.2	14.2 ± 16.8	36.6 ± 8.6

Evolution of AChE (biochemical determinations) in frontoparietal neocortex after nbM lesioning. Data obtained at different times in three series of lesioned animals (for methods and data expression, see Table 1).

to study individual variations.

In this series of animals, cortical recovery of average values is shown, but AChE and CAT values of selected individuals with greater and lesser impairments than the normal recovery group present (Fig. 4). It is important to point out that the higher recovery is accompanied, in most cases, by an increase in contralateral values. In the recovered animals other features are to be pointed out:

(1) Intrinsic cortical cholinergic neurons increase their activity, and a part of the cortical cholinergic fibers are thicker than the normal ones (Fig. 3)

(2) A variable number of cholinergic neurons can be demonstrated in the lesioned area. The number of these cells is not in close agreement with the level of the recovery. Simultaneously, a variable number of noncholinergic neurons are observed in the lesioned area. Animals with

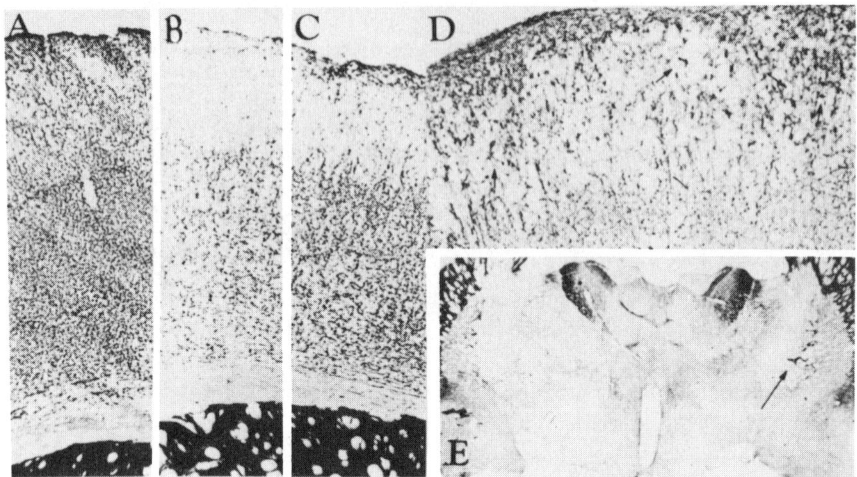

Figure 3. AChE in different cases of unilateral (right) lesioned animals with 30 nmols of quisqualic acid. A, B = AChE in frontoparietal neocortex 14 days after lesioning. Loss of cholinergic fibers in the lesioned side: A = left unlesioned side; B = right lesioned side. C, D = AChE in frontoparietal neocortex after 30 days. Partial recovery of the cholinergic network (higher density of fibers in C than in B) and hyperactivity of fibers and cortical cholinergic neurons (thin and thick cholinergic fibers and more intense positive reaction in cortical cholinergic neurons in D). E = hyperactive cholinergic neurons in posterior nbM areas ipsilateral to the lesion (arrow), 21 days after lesioning. The histochemical incubation was time-regulated to prevent the staining of the left nbM normoactive cholinergic neurons. (A, B, C, D: 40-micron slides from perfused animals; E: 20-micron slide of fresh tissue after fixation. Karnovsky and Roots (1964) modified method) (A, B, C = 40x; D = 80x; E = 6x).

bad recovery showed a decrease in noncholinergic cells higher than 55%, and animals with the best recovery showed a decrease in noncholinergic cells lesser than 40%.

(3) At the EM level, at 30 days, several cholinergic-positive but dysplasic cells could be seen in the lesioned basalis area, and also necrotic cells.

Factors Involved in the Evolution of the Experimental nbM Lesions

From these results, there arise factors that influence the extent and the functional implications of the nbM lesion.

The factors involved in the evolution of the experimental nbM lesions could be grouped in two main categories. The first one can be made linking the factors that promote cortical deficits and behavioral impairment.

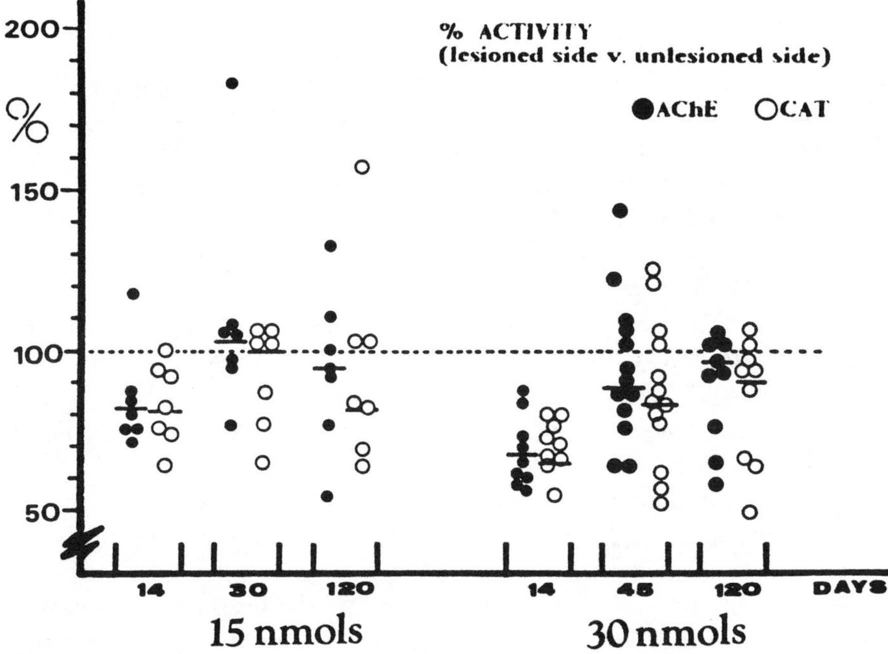

Figure 4. AChE and CAT activities in the ipsilateral frontoparietal neocortex of two series of animals after a single injection of 15 or 30 nmols of quisqualic acid into the nbM. Each dot is the value of each individual in the group. Average values in each group are represented by a line. Enzymatic activities are expressed as per cent of the contralateral unlesioned cortex.

Two of these are the most important causes: the location of the lesion, and the nature of the toxin. Both factors will give the number, type, and dysfunctional level of affected neurons and consecutively will qualify the lesion.

Location of the lesion will be important because the number of cholinergic and noncholinergic corticopetal neurons, and of other neurons belonging to the previously mentioned striatal and amygdalar basal forebrain systems, is a consequence of our manipulation in the extension of the Alzheimer model. Two points should be remembered. On the one hand, there exists a good relationship between several sets of neurons and the cortical regions. In our case, the lesioned posterior part of the nbM innervate the dorsolateral region of the frontoparietal neocortex. On the other hand, many other neurons, both corticopetal noncholinergic and neurons functionally working with the striatal and the amygdalar complexes will be damaged, and, perhaps, side effects will always be present in the models.

The feature of the toxin is the other most important factor in the promotion of the effects. A more diffusible toxin should give a greater forebrain lesion and more elevated cortical deficits. The importance of this factor sharply increases when we consider all together the type of glutamatergic receptor it overstimulates (Zaczek and Coyle, 1982), and the density of this glutamatergic receptor in both the cholinergic and non-cholinergic neurons affected by the toxin. Also, the normal or abnormal situation of the neuron is important for the response (Balazs et al., 1988). Unfortunately, we only have a general idea about the chemical response of the different neurons, and selective lesions cannot be outlined now. In our experience a very variable general neuronal decrease in the lesioned area is observed. Values obtained come between 22 and 87%.

The second category could analyze factors that promote recovery of the deficits and/or lead to lesser long-term alterations.

In the lesioned nbM, remaining cholinergic neurons are probably the most important factor for recovery. Two sets of these cholinergic neurons could be considered: the surviving neurons in the lesioned area, and the neighboring nbM neurons surrounding the lesion. In the cortex, a variable time after lesioning, an increase was observed in the number of AChE-positive fibers and also, in several cases, in the diameter of the axons. Starting in the cortical remnant axons, an outgrowth of axon terminals occurs, and deinnervated cortical fields become innervated. Moreover, cholinergic fibers can increase their functionality as histochemical activities are higher in the cortical slices, which display a less intense but normal pattern. The number and kind of the affected non-cholinergic neurons induced different evolutions of the lesions, because different models produced with diverse toxins or places of injection have a different recovery.

Contralateral nbM participate in the recovery as has been proposed (Pearson et al., 1986; Pedata et al., 1982). We have observed a cholinergic contralateral hyperactivity in selected cases, and a change in the AChE histochemical pattern, increasing reactivity for AChE in cortical layers (II and III) with small activity.

In the neocortex, adaptive changes in cortical neurons are observed. AChE and CAT positive neurons increase their enzymatic activities. M receptors in cortical neurons are maintained.

Neuroactive factors, considering under this denomination several well known and other nonisolated but putative factors, must play an important role (Gage et al., 1989; Haroutunian et al., 1986a; Hefti, 1986; Hefti et al., 1984, 1989; Kiss et al., 1988; Maysinger et al., 1989; Morgan, 1989; Snider and Johnson, 1989). NGF is demonstrated as an essential factor for maintaining the functionality of the cholinergic forebrain neurons. Increased immunoreactivity for NGF receptors has been observed by us in selected cortical neurons and fibers four days after lesioning. On the other hand, different factors induced by the modified cholinergic and noncholinergic neurons in the lesioned nbM, are, probably, the real reason different models, or individuals of a model, have a good or bad recovery, or, to the contrary, a slow impairment of the cholinergic innervation. Involutive neurons have been observed by us 30 days after lesioning, at which time survival neurons reached a normal appearance. This fact has been interpreted by us as a new wave of necrosis not directly related to the neurotoxin, probably promoted by necrotic factors.

Conclusions

Logically, the most important practical implications and/or conclusions that we can extract from the results obtained in the study of the experimental Alzheimer models will be the most useful ones to explain the etiopathogeny of AD and to open new pathways for diagnosis and treatment of this neurodegenerative disorder.

At first, it is necessary to mention that Alzheimer nbM models give a confirmation of the cholinergic theory of AD. An anatomofunctional correlation among basal forebrain cholinergic deficits in the neocortex has been demonstrated. Cortical cholinergic markers AChE, CAT, and a high affinity system of acetylcholine uptake are presynaptic, and in rodents, about 70% of the total value corresponds to the forebrain corticopetal complex. Of the total values, 30% are from intrinsic neurons. M1 receptors and the greater percentage of M2 receptors are postsynaptic.

The free evolution of the cortical cholinergic deficits induced after nbM lesioning is very variable, depending on intrinsic and extrinsic factors. Following different experimental procedures, it is possible to obtain models with permanent cortical deficits, models with a recovery of their deficits and, also, individuals that present long-term increased impairments after a single lesion whose induced deficits are expected to recover. The theoretical and practical meaning of these results is different in each case.

The last-mentioned group of animals could represent the progressive involution arising in AD. By individual but unknown causes, a long-term necrosis and dysfunction of cholinergic neurons, in the location of the lesion and other nbM surrounding areas, has been produced after the nbM disturbance that was induced by lesioning. Several models displaying permanent deficits (after multiple lesions or high doses of toxin) could be the animal reflex of advanced AD. I consider it very interesting that several models or individuals of a model tend to a deep involution when similar ones can recover on the basis of the abilities of the survival neurons after lesioning. Buell and Coleman (1979) pointed out that the most important characteristic to differentiate normal senile involution and AD neurodegeneration was the loss of plasticity (or adaptive ability) to compensate the involutive disappearance of nerve endings with newly created terminals. A very important research line will be the characterization of involutive factors or inhibitors of neuronal plasticity.

A common feature of deep and persistent cholinergic impaired models is the maintenance of higher values of postsynaptic markers, mainly M1 receptors. This fact is in close agreement with the Palacios interpretation of the evolution of the M receptors in AD: M receptor density is maintained up to the moment at which a large decrease of the cortical neurons is reached (Palacios et al., 1987).

Several studies have displayed a high persistence of M receptors and their plastic features (Mckinney and Coyle, 1982; Norman et al., 1986; Palacios, 1982)

The practical consequences of M postsynaptic maintenance are very clear. Many laboratories are looking for new cholinergic drugs, especially chemical agonists of M1 receptors without the classical side effects of the classical cholinergic drugs (Bartus et al., 1987; Haroutunian et al., 1985; Palacios and Spiegel, 1986). These have not been found yet, and other laboratories try to outline new therapies such as cholinergic implants, overproduction of acetylcholine, and so on (Goyal , 1989; Haroutunian et al., 1985, 1986b). In this item, the existence of different classes of M receptors is to be considered (Noronha-Blob et al., 1987).

Several other studies have displayed a high persistence of M receptors and their plastic features (McKinney and Coyle, 1982; Norman et al., 1986; Palacios, 1982; Pintor et al., 1988; Vige and Briley, 1989).

It is necessary to emphasize the models showing a recovery of the deficits because they demonstrate the existence of (1) several plastic possibilities of nervous structures, and (2) factors promoting recovery. Several of our models have shown a clear recovery. Similar results have been pointed out in other studies (Casamenti et al., 1988; Dunnett et al., 1987; Gage et al., 1983; Gardiner et al., 1987; Pedata et al., 1982; Santos Benito et al., 1988; Vige and Briley, 1989; Wenk and Olton, 1984). These findings could not be of application for advanced steps in AD, which are characterized by a loss of plasticity. But we can suppose that if an early diagnosis of AD can be obtained, the treatment of the first levels of this disorder must take advantage of the results here presented. Three observed findings on cholinergic nbM neurons are of great interest in this field: (1) the ability of neurons to survive in the lesioned area, (2) the hyperactivity as a response of these neurons and other cholinergic ones located in neighboring nbM places, and (3) the ability for terminal outgrowth of all these neurons. It has been demonstrated that factors such as NGF promote survival and recovery (Bartus et al., 1986; Dreyfus et al., 1989; Gage et al., 1989; Haroutunian et al., 1986; Hefti, 1986; Hefti et al., 1984), and we have found an increase in the production and in the content of a specific receptor for this factor.

The plastic response of the cortical cells is another support for recovery. Pyramidal and short-axon cells (Stichel et al., 1987) should intervene in the reconstruction of the normal cortical cholinergic pattern. In a similar way, intrinsic cortical cholinergic cells display hyperactivity, but, up to now, the meaning of this ability has not been explored. These responses can be related to the recovery of cholinergic extrinsic innervation in an agonist way, or to the contrary, the response could be directed to maintain other functions such as vascular regulation.

It is very important to study intrinsic and extrinsic factors that actually could promote recovery in such a way as NGF do (Maysinger et al., 1989; Morgan, 1989). Many laboratories are looking for natural factors or drugs that could enhance plastic abilities of the neurons, especially the cholinergic ones. We have tested a nootropic, pyritinol, in a quisqualic model in which physiological recovery is obtained after 45 days of lesioning. Under pyritinol treatment, an induction of recovery was achieved in the first period, in which a cholinergic deficit is a common feature (Fig. 5). This is a promising result that could explain the success obtained in clinical trials with intermediately impaired AD patients (Herrman, 1988). Pyritinol has demonstrated several positive biochemical and behavioral effects (Greiner et al., 1988; Marston et al., 1987; Martin and Widdowson, 1989) especially in deficit situations of brain metabolism. All of these results demonstrate the value of the Alzheimer experimental models in preclinical research.

Finally, it is not possible to avoid the discussion of a controversial topic concerning Alzheimer models. Although the problems of the planning and the interpretation of the results in behavioral trials will not be considered here (because it is far removed from the goals of this review), it is necessary to point out the striking differences found by several authors between the cortical deficits, the behavioral impairment, and the recovery of the cholinergic function. Selected papers showed persistent cholinergic deficits coexisting with improved behavior (Bartus

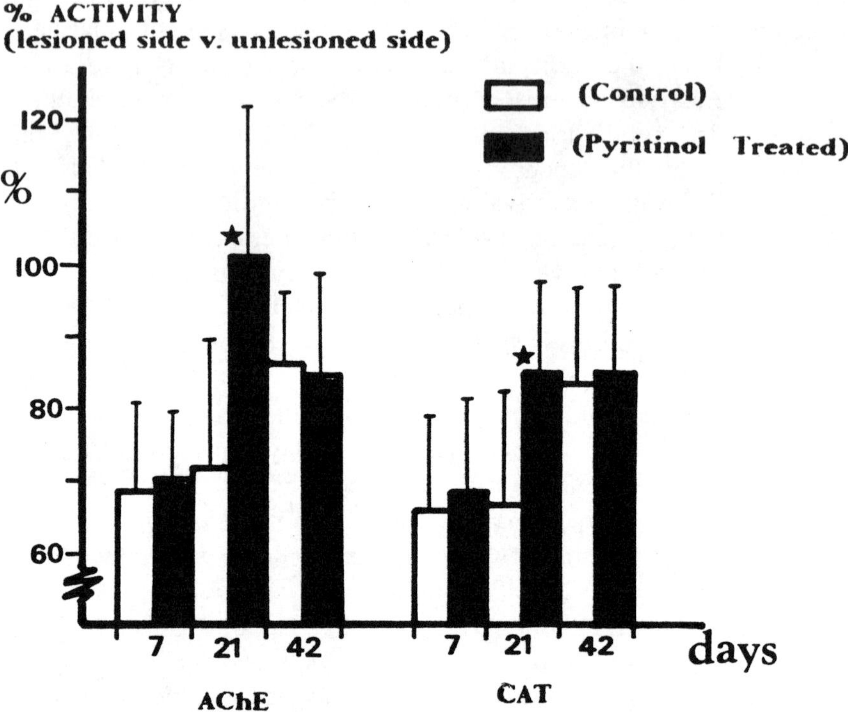

Figure 5. Evolution of AChE and CAT in the ipsilateral frontoparietal neocortex after unilateral (right) nbM lesioning with 30 nmols of quisqualic acid and a treatment with pyritinol dihydrocloride (600 mg/kg/day in a single dosage). Free recovery of the cortical cholinergic markers is observed in this model at 42 days, but the recovery is produced after 21 days. The pyritinol-treated animals showed an early recovery at 21 days, accompanied by a clear AChE hyperactivity (n = 12 animals per group).

et al., 1985). Although a cholinergic perspective is a common starting point for models of age-related cognitive and neurodegenerative disorders (Bartus, 1988), many authors are beginning to claim the importance of the noncholinergic neurons of the nbM in the promotion and maintenance of the behavioral impairments (Robbins et al., 1989; Wenk et al., 1989). Neurotensin, somatostatin, and GABA neurons are the most important of these neurons. These results have a strong anatomofunctional basis. Several times in this review the new conception of the structure of the nbM has been mentioned, emphasizing that the classically described cholinergic forebrain nuclei need now to be considered a part of a corticopetal system, consisting of cholinergic and several kinds of noncholinergic neurons (Alheid and Heimer, 1988).

Acknowledgments. Work described in this chapter was supported in large part by grants from the Spanish Fondo de Investigaciones Sanitarias (FISSS 88/1920) and DIGYCYT (PM 88/0016), and from the German Merck Darmstadt Laboratories. The author thanks Drs. M.L. Bentura, R. Vilariño, C. Zahonero, and L. Rivas for collaboration in the research on AD models. The excellent technical assistances of M.J. Chulia, B. Sanchez, and A. Simon is greatly appreciated.

References

Alheid GF, Heimer L (1988): New perspectives in basal forebrain organization of special relevance for neuropsichiatric disorders: the striatopallidal, amygdaloid and corticopetal components of substantia innominata. *Neuroscience* 27:1–39

Armstrong DM, Benzing WC, Evans J, Terry RD, Shields D, Hansen LA (1989): Substance P and somatostatin coexists within neuritic plaques. Implications for the pathogenesis of Alzheimer's disease. *Neuroscience* 31:663–371

Balazs R, Jorgensen OS, Hack N (1988) N-methyl-D-aspartate promotes the survival of cerebellar granule cells in culture. *Neuroscience* 27:437–451

Bartus RT (1988): The need for common perspectives in the development and use of animal models for age-related cognitive and neurodegenerative disorders *Neurobiol Aging* 9:445–451

Bartus RT, Dean RL, Beer B, Lippa AS (1982): The cholinergic hypothesis of geriatric memory dysfunction. *Science* 217:408–417

Bartus RT, Dean RL, Pontecorvo MJ, Flicker C (1985) a: The cholinergic hypothesis: a historical overview, current perspective, and future directions. In: *Memory Dysfunctions: An Integration of Animal and Human Research from Preclinical and Clinical Perspective* Olton DS, Gamzu E, Corkin S, eds. (*Ann NY Acad Sci* Vol. 444) New York: The New York Academy of Sciences Press, pp. 332–358

Bartus RT, Dean RL, Flicker C (1987): Cholinergic Psychopharmacology: an integration of human and animal research on memory. In: *Psychopharmacology: The Third Generation of Progress*, Metzler HY, ed. New York: Raven Press, pp. 219–232

Bartus RT, Flicker C, Dean RL, Pontecorvo M, Figueiredo JC, Fisher S k (1985)b: Selective memory loss following nucleus basalis lesions: long term behavioral recovery despite persistent cholinergic deficiencies. *Pharmacol Biochem Behav* 23:125–135

Bartus RT, Pontecorco MJ, Flicker C, Dean RL, Figueiredo JC (1986): Behavioral recovery following bilateral lesions of the nucleus basalis does not occur spontaneously. *Pharmacol Biochem Behav* 24:1287–1292

Bear MF, Carnes KM, Ebner FF (1985): Investigation of cholinergic circuitry in cat striate cortex using acetylcholinesterase histochemistry. *J Comp Neurol* 234:411–430

Bennett-Clarke CA (1988): Cortical projections from somatostatin neurons of the basal forebrain in the rat. *Soc Neurosci Abst* 81:8–8

Bertoni-Freddari C, Guiuli C, Pieri C, Paci D (1986): Quantitative investigation of the morphological plasticity of synaptic function in rat dentate gyrus during aging. *Brain Res* 366:187–192

Bigil V, Woolf NJ (1982): Cholinergic projections from the basal forebrain to frontal, parietal, temporal, occipital and cingulate cortices: a combined fluorescent tracer and acetylcholinesterase analysis. *Brain Res Bull* 8:727–749

Briegon A, Greenberger V, Segal M (1984): Quantitative histochemistry of brain acetylcholinesterase and learning rate in the aged rat. *Neurobiol Aging* 7:215–217

Buell SJ, Coleman PD (1979): Dendritic growth in the aged human brain and failure of growth in senile dementia. *Science* 206:854–856

Butcher LL, Woolf NJ (1986): Central cholinergic systems: synopsis of anatomy and overview of physiology and pathology. In: *The Biological Substrates of Alzheimer's Disease*, Scheibel AB,Weschler AF, Brazier MAB, eds. Orlando, San Diego: Academic Press, pp. 73–86

Casamenti F, Di Patre PL, Bartolini L, Pepeu B (1988): Unilateral and bilateral nucleus basalis lesions: differences in Neurochemical and behavioural recovery. *Neuroscience* 24:209–215

Chrobak JJ, Hanin I, Schmechel DE, Walsh Th J (1988): AFG4A-induced working memory impairment: behavioral, neurochemical and histological correlates *Brain Res* 463:107–117

Chui HCW, Bondareff CZ, Slager U (1984): Stability of neuronal number in the human nucleus basalis of Meynert with age. *Neurobiol Aging* 5:83–88

Collerton D (1986): Cholinergic function and intellectual decline in Alzheimer's disease. *Neuroscience* 19:1–28

Coyle JT, Price DL, Delong MR (1983): Alzheimer's disease: A disorder of cortical cholinergic innervation. *Science* 219:1184–1190

Crawley JN, Wenk GL (1989): Co-existence of galanin involved in memory processes and dementia? *Trends Neurosci* 12:278–281

Davies P (1979): Neurotransmitter-related enzymes in senile dementia of the Alzheimer type. *Brain Res* 171:319–327

De Belleroche J, Gardiner IM, Hamilton MH, Birdsall JM (1985): Analysis of muscarinic receptor concentration and subtypes following lesion of rat substantia innominata. *Brain Res* 340:201–209

Deutsch JA (1971): The cholinergic synapse and the site of memory. *Science* 174:788–794

Dohanich GP, McEven BS (1986): Cholinergic limbic projections and behavioral role of basal forebrain nuclei in the rat. *Brain Res Bull* 26:477–482

Drachman DA, Leavitt J (1974): Human memory and the cholinergic system. *Arch Neurol* 30:113–121

Dreyfus CF, Bernd P, Martinez HJ, Rubin SJ, Black IB (1989): GABergic and cholinergic neurons exhibit high-affinity nerve growth factor binding in rat basal forebrain. *Exp Neurol* 104:181–185

Dunnett SB, Whisham IO, Jones GH, Bunch SF (1987): Behavioral, biochemical and histochemical effects of different neurotoxic amino acids injected into nucleus basalis magnocellularis of rats. *Neuroscience* 20:653–669

El-Defrawy SR, Coloma F, Jhamandas K, Boegman RJ, Beninger RJ, Wirxhing BA (1985): Functional and neurochemical cortical cholinergic impairment following neurotoxic lesions of the nucleus basalis magnocellularis in the rat. *Neurobiol Aging* 6:325–330

El-Defrawy SR, Coloma F, Jhamandas K, Boegman RJ, Shipton L (1986): Lack of recovery of cortical cholinergic function following quinolic or ibotenic acid injections into the nucleus basalis magnocellularis in rats. *Exp Neurol* 91:628–633

Ellmans GL, Courtney KD, Andres V, Featherstone RM (1961): A new and rapid colerimetric determination of acetylcholinesterase activity. *Biochem Pharmacol* 7:88–95

Fisher RS, Buchwald NA, Hull CD, Levine MS (1988): GABAergic basal forebrain neurons project to the neocortex: The localization of glutamic acid decarboxylase and choline acetyltransferase in feline corticopetal neurons. *J Com Neurol* 272:489–502

Fonnum F (1975): A rapid radiochemical method for the determination of choline acetyltransferase. *J Neurochem* 24:407–409

Freund TF, Antal M (1988): GABA-containing neurons in the septum control inhibitory interneurons in the hippocampus. *Nature* 36:170–173

Gage FH, Batchelor P, Chen KS, Chin D, Higgins GA, Koh S, Deputy S, Rosenberg MB, Fisher W, Bojörklund A (1989): NGF receptor reexpresion and NGF-mediated cholinergic neuronal hypertrophy in the damaged adult neostriatum. *Neuron* 2:1177–1184

Gage FH, Bojörklund A, Stenevi U (1983): Reinnervation of the partially deafferented hippocampus by compensatory collateral sprouting from aparent cholinergic and adrenergic afferents. *Brain Res* 268:27–37

Gajdusek DC (1987): A new recognized mechanism of pathogenesis in Alzheimer's disease, Amyotrophic Lateral Sclerosis, and other degnerative neurological diseases: the beta-fibriloses of brain. In: *Nutrition, Health and Peace*, Jarillawa RJ, Schowoebel SL, eds. Palo Alto: Linus Pauling Institute, pp. 21–55

Gardiner IM, De Belleroche J, Prom BK, Hamilton MH (1987): Effect of lesion of the nucleus basalis of rat on acetylcholine release in cerebral cortex. Time course of compensatory events. *Brain Res* 407:263–271

Gower AJ (1986): Lesioning of the nucleus basalis in the rat as a model of Alzheimer's disease. *Trends Pharmacol Sci* 12:432–434

Goyal RK (1989): Muscarinic receptor subtypes: physiology and clinical implications. *New Engl J Med* 321:1022–1029

Greiner HE, Haase AF, Seyfried CA (1988): Neurochemical studies on the mechanism of action of pyritinol. *Pharmacopsychiat* 21:26–32

Haroutunian V, Davis KL, Davis B m, Horvath PP, Johns CA, Morhs RC (1985): The study of cholinomimetics in Alzheimer's disease and animal models. In: *Psychopharmacology: Recent Advances and Future Prospects*, Iversen SD, ed. Oxford, New York: Oxford University Press, pp. 218–259

Haroutunian V, Kanof PD, Davis KL (1986a): Partial reversal of lesion induced deficits in cortical cholinergic markers by nerve growth factor *Brain Res* 386:397–399

Haroutunian V, Kanof PD, Tsuboyama GK, Campbell GA, Davis KL (1986b): Animal models of Alzheimer's disease: behavior, pharmacology, transplants. *Can J Neurol Soc* 13:385–393

Hefti F (1986): Nerve growth factor promotes survival of septal cholinergic neurons after fimbrial transections. *J Neurosci* 6:2155–2162

Hefti F, Dravid A, Hartikka J (1984): Chronic intraventricular injections of nerve growth factor elevate hippocampal choline acetyltransferase in adult rats with partial septo-hippocampal lesions. *Brain Res* 293:305–311

Hefti F, Hartikka J, Krusel B (1989): Function of neurotrophic factors in the adult and aging brain and their possible use in the treatment of neurodegenerative diseases. *Neurobiol Aging* 10:515–533

Hepler DJ, Olton DS, Wenk GL, Coyle JT (1985): Lesions in nucleus basalis magnocellularis and medial septal area of rats produce qualitatively similar memory impairment. *J Neurosci* 5:866–873

Herrmann WM, Kubicki SK, Rohmel J (1988): Vigilance classification system: Development, sample values, and applications in nootropic drug research. *Z Gerontopychol Psychiat* 1(s):1–33.

Johnston MV, McKinney M, Coyle JT (1981): Neocortical cholinergic innervation: a description of extrinsic and intrinsic components in the rat. *Exp Brain Res* 43:159–172

Karnovsky MJ, Roots LA (1964): Direct-coloring thiocholine method for cholinesterase. *J Histochem Cytochem* 12:219–221

Kiss J, McGovern J, Patel AJ (1988): Immunohistochemical localizaton of cells containing nerve growth factor receptors in the different regions of the adult rat forebrain. *Neuroscience* 27:731–748

Knežević, S, Mubrin Z, Risberg J, Vucinic G, Spilich G, Gubarev N, and Wannenmacher W (1989). Pyritinol treatment of SDAT patients: Evaluation of psychiatric and neurological examination, psychometric testing, and rCBF measurements. *International Journal of Clinical Psychopharmacology* 4(1):25–28

Koelle GB, Friedenwald JS (1949): A histochemical method for localizing cholinesterase activity. *Proc Soc Exp Biol Med* 70:617–622

Krnjević K, Silver A (1965): A histochemical study of cholinergic fibers in the cerebral cortex. *J Anat* 99:711–759

Lamour Y, Dutar P, Jobert A (1982): Topographic organization of basal forebrain neurons projecting to the rat cerebral cortex. *Neurosci Lett* 34:117–122

Mann DMA, Yates PO, Marcyniuk B (1984a): Monoaminergic neurotransmitter systems in presenile Alzheimer's disease and in senile dementia of Alzheimer type. *Clin Neurophatol* 3:199–205

Mann DM A, Yates PO, Marcyniuk B (1984b): Changes in nerve cells of the nucleus basalis of Meynert in Alzheimer's disease and their relationship to aging and the the accumulation of lipofuscin pigment. *Mech Aging Dev* 25:189–204

Marston HM, Martin KJ, Robbins TW (1987): Effect of the cholinergic administration of pyrithioxin on behavior and cholinergic function in young and aged rats. *J Psycho Pharmacol* 1(4):237–243

Martin KJ, Widdowson L (1989): The role of membrane phospholipids in acetylcholine synthesis: Observations with pyritinol. In: *Pharmacological Interventions on Central Cholinergic Mechanism in Senile Dementia (Alzheimer's Disease)*, Kewitz, Thompson, Bickel, eds. München:W Zuchschwerlt, Vorlag, pp. 129–132

Mash DC, Flynn DD, Potter LT (1985): Loss of M2 muscarinic receptors in the cerebral cortex in Alzheimer's disease and experimental cholinergic denervation. *Science* 228:1115–1117

Maysinger D, Garofolo L, Jalsenjak I, Cuello AC (1989): Effects of microencapsulated monosialoganglioside GM1 on cholinergic neurons. *Brain Res* 496:165–172

McKinney M, Coyle JT (1982): Regulation of neocortical muscarinic receptors effects of drug treatment and lesions. *J Neurosci* 2:97–105

Mesulam MM, Mufson EJ, Levey AI, Weiner EJ (1983a): Cholinergic innervation of cortex by the basal forebrain: Cytochemistry and cortical connections of the septal area, diagonal band nuclei, nucleus basalis (substantia innnominata) and hypothalamus in the rhesus monkey. *J Comp Neurol* 214:170–197

Mesulam MM, Mufson Ej, Wainer BH, Levey AI (1983b): Central cholinergic pathways in the rat: an overview based on an alternative nomenclature (Ch1-Ch6). *Neuroscience* 10:1185–1201

Miquel J, Johnson JE, Cervos-Navarro J (1983): Comparison of CNS aging in human and experimental animals In: *Brain Aging: Neuropsychology and Neuropharmacology (Aging, Vol 21)* Cervos-Navarro J, Sarcander HI, eds. New York: Raven Press, pp. 231–258

Morgan DG (1989): Considerations in the treatment of neurological disorders with trophic factors. *Neurobio Aging,* 10:547–549

Norman AB, Blaker SN, Thal L, Creese J (1986): Effects of aging and cholinergic deafferentation on putative muscarinic cholinergic receptor subtypes in rat cerebral cortex. *Neurosci Lett* 70:289–294

Noronha-Blob L, Lowe VC, Hanson RC (1987): Heterogeneity of muscarinic receptors coupled to phosphoiinositide breakdown in guinea pig brain and peripheral tissues. *Life Sci* 41:967–975

Ordy JM, Thomas GJ, Volpe BT, Dunlop WP, Colombo PM (1988): An animal model of human-type memory loss based on aging, lesion, forebrain ischemia, and drug studies with the rat. *Neurobiol Aging* 9:667–683

Orzi F, Diana G, Casamenti F, Palombo E, Fieschi C (1988): Local cerebral glucose utilization following unilateral and bilateral lesions of the nucleus basalis magnocellularis in the rat. *Brain Res* 462:99–103

Palacios JM (1982): Autoradiografic localization of muscarinic cholinergic receptors in the hippocampus of patients with senile dementia. *Brain Res* 243:173–175

Palacios JM, Cortes R, Probst A (1987): Receptor plasticity in the human brain: some autoradiographic studies. *J Recep Res* 7:581–597

Palacios JM, Spiegel R (1986): Muscarinic cholinergic agonist: pharmacological and clinical perspectives. *Prog Brain Res* 70:485–497

Pearson RC A, Weal JW, Powell TP S (1986): Hypertrophy of cholinergic neurones of the basal nucleus in the rat following damage of the contralateral nucleus. *Brain Res* 382:149–152

Pedata F, Lo Conte G, Sorbi S, Marconcini-Pepeu I, Pepeu G (1982): Changes in high affinity choline uptake in rat cortex following lesions of the magnocellular forebrain nuclei. *Brain Res* 233:359–367

Pintor A, Fortuna S, Volpe MT, Michalek H (1988): Muscarinic plasticity in the brain of senescent rats: down-regulation after repeated administration of diisopropyl fluorophosphate. *Life Sci* 42:2113–2121

Robbins TW, Everitt BJ, Marston HM, Wilkinson J, Jones GH, Page KJ (1989): Comparative effects of ibotenic acid- and quisqualic acid-induced lesions of the substantia innominata on attentional function in the rat: further implications for the role of the cholinergic neurons of the nucleus basalis in cognitive processes. *Behav Brain Res* 35:221–240

Rye DB, Wainer BH, Mesulam MM, Mufson EJ, Saper CB (1984): Cortical projections arising from the basal forebrain: a study of cholinergic components employing combined retrograde tracing and immunohistochemical localizations of choline acetyltransferase. *Neuroscience* 13:627–643

Santos-Benito FF, Gonzalez JL, de la Torre F (1988): Choline acetyltransferase activity in the rat brain cortex, homogenate, synaptosomes, and capillaries after lesioning the nucleus basalis magnocellularis. *J Neurochem* 50:395–399

Schwaber JS, Rogers WT, Satoh K, Fibiger HC (1987): Distribution and organization of cholinergic neurons of the rat forebrain demonstrated by computer-aided data acquisition and three dimensional reconstruction. *J Comp Neurol* 263:309–325

Senut MC, Menetrey D, Lamour Y (1989): Cholinergic and peptidergic projections from the medial septum and the nucleus of the diagonal band of Broca to dorsal hippocampus, angulate cortex and olfactory bulb: a combined wheat-germ agglutinin-apohorseradish peroxidase-gold immunohistochemical study. *Neuroscience* 30:385–403

Shute CD, Lewis PR (1963): Cholinergic neurons pathways in the brain. *Nature* 199:1160–1164

Shute CD, Lewis PR (1967): The ascending cholinergic reticular system: neocortical, olfactory and subcortical projections. *Brain* 90:497–520

Smith G (1988): Animal models of Alzheimer's disease: experimental cholinergic denervation. *Brain Res Rev* 13:103–118

Snider WD, Johnson EM (1989): Neurotrophic molecules. *Ann Neurol* 26:489–506

Sofroniew MV, Pearson RC A, Powell TP S (1987): The cholinergic nuclei of the basal forebrain of the rat: normal structure, development and experimentally induced degeneration. *Brain Res* 411:310–331

Springer JE, Tayrien MW, Loy R (1987): Regional analysis of age-related changes in the cholinergic system of the hippocampal formation and basal forebrain of the rat. *Brain Res* 407:180–184

Stichel CC, Dolabela de Lima A, Singer W (1987): A search for choline acetyltransferase-like immunoreactivity in neurons of cat striate cortex. *J Comp Neurol* 258:99–111

Toledano A (1988): Hypotheses concerning the aetiology of Alzheimer's disease. *Pharmacopsychiat* 21:17–25

Toledano A (1989): Alzheimer experimental. Modelos biologicos y su utilidad en la caracterizacion de medicamentos encefalotropicos. In: *Trastornos cerebrales organicos.* Toledano A, Maurer K and Wurtman RJ, eds. Braunschweig: Vieweg, pp. 15–43

Toledano A, Barca MA, Moradillo I (1988): Mitochondrial dehydrogenases during sensescence of the cerebellar cortex. A comparative electron microscope study. *Cell Molec Biol* 34:175–189

Vige X, Briley M (1989): Muscarinic receptor plasticity in rats lesioned in the nucleus basalis of Meynert *Neuropharmacol* 28:727–732

Walker LC, Koliatsos VE, Kitt CA, Richardson RT, Rokaeus APrice D L (1989): Peptidergic neurons in the basal forebrain magnocellular complex of the rhesus monkey. *J Comp Neurol* 280:272–282

Watson M, Vickroy TW, Fibiger HC, Roeske WR, Yamamura HI (1985): Effects of bilateral iobotenate induced lesions of the nucleus basalis magnocellularis upon selective cholinergic biochemical markers in the rat anterior cerebral cortex. *Brain Res* 346:387–391

Wenk GL, Markowska AL, Olton DS (1989): Basal forebrain lesions and memory: alterations in neurotensin not acetylcholine, may cause amnesia. *Behav Neurosci* 103:765–769

Wenk H, Bigl V, Meyer U (1980): Cholinergic projections from magnolcellular nuclei of the basal forebrain to cortical areas in rats. *Brain Res Rev* 2:295–361

Wenk GL, Olton DS (1984): Recovery of neocortical choline acetyltransferase activity following iobotenic acid injection into the nucleus basalis of Meynert in rats. *Brain Res* 293:184–186

Wisniewsky HM, Sinatra RS, Iqbal K, Grundke-Iqbal I (1981): Neurofibrillary and synaptic pathology in the aged brain. In: *Aging and Cell Structures* Vol. 1, Johnson JE, ed. New York: Plenum Press, pp. 105–142

Zaczek R, Coyle JT (1982): Excitatory amino acids analogues: neurotoxicity and seizures. *Neuropharmacol* 21:15–26

Neurodegenerative Changes in Aging and Dementia: A Comparison of Alzheimer and Lewy Body Type Pathology

Elaine K. Perry, Elizabeth Marshall, Anthony Cheng, James A. Edwardson, and Robert H. Perry

Whereas cortical senile plaques are evident in the vast majority of elderly people and as such can be considered part of the "normal" aging process, Lewy bodies are found in the cortex of a very small proportion (<5%) of normal aged (over 70 yr) individuals and may be more specifically disease-associated. Lewy bodies invariably appear in the cortex of elderly patients suffering from a form of dementia that is clinically distinct from Alzheimer's disease (AD)—Senile Dementia of Lewy Body Type (SDLT). SDLT is part of the spectrum of Lewy body disease (which also includes diffuse Lewy body disease) and is distinguished by the relatively high incidence of mainly visual hallucinations. Although memory impairment (the cardinal feature of AD, also evident in SDLT) may to some degree be an inevitable concomitant of the aging process, hallucinations are not generally associated with normal aging and presumably relate in SDLT to specific disease-related alterations in brain function. At the histological level there is, however, no obvious correlation between the distribution or density of cortical Lewy bodies and the incidence of hallucinations, although at the biochemical level an interesting relationship is apparent. Distinct profiles of cholinergic and monoaminergic transmitter activities were evident in AD and SDLT, particularly with respect to the occurrence of hallucinations in the latter. Thus the presynaptic cholinergic enzyme choline acetyltransferase was significantly lower, whereas pre- and post-synaptic serotonergic activities (5HIAA and $5HT_2$ receptor binding) were relatively spared in the hallucinating SDLT group compared to AD and the non-hallucinating SDLT group. The concept of an imbalance between serotonin and acetylcholine with relatively hyperactive serotonergic and hypoactive cholinergic activities being associated with hallucinating experiences is consistent with the psychopharmacology of drugs affecting the respective systems. Thus, cholinergic antagonists and selective serotonergic agonists tend to be hallucinogenic, although whether cholinergic agonist or serotonergic (S_2) antagonist drug treatments would alleviate hallucinations in SDLT remains to be determined.

Neurodegenerative Changes in Aging

Multiple changes occur in the human brain during development and aging, some of these being common to both processes. Thus, cell death appears to be an essential part of development and an inevitable (and perhaps also essential) concomitant of old age (Perry and Perry, 1988). Certain morphological features are, however, confined to aging, being absent from the young brain and only apparent in some normal individuals around the fifth or sixth decade of life. Such features include senile (neuritic or A4) plaques, neurofibrillary tangles, granulovacuolar degeneration, Hirano bodies, and vascular amyloid. The extent to which one of the hallmarks of Parkinson's disease—intraneuronal Lewy bodies—develop as a concomitant of normal aging is controversial. Some authors have suggested that a substantial minority of the elderly develop Lewy bodies, although others (Perry et al., 1990b) argue that they are rare and occur only in preclinical cases of Lewy body disease (Parkinson's disease or Lewy body dementia, for example).

In a recent survey of 131 neurologically and psychiatrically normal individuals aged between 50 and 100 years, the prevalence of plaques, tangles, and Lewy bodies in vulnerable brain areas was examined. Figure 1 illustrates that whereas plaques and to a lesser extent tangles are present in the majority of elderly individuals, Lewy bodies are extremely rare, and even in the tenth decade of life are apparent in less than 5% of the population. It is notable that if the occurrence of plaques or tangles, or both, is considered together in both neo- and archicortical structures (such as hippocampus and entorhinal cortex), all individuals over the age of 90 years are found to be affected to some degree.

It seems reasonable to conclude from this and other studies (reviewed in Perry et al., 1990b) that so-called Alzheimer-type pathology (plaques and tangles) is to a certain extent an inevitable consequence of normal aging, whereas Lewy bodies may be more specifically related to disease. Whether all individuals are ultimately destined—if they live long enough—to develop Alzheimer's disease is debatable (for detailed arguments on this issue, see Hauw et al., 1988; Perry and Perry, 1988; Storandt, 1988). The arguments here are not just academic; if a pathological process can be shown to be specifically disease-related, then there is

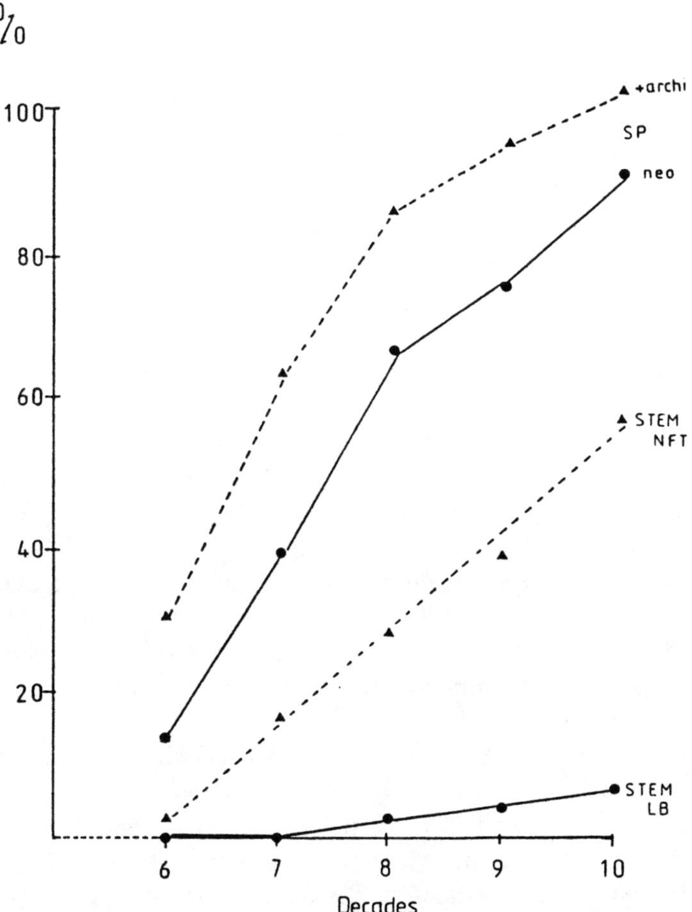

Figure 1. Age-related prevalence of senile plaques (SP) in the neocortex (neo) and combined neo- and archicortex (+ archi) and prevalence of neurofibrillary tangles (NFT) and Lewy bodies (LB) in the brain stem (substantia nigra and locus coeruleus, for example). The number of individuals in the 6th to 10th decades were, respectively: 22, 31, 36, 28, and 14. Note that the vertical axis denotes the percentage of individuals affected in each decade and bears no relation to the absolute numbers of the histological features identified.

hope for a treatment or cure, but if it is invariably associated with increasing age, then, like aging itself, the therapeutic outlook is less promising. With this in mind the nature of three major disorders—Alzheimer's and Parkinson's disease and senile dementia of Lewy body type (SDLT), in which the histological features demonstrated in Fig. 1 occur—is compared from the clinical, pathological, and, most extensively, neurochemical viewpoints.

Contrast of Alzheimer and Lewy Body Type Dementia

Clinical and neuropathological aspects. The combined features of numerous neocortical plaques and tangles are diagnostic of Alzheimer's disease and invariably evident in middle-aged cases of Down's syndrome. Plaques and tangles may occur individually or in combination, to a variable extent, in other diseases in which the presence of these histopathological features is not diagnostic of the disease. A4 plaques, for example, are seen in some but not all cases of Parkinson's and diffuse Lewy body disease, in SDLT (see below), progressive supranuclear palsy, dementia pugilistica, motoneuron disease, renal dialysis, and Guam Parkinsonism dementia. Diseases in which Lewy bodies invariably occur include Parkinson's disease, diffuse Lewy body disease, and senile dementia of Lewy body type. These three diseases can be considered part of the spectrum of Lewy body disease in the elderly, depending upon the relative extent of Lewy body formation and neuronal loss in different brain regions. In other diseases such as Hallervorden Spatz, subacute sclerosing panencephalitis, motoneuron disease, and corticobasal degeneration, Lewy bodies are evident in a minority of cases; in the geographically distinct Parkinsonian dementia complex of Guam, the majority are affected (see also Gibb and Lees, 1988).

There is currently some controversy concerning the classification of elderly demented individuals in which moderate numbers of cortical Lewy bodies are evident at autopsy. Some argue that the presence of plaques places them in the Alzheimer's disease category; however not all SDLT cases have significant numbers of neocortical plaques (and indeed many other diseases are variably associated with the appearance of plaques; see above). Others (Hansen et al., 1989) maintain that the atypical dementia associated with Lewy bodies places such cases in an Alzheimer's disease subgroup, although in our series both the clinical and pathological features were generally so distinct as to exclude a diagnosis of Alzheimer's disease. Yet others (Byrne et al., 1989) maintain these cases should be categorized as diffuse Lewy body disease (DLBD); although in the SDLT series the density of neocortical Lewy bodies was an order of magnitude below that seen in DLBD, and indeed such a broad interpretation of the diagnostic criteria for DLBD would include many cases of Parkinson's disease, since cortical Lewy bodies occur in virtually all cases of Parkinson's over 75 yr (Perry et al., unpublished observation). Our classification of SDLT, including comparisons with Alzheimer's and Parkinson's disease, is summarized in Table 1 and the principal clinical and neuropathological features of SDLT outlined in Table 2.

Table 1

Dementia Classification

	Clinical presenting features	Pathological Cortex	Substantia Nigra
Alzheimer's	Progressive memory loss	Plaques and tangles	Normal
Parkinson's	Extrapyramidal movement disorder	Variable (some with sparse Lewy bodies and plaques)	Extensive neuron loss, Lewy bodies
Senile dementia of Lewy body type	Confusional state (often fluctuating with visual)	Lewy bodies (moderate) and plaques (variable)	Moderate neuron loss, Lewy bodies

Senile dementia of Lewy body type, Parkinson's disease, and diffuse Lewy body disease are part of the spectrum of Lewy body disease in the elderly—defined by the presence of Lewy bodies as a diagnostic feature (Perry et al., 1989,1990).

That SDLT and Alzheimer's disease are clinically distinct is emphasized in Table 3, which demonstrates the reduced severity of dementia, increased incidence of extrapyramidal features and hallucinations and, most importantly, higher Hachinski scores, which led to many SDLT cases being initially classified by the clinician as multi-infarct dementia.

The principal neuropathological features of SDLT, distinguishing it from Alzheimer's disease, are the presence in the cerebral cortex of Lewy or Lewy-like bodies, the prevalence of Lewy bodies in several subcortical nuclei, and neuronal loss in the substantia nigra. Cortical Lewy bodies are generally present at a density far below that seen in diffuse Lewy body disease and are concentrated in areas known to receive a dense dopaminergic innervation, such as the cingulate or entorhinal cortex, being absent or rare in the occipital cortex. Since Lewy bodies in the midbrain or basal forebrain are invariably associated with neuronal loss—for example nigral degeneration in both SDLT and Parkinson's disease—it is reasonable to suppose that neuronal loss also occurs in the cortex in SDLT. The nature of the cortical neurons affected by Lewy bodies is as yet unknown, although it is interesting to note that the Lewy bodies are predominantly localized in lower cortical layers. In this respect they parallel the terminations of the dopaminergic innervation of the cortex from the ventral tegmental area in the midbrain and also the distribution of in-

Table 2

Senile Dementia of Lewy Body Type

<u>Clinical Features</u>

Relatively acute presentation

Fluctuating course with congnitive and behavioral impairment (may be misdiagnosed initially as multi-infarct dementia)

Visual hallucinations/aggression/psychotic episodes/depression as presenting features

Progressive underlying dementia

Visuospatial dysfunction

Mild extrapyramidal features such as flat facies (absence of Parkinsonian triad of tremor, rigidity, and akinesia)

<u>Neuropathological Features</u>

Lewy body formation and selective neuronal loss in brainstem and other subcortical nuclei (eg., locus coeruleus, nucleus of Meynert)

Lewy body formation in archi- and limbic cortex, with neocortical densities either below that in diffuse Lewy body disease or not detectable

Numerous neocortical senile plaques in most, but not all cases

Few or no neocortical tangles

No significant ischaemic lesions

trinsic tyrosine hydroxylase immunoreactive neurons recently described in the lowest layers of the human cortex (Gaspar et al., 1987). These neurons have been noted in diffuse Lewy body disease to be affected by Lewy body formation (Kuljis et al., 1989). Whether these intrinsic cortical neurons are in fact dopaminergic is unclear, since they do not apparently contain DOPA-decarboxylase (Gaspar et al., 1987), although they are restricted to the adult primate brain (disappearing in rodents in the course of development). The dopaminergic system clearly merits investigation in SDLT in view of these observations and the neuronal loss in the substantia nigra (Perry et al., 1990a). However, since Lewy bodies are not confined to dopaminergic neurons in the diseased human brain (being apparent in cholinergic, noradrenergic, and serotonergic subcortical nuclei in, for example, Parkinson's disease), cortical neurons affected may not be homogenous in terms of cell type or transmitter systems affected.

Neurochemical aspects. Frozen brain tissue from Alzheimer, Parkinson, and SDLT cases (classified according to Table 2) were subjected to analy-

Table 3

Comparison between Senile Dementias of Lewy Body Type and Alzheimer Type

	SDLT	SDAT
Case numbers	14	14
Age at onset	81±6	81±5
Duration (referral to death)	3±2	6±3
Mental Test Scores		
— presentation	25±7	14±6
— terminal	14±9	5±8
Hachinski scores	6±2	2±2
Hallucinations		
— presentation	8	0
— terminal	8	3
Extrapyramidal features	2.2±1.3	0.9±0

Differences between the two groups reached statistical significance in all respects except age (see Perry et al., 1990).

sis of cholinergic and monoaminergic parameters. These included choline acetyltransferase (ChAT—Fonnum, 1975), muscarinic receptor binding (in particular, the low affinity agonist subtype L, Smith et al., 1988), the nicotinic receptor (as defined by high affinity (3H) nicotine binding; Perry et al., 1987), dopaminergic and serotonergic amines and metabolites (dopamine, homovanillic acid [HVA], 3,4-dihydroxyphenylacetic acid [DOPAC], serotonin and 5-hydroxyindolacetic acid [5HIAA], separated by HPLC according to the method of Wagner et al., 1982) and the serotonergic receptor subtype S_2 (demonstrated autoradiographically in cryostat sections of neocortex, Yates et al., 1990)

The presynaptic cholinergic marker, ChAT, was most extensively reduced in the cortex in SDLT compared with Alzheimer's and Parkinson's disease, although there was a tendency towards increased basal ganglia activity in the caudate nucleus (Fig. 2).

The more extensive ChAT loss in SDLT compared with Alzheimer's disease is consistent with the report of Hansen et al. (1989) that the "Lewy Body Variant" of Alzheimer's disease has lower enzyme activity compared with classical Alzheimer's disease. There were striking differences in the present series between the muscarinic and nicotinic receptors (Fig. 3).

The muscarinic subtype, L, was elevated in Parkinson's disease and

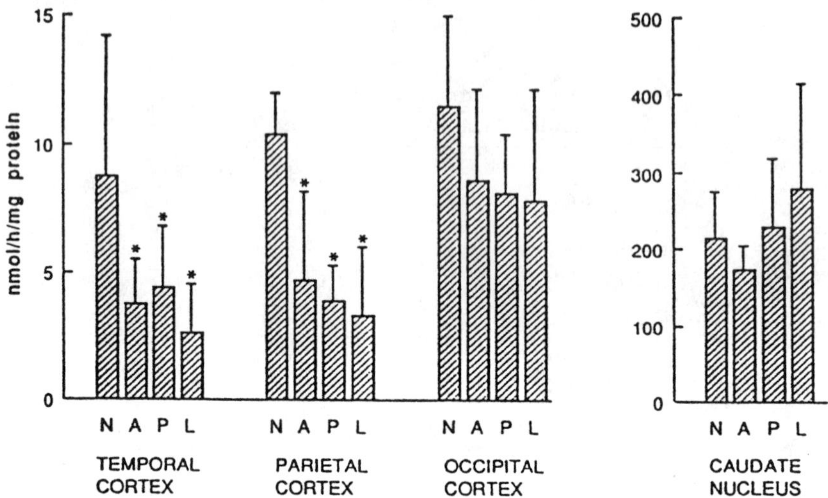

Figure 2. Comparison of choline acetyltransferase in normal (N), Alzheimer's disease (A), Parkinson's disease (P) and senile dementia of Lewy body type (L). Reductions, compared with the normal, marked with an asterisk were significant (Student's t-test, $p < 0.05$ or less).

SDLT compared with Alzheimer's disease, a difference that probably reflects the relative preservation of neocortical neurons with sparse or absent neocortical tangles in the former two disorders. A degree of receptor "supersensitivity" may reflect the functional integrity of cholinoceptive neurons and perhaps augers well for potential cholinotherapy in relation to dementia in Parkinson's disease and SDLT. Fluctuations in cognitive function in SDLT (but not generally in Alzheimer's disease) may incidentally relate to decremental losses of cholinergic input followed by compensatory muscarinic supersensitivity in intact cholinoceptive neurons. In contrast to the muscarinic receptor, nicotinic receptor binding was reduced in all of the diseases investigated (Fig. 3).

In the caudate, reductions in dopamine (Fig. 4) almost exactly paralleled the loss of neurons in the substantia nigra which was around 40% in SDLT, 80% in Parkinson's disease, and insignificant in Alzheimer's disease (Perry et al., 1990a).

Changes in the dopaminergic metabolites, HVA and DOPAC, were dissimilar, indicative of differences in dopamine metabolism in surviving neurons. Caudate serotonin and its metabolite, 5HIAA, were most extensively decreased in Parkinson's disease, although serotonin itself was also affected in the other disorders (Fig. 4). In the neocortex, 5HIAA was significantly reduced compared with the normal in all clinical categories; a trend towards decreased HVA did not reach significance (Fig. 5). It

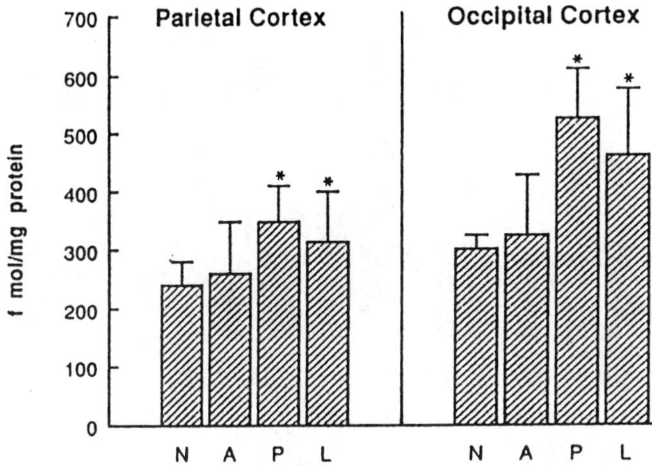

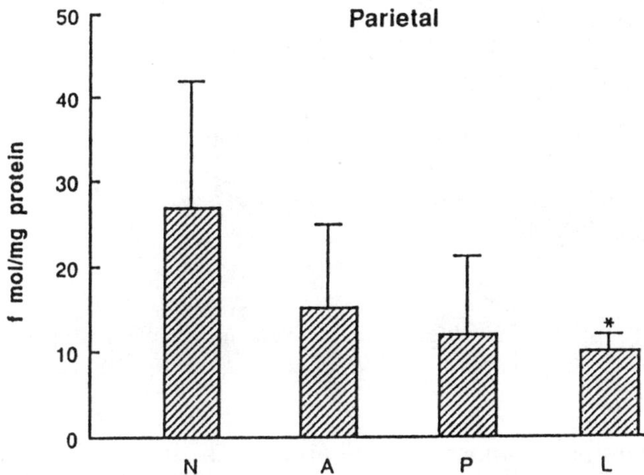

Figure 3. Muscarinic and nicotinic receptor binding in the different clinical groups (as in Fig. 2); significant changes compared with the normal marked by an asterisk ($p<0.05$ or less).

should be noted that measurements of dopamine or its metabolite levels in Parkinson's disease are complicated by the effects of drug therapy.

To summarize the neurochemical data so far obtained, SDLT could be distinguished from Alzheimer's disease by a more extensive presynaptic cholinergic deficit, an increase in muscarinic receptor binding, and a loss

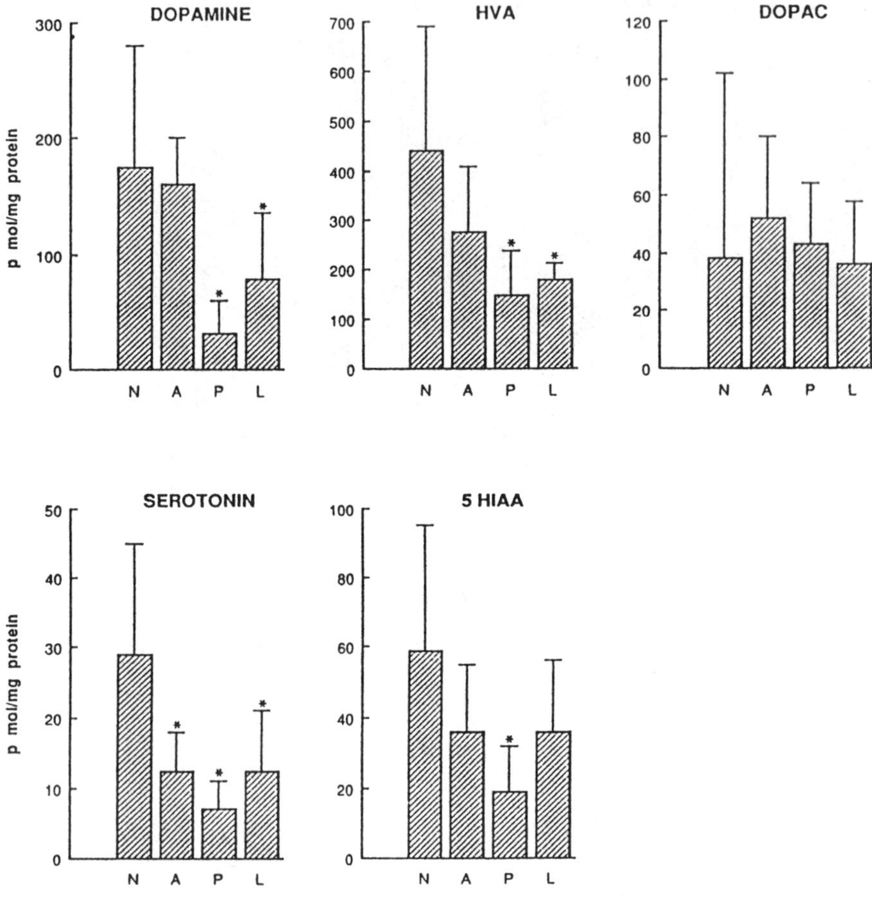

Figure 4. Levels of dopamine, serotonin, and metabolites in the caudate in the different clinical groups (as in Fig. 2). Significant differences compared with the normal ($p < 0.05$ or less) marked by an asterisk.

of striatal dopamine. Compared with Parkinson's disease, SDLT was also more affected by a cholinergic deficit but was less affected by the loss of striatal dopamine. In addition, the postsynaptic serotonergic receptor subtype S_2 and levels of both somatostatin and corticotrophin-releasing factor were decreased in SDLT and Alzheimer's disease but not in Parkinson's disease (Perry et al., in preparation). Thus, there are both qualitative and quantitative neurochemical differences between SDLT and the other two diseases, although apart from the nigral levels of dopamine in Parkinson's disease, these do not obviously account for the distinct clinical features in the different groups. As suggested, extensive presynaptic cholinergic deficits in conjunction with postsynaptic muscarinic supersensitivity

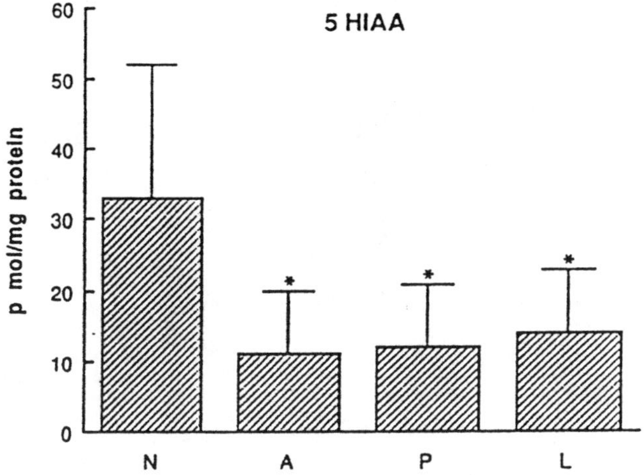

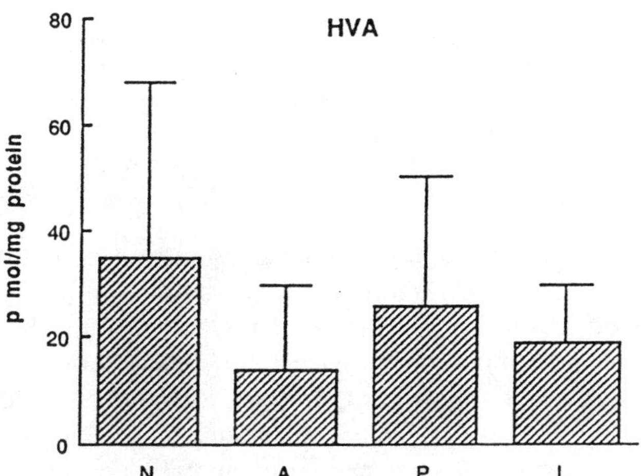

Figure 5. Levels of serotonergic and dopaminergic metabolites, 5HIAA, and HVA in temporal cortex in the different clinical groups (as in Fig. 2). Reductions in 5HIAA were significant ($p < 0.05$ or less) in all pathological groups.

may account for fluctuations in cognitive function in SDLT. A more impressive neurochemical-clinical correlate emerges on closer inspection of the SDLT group in relation to the incidence of one particular clinical symptom—hallucinations.

The majority of SDLT cases reported experiencing hallucinations, mainly of a visual nature (see Table 4), in the course of their disease. The complex scenic images of people or animals indicate that the region

Table 4

Examples of Visual Hallucinations in Senile Dementia of Lewy Body Type

. . . "disturbed by vivid images of animals"

. . . "picking at non-existent objects in her lap"

. . . "gypsies trying to climb in through the windows"

. . . "talks to (non-existent) people and reaches to touch them"

. . . "takes off her slipper and tries to kill the beetles on the floor"

. . . "saw railway trains and racehorses in the house"

. . . "people disguised as bushes, or with petals as clothes, in the garden"

. . . "saw (and touched) strings connecting furniture with walls, ceilings, etc."

. . . "saw men rising from trap door"

. . . "people forcing him to sit down which he resisted"

Examples are derived verbatim from the case notes of different individuals.

of the brain involved is in temporal cortex association areas. In Penfield's classical studies of electrical stimulation of the cortex in conscious patients (Penfield and Jasper, 1954) similar pictures were elicited during stimulation of the temporal lobe whereas stimulation of occipital areas gave rise to relatively simple images (stars, triangles, and moving lines, for example). The SDLT group was divided into subgroups with and without reported hallucinations, and the temporal lobe was examined for cholinergic and monoaminergic activities. Neurochemical comparisons between these subgroups and the normal are provided in Table 5. Striking differences were apparent between the groups in relation to presynaptic cholinergic and serotonergic activities and in binding to the S_2 receptor. Thus, ChAT was most extensively decreased (to under 20% of the normal) in the hallucinating subgroup whereas reductions in 5HIAA and S_2 were only apparent in the nonhallucinating subgroup. This finding is compatible with the concept of a serotonergic–monoaminergic imbalance in the cortex being associated with the occurrence of hallucinations. A relative hyperactivity of serotonin and hypoactivity of acetylcholine in hallucinating individuals is also consistent with the effects of specific drugs. S_2 agonists (e.g., LSD) and anticholinergics (e.g., atropine) can induce hallucinations in normal individuals (Glennon et al., 1984; Goodman and Gilman, 1975). Obviously, other transmitters such as dopamine and glutamate, known to be associated with hallucinations, may be involved in the manifestation of this symptom in SDLT, although present data in the cortex suggest no significant changes in either the dopaminergic metabolite (see Fig. 5) or in NMDA glutamate receptor binding including its modulation by glycine (Bowen and Perry, unpublished observations) in the disease. It is notable that the prevalence of hallucinations in Alzheimer's and Parkinson's diseases is rarer than in SDLT and, interestingly, reductions in cholinergic activities are not as extensive, whereas reductions in serotonergic activities are more extensive, compared with the hallucinating SDLT subgroup. One important conclusion from this observation is that patients with dementia who experience hallucinations may have a relatively selective cholinergic deficit (other transmitters being less affected), and so this symptom might be a clinical indicator of patients who respond more positively to cholinotherapy—a conclusion also supported by the relative lack of tangles in SDLT (Perry et al., 1990a).

A point worth noting regarding hallucinogenesis in neurodegenerative disease is that, unlike memory impairment, which appears to be a common complaint in normal old people, hallucinations are not normally experienced as a function of increasing age. Instead they reflect either a pathological process in the brain or the effects of exogenous drugs or other naturally occurring agents. The neuropathological counterpart of hallucinations in SDLT is not yet clear and does not appear to be the density of Lewy bodies in the temporal cortex (Perry et al., 1990a). It may instead be due (as suggested above) to the relatively rapid and selective degeneration of the cholinergic system—perhaps as a result of Lewy body prevalence in the Meynert. This may also contribute to the acute confusion that marks the initial and many of the ensuing stages of the disease, and fluctuations in the level of confusion may (as suggested earlier) reflect the variable ability of the postsynaptic cholinergic receptor to compensate. In normal aging, by contrast, degeneration of many different neuronal pathways may occur conjointly. Reports on age-associated alterations in transmitter activities are notoriously inconsistent, although there does appear to be agreement concerning an age-related S_2 loss in the human cortex (Marcusson et al., 1984; Cheetham et al., 1988) which may render hyperserotonergic activity an unlikely consequence of the normal aging process. Clearly, further studies of normal aging itself are now warranted to unravel the significance of neurodegenerative changes

Table 5

Neurotransmitter Activities in Temporal Cortex

	Normal	Senile dementia of Lewy body type without hallucinations.	Senile dementia of Lewy body type with hallucinations.
Cholinergic			
Choline acetyltransferase	8.7 ±5.6	* 4.0 ±1.7	** 1.5 ±1.6
Muscarinic receptor	158 ±29	* 192 ±29	* 201 ±37
Nicotinic receptor	12.6 ±7.0	* 5.2 ±1.5	** 3.7 ±2.2
Serotonergic			
5HIAA	32.6 ±18.5	** 9.7 ±6.0	17.3 ±11.2
S_2 receptor	42.0 ±16.5	** 16.4 ±2.4	33.1 ±20.1
Dopaminergic			
HVA	36.5 ±32.8	* 13.2 ±9.1	24.6 ±14.5

For units, see Figs. 2–5; S_2 receptor binding, measured by Magiscan from autoradiographic images, as fmol/mg tissue in the third layer of the cortex where a distinct band of intense (^{3}H)-Ketanserin binding was apparent in the normal.

*, **, significantly different from the normal, $p<0.05$ and <0.01 (Student's t-test). Numbers in the normal, nonhallucinating and hallucinating groups were 6–12, 4–5, and 4–6, respectively.

in age-associated diseases such as SDLT and Alzheimer's disease. Many of the clinical symptoms in such diseases may depend critically on preceding or accompanying age-associated deficits.

Conclusion

Returning, finally, to the question raised at the beginning of this chapter—whether pathological features such as plaques and Lewy bodies are specifically disease associated—the importance of studying the normal aging process is further highlighted. Thus if all the features of Alzheimer's disease, including for example plaques and tangles, occur to a lesser extent with increasing age, then examination of normal (postmaturity) aging is likely to reveal the earliest and most important events on the pathological cascade that lead to the disease. Understanding those that are primary should lead to the most successful therapeutic strategies, albeit not necessarily preventative. In contrast, investigations into the nature of diseases such as Parkinson's and SDLT, in which the key pathological feature—the Lewy body—is not evident in normal aging, should be based on a different premise. Here there is every reason to expect that a specific factor or factors (genetic and/or environmental) is responsible for the disease and that the ultimate goal of prevention is much more realistic.

Acknowledgment. Many thanks to Dawn Hinds for manuscript preparation.

References

Byrne EJ, Lennox G, Lowe J, Godwin-Austen R (1989): Diffuse Lewy body disease: clinical features in 15 cases. *J Neurol Neurosurg Psychiat* 52:709–717

Cheetham SC, Crompton MR, Katona CLE, Horton RW (1988): Brain 5HT2 receptor binding in depressed suicide victims. *Brain Res* 443:272–280

Fonnum F (1975): A rapid radiochemical method for the determination of choline acetyltransferase. *J Neurochem* 24:407–409

Gaspar P, Berger B, Febvret A, Vigny A, Krieger-Poulet M, Borri-Vottattorni C (1987): Tyrosine hydroxylase-immunoreactive neurons in the human cerebral cortex: a novel catecholaminergic group? *Neurosci Lett* 80:257–262

Gibb WRG, Lees AJ (1988): The relevance of the Lewy body to the pathogenesis of idiopathic Parkinson's disease. *J Neurol Neurosurg Psychiat* 51:745–752

Glennon RA, Titeler M, McKenney JD (1984): Evidence for 5HT2 involvement in the mechanism of action of hallucinogenic agents. *Life Sci* 35:2505–2511

Goodman LS, Gilman A (1975): *The Pharmacological Basis of Therapeutics.* New York: MacMillan Publishing Company

Hansen L, Salmon D, Galaska D, Masliah E, Katzman R, De Teresa R, Thal L, Bay MM, Hofstelter R, Klanber M, Rice V, Butters N, Alford M (1989): The Lewy body variant of Alzheimer's disease. *Neurol* 40:1–8

Hauw JJ, Duyckaerts C, Delaere P (1988): Neuropathology of aging and DAT: how can age-related changes be distinguished from those due to disease processes. In: *Etiology of Dementia of Alzheimer's Type*, Henderson AS, Henderson JH, eds. Chichester: John Wiley and Sons, pp. 195–212

Kuljis R, Martin-Vasallo P, Peress NS (1989): Lewy bodies in tyrosine hydroxylase synthesizing neurons of the human cerebral cortex. *Neurosci Lett* 106:49–54

Marcusson JO, Norgan DG, Winblad B, Finch CG (1984): Serotonin-2 binding sites in human frontal cortex and hippocampus. Selective loss of 2-A sites with age. *Brain Res* 311:51–56

Penfield W, Jasper H (1954): *Epilepsy and the Functional Anatomy of the Human Brain.* London: J and A Churchill

Perry EK, Perry RH (1988): Aging and dementia: neurochemical and neuropathological comparisons. In: *Etiology of Dementia of Alzheimer's Type.* Henderson AS, Henderson JH, eds. Chichester: John Wiley and Sons, pp. 214–228

Perry EK, Perry RH, Smith CJ, Dick DJ, Fairbairn A, Blessed G, Candy JM, Johnson M, Edwardson JA (1987): Nicotinic receptor abnormalities in Alzheimer's and Parkinson's diseases. *J Neurol Neurosurg Psychiat* 50: pp. 805–809

Perry RH, Irving D, Blessed G, Fairbairn AF, Perry EK (1990a): Senile dementia of Lewy body type: a clinically and histopathologically distinct form of Lewy body dementia in the elderly. *J Neurol Sci* 95:119–139

Perry RH, Irving D, Tomlinson BE (1990b): Lewy body prevalence in the elderly: relationship to psychiatric status, Alzheimer pathology and catecholamine neuronal density. *J Neurol Sci* 100:223–233

Smith CJ, Perry EK, Perry RH, Candy JM, Johnson M, Bonham JR, Dick DJ, Fairbairn A, Blessed G, Birdsal NJM (1988): Muscarinic cholinergic receptor subtypes in hippocampus in human cognitive disorders. *J Neurochem* 50:847–856

Storandt M (1988): Relationship of normal aging and dementing diseases in later life. In: *Etiology of Dementia of Alzheimer's Type.* Henderson AS, Henderson JH, eds. Chichester: John Wiley and Sons, pp. 231–239

Wagner J, Vitali P, Palfreyman MG, Zraika M, Huot S (1982): Simultaneous determination of 3,4-dihydroxyphenylalanine, 5-hydroxytryptophan, dopamine, 4-hydroxy-3-methoxyphenylalanine, norepinephrine, 3,4-dehydroxyphenylacetic acid, homovanillic acid, serotonin and 3,4-dehydroxyindoacetic acid in rat cerebrospinal fluid and brain by high performance liquid chromatography with electrochemical detection. *J Neurochem* 38:1241–1254

Yates M, Leake A, Candy JM, Fairbairn AF, McKeith, IG, and Ferrier IN (1990): 5HT2 receptor changes in major depression. *Biol Psychiat* 27:489–496

Cortical Morphological Changes in Dementia

Heiko Braak and Eva Braak

Many diseases of the brain leading to impairment of intellectual capacities are associated with morphological changes in the antero-medial portions of the temporal lobe. Among these are Alzheimer's disease, Parkinson's disease, and the syndrome of dementia with argyrophilic grains.

Alzheimer's disease is the most important of these disorders and accounts for approximately 60% of all of the cases of presenile and senile dementia (Mann, 1985; Price, 1986; Von Braunmühl, 1957). Parkinson's disease, on the other hand, should not be considered to be purely a disorder of the extrapyramidal system, since it is frequently associated with a more or less obvious intellectual decline (Cummings, 1988). Finally, it is the experience of many neuropathologists that a surprisingly large number of cases of clinically proven dementia fail to reveal the features of known diseases associated with progressive dementia (Ulrich et al., 1986). Application of newly developed silver techniques revealed in quite a number of these cases unusual pathological changes in the form of small argyrophilic grains in the antero-medial portions of the temporal lobe (Braak and Braak, 1987, 1989; Itagaki et al., (1989).)

Before going into details it appears appropriate to recapitulate that the cerebral cortex in the mammalian brain is formed of two fundamentally different types of gray matter: the isocortex and the allocortex. In the human brain the extended and more or less uniformly built isocortex represents the dominating structure. The heterogeneous allocortex is small in comparison and comprises, among other areas, the hippocampal formation and the entorhinal region. The entorhinal cortex spreads over the anterior portions of the parahippocampal gyrus and gyrus ambiens and exhibits a particularly complex lamination pattern with up to 12 layers. None of these layers corresponds to laminae of the isocortex.

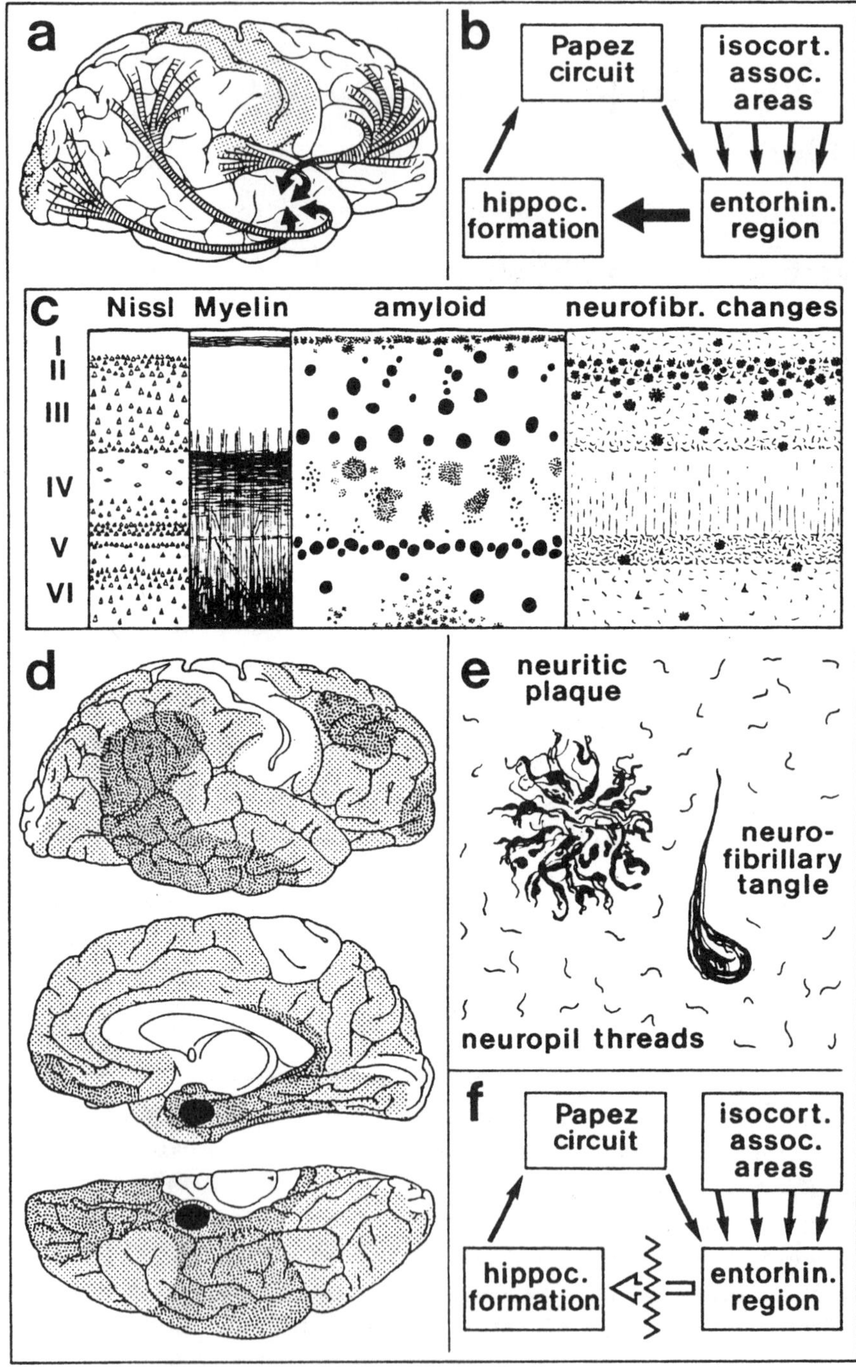
a
b
Papez circuit
isocort. assoc. areas
hippoc. formation
entorhin. region
c
Nissl
Myelin
amyloid
neurofibr. changes
I
II
III
IV
V
VI
d
e
neuritic plaque
neuro-fibrillary tangle
neuropil threads
f
Papez circuit
isocort. assoc. areas
hippoc. formation
entorhin. region

Figure 1. (a) and (b): Fiber bundles from isocortical association areas converge upon the entorhinal cortex which in turn projects via the perforant path (thick arrow) to the hippocampal formation. (c): Schematic lamination pattern of an isocortical area (striate area, Brodmann field 17) and its Alzheimer-related pathology as seen in preparations stained for amyloid and neurofibrillary changes. Note the difference in the distribution pattern of amyloid plaques and neuritic plaques. (d): Distribution of Alzheimer-related cortical changes. The most severe changes are seen in the entorhinal region (black spot). (e): Neurofibrillary changes consist of three kinds of lesions, namely the neuritic plaques, neurofibrillary tangles, and neuropil threads. (f): The transport of information from isocortical association areas to the hippocampal formation is bilaterally interrupted by destruction of the cells of origin of the perforant tract. This also hampers the flow of information within the Papez circuit (with kind permission from Braak and Braak, 1990b).

The "transentorhinal region" is squeezed in between the proper entorhinal territory and the adjoining temporal cortex and is easily recognized on account of the conspicuous descent of the superficial entorhinal cellular layer (Braak, 1980; Braak and Braak, 1985).

Recent investigations in the primate brain have shown that the entorhinal region receives a particularly dense input from all association areas of the isocortex providing it with abundant somato-motor, somato-sensory, acoustic, and visual information (Fig. 1a; Van Hoesen, 1982; Van Hoesen et al., 1972). The outer entorhinal layers generate the perforant path and via this fiber tract the information is transported to the hippocampal formation (Fig. 1b, thick arrow). This information transfer from isocortical association areas via the entorhinal region to the hippocampal formation is of significance for mnestic functions.

The parvocellular layer of the presubiculum—an important relay station of the Papez circuit—is another major source of input to the entorhinal region. The Papez circuit carries information from the subiculum directly or via the mamillary body to the anterior nuclear complex of the thalamus and from here via the retrosplenial region and presubiculum to the entorhinal cortex. Again, the perforant path ultimately transmits the information to the hippocampus (Fig. 1b).

All of the territories under consideration—the isocortical association areas, the entorhinal region, the hippocampal formation, and the relay stations of the Papez circuit—exhibit characteristic and thoroughgoing structural changes during the course of Alzheimer's disease. In Parkinson's disease, the cortical change is far more restricted and virtually confined to the entorhinal region. This territory is also found to be severely affected by dementia with argyrophilic grains.

Alzheimer's Disease

The hallmark of Alzheimer's disease is a progressive accumulation of insoluble fibrous material that normally does not occur in the central nervous system. This material is deposited in both extracellular and intraneuronal locations. The extracellular deposits are formed of amyloid while the intraneuronal components make up the neurofibrillary changes of the Alzheimer's type. Modern silver techniques, which take advantage of physical development of the latent image, are available for both the amyloid (Campbell et al., 1987) and neurofibrillary changes (Gallyas and Wolff, 1986). The sensitivity and specificity of these advanced stains is comparable to that seen in immunostained preparations (Braak et al., 1989).

Amyloid deposits. Large numbers of amyloid deposits can be seen in virtually all subdivisions of the cerebral cortex in Alzheimer's disease. Most of these patchy deposits remain unstained in Congo red and thioflavin preparations—the conventional techniques used for the demonstration of amyloid—but they can consistently be recognized in immunostains for β-protein and its precursors, or in the silver stains under consideration. In general, the amyloid patches turn out to be devoid of pathologically changed argyrophilic neurites. Furthermore, they do not reveal glial cell accumulations or neuronal distortions in Nissl stained material. It has to be emphasized that large numbers of amyloid deposits occur in layers that never or only occasionally harbor neuritic plaques; among these are the molecular layer and the multiform layer of the isocortex (Fig. 1c). Many amyloid deposits are even present in the white substance. Thus, it appears inappropriate to consider the amyloid patches as forerunners of neuritic plaques or as structures that inevitably transform into neuritic plaques.

Within the hippocampal formation, the amyloid patches exhibit a rather specific distribution. Besides the patches restricted to the pyramidal layer and some additional fluffy material close to the free surface of the fascia dentata, there are two conspicuous rows of densely packed amyloid deposits. These rows closely follow the course of the perforant path and are thus confined to narrow zones of the molecular layer of CA1 and the fascia dentata. These zones correspond to the termination field of the perforant path. The parvopyramidal layer of the presubiculum—the last relay station of the Papez circuit—is infested with amyloid (similar to the distribution of amyloid seen in cases of Parkinson's disease with cognitive decline; see Fig. 2a; Kalus et al., 1989). The entorhinal region also contains quite a number of amyloid patches in a layer specific manner (similar to Fig. 2a). It should be noted that the second layer—harboring most of the neurons generating the perforant tract—remains devoid of amyloid.

Neurofibrillary changes. Preparations stained for neurofibrillary changes show three kinds of lesions: neuritic plaques, neurofibrillary tangles, and neuropil threads (Fig. 1e; Braak and Braak, 1988a). All of these structures develop by intraneuronal accumulation of insoluble and intensely argyrophilic material.

The globular neuritic plaques are particularly enigmatic structures. They are formed of an intricate feltwork of pathologically changed and often ballooned processes of nerve cells, dendrites and axons (Mann, 1985; Price, 1986; Von Braunmühl, 1957). It is important to note that only plaques showing distended argyrophilic processes should be referred to as neuritic plaques.

Neurofibrillary tangles develop within the somata of nerve cells and may extend into the dendrites. The proximal axon, however, always remains free of these changes. Initial stages of tangles can generally be found close to lipofuscin deposits. The amount of pathological material quickly increases and sooner or later the cell body becomes filled by a tangle that frequently seems to push the nucleus aside. Tangle-bearing neurons will ultimately die. The remnants of tangles—often referred to as "ghost"-tangles—will then become engulfed by astrocytes (Probst et al., 1982). Within the isocortex, tangle-bearing neurons all belong to the class of pyramidal cells. Interneurons appear to resist the development of tangles.

Neuropil threads contribute considerably to the total amount of pathological material accumulating in the cortex although they appear—at first glance—as relatively inconspicuous structures (Braak et al., 1986). With the use of transparent Golgi impregnations counterstained for neurofibrillary changes, it has been shown that cortical neuropil threads occur in distal portions of spiny dendrites, which belong to tangle pyramidal cells (Braak and Braak, 1988b). Interneurons as well as glial cells remain devoid of these specific alterations. In some areas and layers of the cortex—such as in layer V of the striate area—a particularly pronounced thread density is noted (Fig. 1c; Braak et al., 1989).

Distribution patterns. The neurofibrillary changes exhibit a distinct but varying distribution in different areas of the cerebral cortex. The highly specific silver techniques enable one to use whole hemisphere sections, or even double hemisphere sections, which immediately reveal—even when looked at with the unaided eye—striking differences in the severity

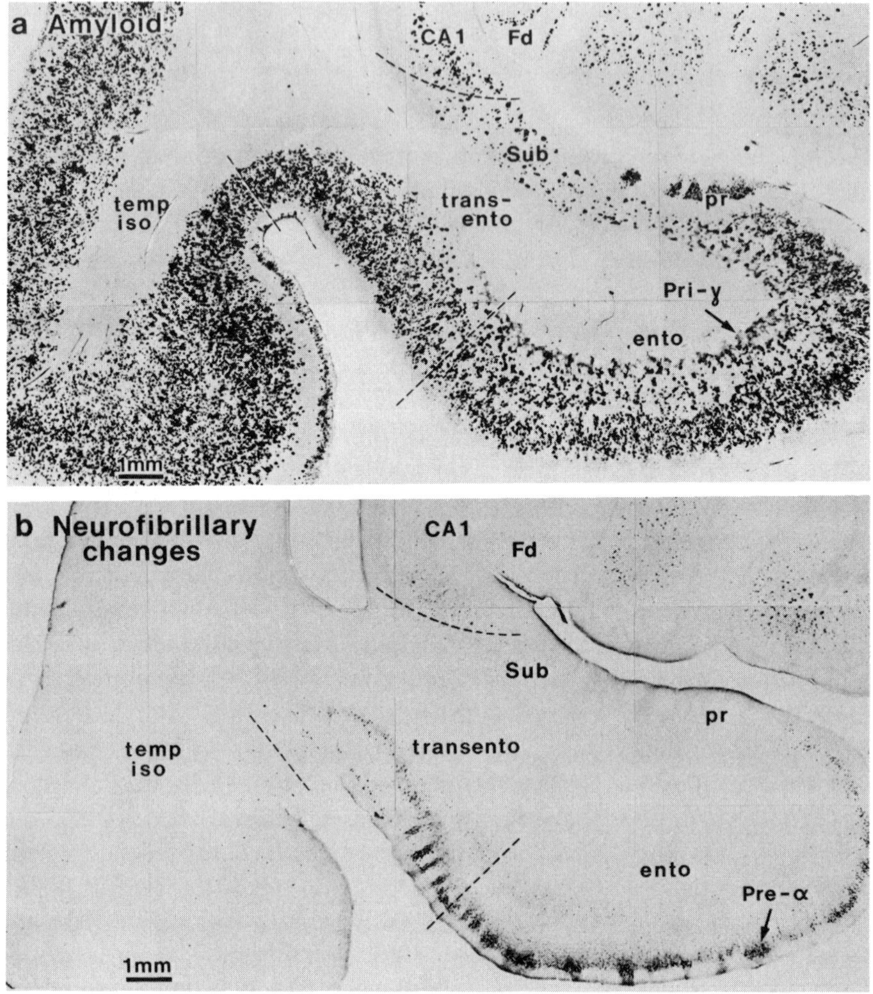

Figure 2. Characteristic cortical changes seen in cases of Parkinson's disease with cognitive decline. (a): Amyloid patches. The hippocampal formation contains only a small amount of amyloid while the presubiculum, the entorhinal region, and the isocortex are infested with plaque-like amyloid deposits. (b): Neurofibrillary changes. Both the hippocampal formation and isocortical areas display only a mild affection. Note the conspicuous changes in the outer cellular layer of the entorhinal region. Virtually all projection cells of this layer contain a tangle. CA1: first sector of the Ammon's horn; ento: entorhinal region; Fd: fascia dentata; pr: parvocellular layer of the presubiculum proper; Pre-alpha: superficial entorhinal layer; Pre-gamma: the deep entorhinal layer; Sub: subiculum; temp iso: temporal isocortex; transento: transentorhinal region. Successive sections, 100 μm. Silver stains for amyloid (a) and neurofibrillary changes (b). (With kind permission from Braak and Braak, 1990a.)

of the pathological change. Figure 1d shows the distribution of cortical pathological changes, as compiled from data available in the literature (Brun, 1983; Brun and Gustafson, 1976). In general, the isocortical association areas are heavily involved. Nevertheless, the brunt of the pathology is borne by the allocortex with severe destruction of the Ammon's horn and, in particular, the entorhinal region (Fig. 1d, black spot). The pyramidal cells within the superficial layers of the entorhinal cortex are extremely susceptible to the development of tangles (Braak and Braak, 1985). Advanced stages of Alzheimer's disease show that virtually all projection cells of the second entorhinal layer are affected. Since the same cells generate important portions of the perforant path, the pathological process will almost certainly lead to an impairment of the information transport from the entorhinal region to the hippocampal formation (Fig. 1f). The bilateral lamina-specific destructive process ultimately generates a "disconnection syndrome" in the sense of the late American neurologist Geschwind (Hyman et al., 1984, 1986); the isocortex and allocortex become disconnected from each other. In addition, the lesion severely hampers the flow of information in the Papez circuit. Therefore, the characteristic pattern of pathology may explain the early and severe impairment of intellectual capacities shown by individuals suffering from Alzheimer's disease.

Parkinson's Disease

Similar considerations can be applied to cases of Parkinson's disease with progressive cognitive decline. Major lesions in the cerebral cortex have as yet not been reported (Agid et al., 1986). Application of advanced silver techniques, nevertheless, enabled us to demonstrate a characteristic set of cortical alterations (Braak and Braak, 1990a).

Astonishingly large numbers of amyloid patches can be encountered in almost all areas of the cerebral cortex in the Parkinsonian brain. The density of amyloid patches is clearly above the level seen in the normally aged brain and is similar to that seen in Alzheimer's disease.

However, when focusing on neurofibrillary changes it becomes obvious that the Parkinsonian isocortex remains virtually devoid of intraneuronal alterations. Neuritic plaques are extremely rare. Neurofibrillary tangles and neuropil threads can—in the isocortex—only occasionally be found, but their numbers are by no means high enough to permit the diagnosis of a concomitant Alzheimer's disease. As regards the allocortex, the Ammon's horn is found to be only mildly affected. The density

Figure 3. Summary diagram of neurofibrillary changes seen in cases of Alzheimer's disease (upper row), Parkinson's disease with cognitive decline (middle row), and the dementia with argyrophilic grains (lower row). The distribution pattern of neurofibrillary changes seen in Parkinson's disease differs markedly from that found in fully developed Alzheimer's disease. Neuritic plaques occur only rarely in Parkinson's disease. Both the hippocampal formation and the isocortex display only a few tangles and neuropil threads. Severe destruction is virtually confined to the superficial cellular layer of the entorhinal region. Similarly, in cases of dementia with argyrophilic grains the pathologically changed processes are predominantly found within the superficial layers of the entorhinal cortex. a: alveus; CA1: first sector of the Ammon's horn; Diss: lamina dissecans; entorhin.region: entorhinal region; m: molecular layer of CA1; p: pyramidal layers of CA1; parasub: parasubiculum; Pre-alpha: superficial cellular layer of the entorhinal region (layer II); Pre-alpha, beta, gamma: sublayers of the lamina principalis externa; Pri-alpha, beta, gamma: sublayers of the lamina principalis interna; presubic: presubiculum proper; temp. iso: temporal isocortex; transentorhin. region: transentorhinal region; I–VI: isocortical layers.

of tangles is only slightly above that seen in nondemented age-matched controls. Only the entorhinal region shows severe changes and these are virtually confined to the outer cellular layer, which appears to be infested with tangles and threads (Figs. 2 and 3).

The Parkinsonian brain is obviously prone to develop particularly severe, but circumscribed, area-specific, layer-specific, and cell-type-specific affection of the entorhinal cortex. Initially, the change seems to begin in the transentorhinal region from where it spreads to the proper entorhinal cortex. Both the isocortex and the allocortical hippocampal formation remain almost devoid of neurofibrillary changes. The bilateral circumscribed lesion gradually destroys the perforant path and eventually impairs the transport of isocortical and Papez circuit information from the entorhinal region to the hippocampus (Braak and Braak, 1990a).

Dementia with Argyrophilic Grains

It is the experience of many neuropathologists that a considerable number of clinically well-documented cases of dementia remain uncorrelated with any of the known neuropathological changes even after thoroughgoing post-mortem examination (Ulrich et al., 1986). In a considerable number of such cases it has been possible to demonstrate with the use of advanced silver techniques the presence of conspicuous spindle-shaped grains scattered throughout the neuropil of a restricted number of cortical areas and subcortical nuclei (Braak and Braak, 1987, 1989; Itagaki et al., 1989). The grains are predominantly encountered in sector CA1 of the

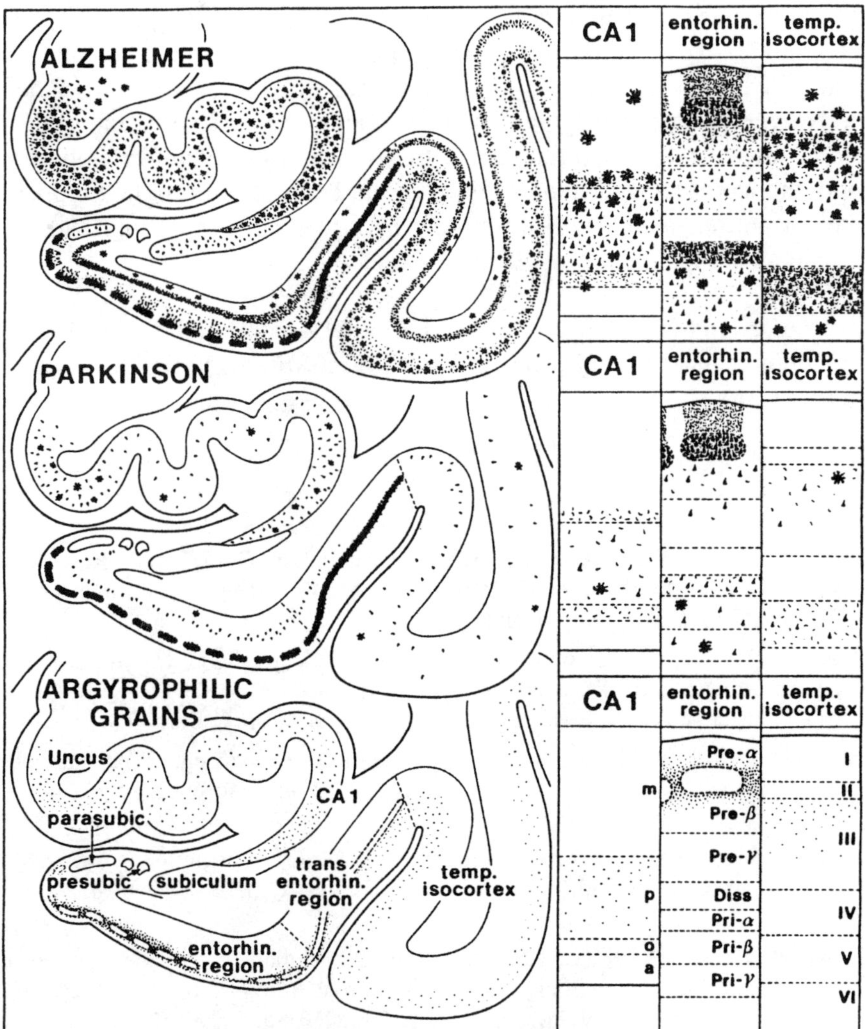

Figure 3. (See description on adjacent page.)

Ammon's horn and in the entorhinal region. A less dense scattering of grains is recognized in the adjoining temporal isocortex, insular, and orbitofrontal cortex. The grains show an elongated body with cone-shaped poles frequently giving off filiform appendages. Size, shape, and pattern of distribution suggest the grains to be pathologically changed nerve cell processes, most probably terminal ramifications of axons. Grains of this type have not been found to occur in the cortex of nondemented individuals of old age.

The antero-medial portion of the temporal lobe is most severely affected in cases of dementia with argyrophilic grains. The highest density of grains is found in the superficial cellular layers of the entorhinal cortex. Their distinct distribution again suggests an impairment of nerve cells generating the perforant path. It may well be that even subtle lesions cause severe disturbances in the transmission of information from both the isocortex and relay stations of the Papez circuit to the hippocampal formation, which will ultimately lead to a corresponding decline in intellectual capacities.

Acknowledgments. This work was kindly supported by the Deutsche Forschungsgemeinschaft, the Friedrich Merz Stiftung, Frankfurt and Degussa, Hanau.

References

Agid Y, Ruberg M, Dubois B, Pillon B, Cusimano G, Raisman R, Cash R, Lhermitte F, Javoy-Agid F (1986): Parkinson's disease and dementia. *Clin Neuropharmacol* 9 (Suppl 2):22–36

Braak H (1980): Architectonics of the human telencephalic cortex. In: *Studies of brain function*, Vol IV, Braitenberg V, Barlow HB, Bizzi E, Florey E, Grüsser OJ, van der Loos H, eds. Berlin Heidelberg New York: Springer, pp. 1–147

Braak H, Braak E (1985): On areas of transition between entorhinal allocortex and temporal isocortex in the human brain. Normal morphology and lamina-specific pathology in Alzheimer's disease. *Acta Neuropathol* 68:325–332

Braak H, Braak E (1987): Argyrophilic grains: characteristic pathology of cerebral cortex in cases of adult onset dementia without Alzheimer changes. *Neurosci Lett* 76:124–127

Braak H, Braak E (1988a): Morphology of the human isocortex in young and aged individuals: Qualitative and quantitative findings. In: *Interdisciplinary Topics in Gerontology,* Vol XXV, von Hahn HP, ed. Basel: Karger, pp. 1–15

Braak H, Braak E (1988b): Neuropil threads occur in dendrites of tangle-bearing nerve cells. *Neuropathol Appl Neurobiol* 14:39–44

Braak H, Braak E (1989): Cortical and subcortical argyrophilic grains characterize a disease associated with adult onset dementia. *Neuropathol Appl Neurobiol* 15:13–26

Braak H, Braak E (1990a): Cognitive impairment in Parkinson's disease: amyloid plaques, neurofibrillary tangles, and neuropil threads in the cerebral cortex. *J Neural Transm* (P-D Sect) 2:45–57

Braak H, Braak E (1990b): Morphology of the cerebral cortex in relation to Alzheimer's dementia. In: *Alzheimer's Disease. Epidemiology, Neuropathology, Neurochemistry and Clinics. (Key Topics in Brain Research)*, Maurer K, Riederer P, Beckmann L, eds. Wien: Springer, pp. 164–170

Braak H, Braak E, Grundke-Iqbal I, Iqbal K (1986): Occurrence of neuropil threads in the senile human brain and in Alzheimer's disease: A third location of paired helical filaments outside of neurofibrillary tangles and neuritic plaques. *Neurosci Lett* 65:351–355

Braak H, Braak E, Kalus P (1989): Alzheimer's disease: areal and laminar pathology in the occipital isocortex. *Acta Neuropathol* 77:494–506

Brun A (1983): An overview of light and electron microscopic changes. In: *Alzheimer's Disease. The Standard Reference*, Reisberg B, ed. New York: Free Press, pp. 37–47

Brun A, Gustafson L (1976): Distribution of cerebral degeneration in Alzheimer's disease. *Arch Psychiat Nervenkr* 223:15–33

Campbell SK, Switzer RC, Martin TL (1987): Alzheimer's plaques and tangles: A controlled and enhanced silver staining method. *Soc Neurosci* (Abstr) 13:678

Cummings JL (1988): The dementias of Parkinson's disease: prevalence, characteristics, neurobiology, and comparison with dementia of the Alzheimer type. *Eur Neurol* 28(Suppl 1):15–23

Gallyas F, Wolff JR (1986): Metal-catalyzed oxidation renders silver intensification selective. Applications for the histochemistry of diaminobenzidine and neurofibrillary changes. *J Histochem Cytochem* 34: 1667–1672

Hyman BT, Van Hoesen GW, Damasio AR, Barnes CL (1984): Alzheimer's disease: Cell-specific pathology isolates the hippocampal formation. *Science* 225:1168–1170

Hyman BT, Van Hoesen GW, Kromer LJ, Damasio AR (1986): Perforant pathway changes and the memory impairment of Alzheimer's disease. *Ann Neurol* 20:472–481

Itagaki S, McGeer PL, Akiyama N, Beattie BL, Walker DG, Moore RW, McGeer EG (1989): A case of adult-onset dementia with argyrophilic grains. *Ann Neurol* 26:685–689

Kalus P, Braak H, Braak E, Bohl J (1989): The presubicular region in Alzheimer's disease: topography of amyloid deposits and neurofibrillary changes. *Brain Res* 494:198–203

Mann DMA (1985): The neuropathology of Alzheimer's disease: A review with pathogenetic, aetiological and therapeutic considerations. *Mech Ageing Dev* 31:213–255

Price DL (1986): New perspectives of Alzheimer's disease. *Ann Rev Neurosci* 9:489–512

Probst A, Ulrich J, Heitz PU (1982): Senile dementia of Alzheimer type: Astroglial reaction to extracellular neurofibrillary tangles in the hippocampus. An immunocytochemical and electron-microscopic study. *Acta Neuropathol* 57:75–79

Ulrich J, Probst A, Wüest M (1986): The brain diseases causing senile dementia. A morphological study on 54 consecutive autopsy cases. *J Neurol* 233:118–122

Van Hoesen GW (1982): The parahippocampal gyrus. New observations regarding its cortical connections in the monkey. *Trends Neurosci* 5:345–350

Van Hoesen GW, Pandya DN, Butters N (1972): Cortical afferents to the entorhinal cortex of the rhesus monkey. *Science* 175:1471–1473

Von Braunmühl A (1957): Alterserkrankungen des Zentralnervensystems. Senile Involution. Senile Demenz Alzheimersche Krankheit. In: *Handbuch der speziellen pathologischen Anatomie und Histologie*, Vol XIII/1A, Lubarsch O, Henke F, Rössle R, eds. Berlin: Springer, pp. 337–539

Neuroendocrine Changes in Aging and Alzheimer's Disease

D. F. Swaab, E. Goudsmit, H. P. H. Kremer,
R. Ravid, and W. Witting

> Here in this well-concealed spot, almost to be covered with a thumbnail lies the very mainspring of primitive existence—vegetative, emotional, reproductive—on which with more or less success, man has come to superimpose a cortex of inhibitions.
>
> — H. Cushing, 1929
> (from Plum and Van Uitert, 1978)

Alterations in hypothalamic structures and functions are thought to be causal in diseases such as anorexia nervosa, depression, Cushing's disease, diabetes insipidus, Prader-Willi-syndrome, polycystic ovaries syndrome, and the malignant neuroleptic syndrome, as well as in disturbances in sleep and temperature regulation. In addition, the hypothalamus is affected in neurodegenerative diseases and might be responsible for particular symptoms, e.g., in Alzheimer's, Parkinson's and Huntington's diseases and possibly in multiple sclerosis. Moreover, this brain region is presumed to be altered in adrenogenital syndrome, by hormones given during development (e.g., diethylstibestrol (DES)), and in Turner's, Klinefelter's and Kallman's syndromes.

In spite of this impressive list, knowledge on the normal development, sexual differentiation, and aging of the human hypothalamus is hardly available. Data on how the pathophysiology of the conditions mentioned above is reflected in the morphology of the human hypothalamus are essentially nonexistent. In the present chapter, recent data on morphological alterations of several hypothalamic structures in normal aging and Alzheimer's disease are presented. The sexually dimorphic nucleus (SDN), the suprachiasmatic nucleus (SCN), the supraoptic nucleus (SON), the paraventricular nucleus (PVN), and the nucleus tuberalis lateralis (NTL) were selected in order to illustrate the variety in changes occurring in brain-endocrine interactions in the human hypothalamus during these conditions. Alzheimer's disease can in many respects be considered as an advanced, accelerated form of aging.

Sexually Dimorphic Nucleus

The sexually dimorphic nucleus of the preoptic area (SDN) was first described in the rat brain by Gorski et al. (1978). The SDN in the male rat is 3–8 times larger than in the female rat, due to differences in perinatal steroid levels (Jacobson et al., 1980). The SDN is involved in aspects of male sexual behavior (i.e., mounting, intromission, and ejaculation). Mounting behavior is not only affected by SDN lesions in male rats, but also in female rats (De Jonge et al., 1989; Turkenburg et al., 1988). The SDN in the young adult human brain is twice as large in males as in females and contains twice as many cells (Swaab and Fliers, 1985; Fig. 1). The SDN is identical to the "intermediate nucleus" described by Braak and Braak (1987). Sexual dimorphism in the human brain is not present at birth (Fig. 1). At that moment, the SDN contains only some 20% of the cell number found around 2–4 years of age. From birth up to this age, cell numbers increase rapidly in a similar way in both sexes. A sex difference does not occur until about the fourth year postnatally, when cell numbers start to decrease in girls, whereas the cell numbers in the SDN remain stable until approximately 50 years of age in males (Swaab and Hofman, 1988). Cell numbers in the male SDN decrease sharply after this age and in females a second phase of marked cell loss sets in after the age of 80 (Fig. 1; Hofman and Swaab, 1989). The sharp decrease in cell numbers in this nucleus later in life might be related to the dramatic hormonal changes which accompany both male and female senescence (Hofman and Swaab, 1989), and to the decrease in male sexual activity. It is not clear whether the hormonal changes would be cause or effect of the observed cell loss in this nucleus. Cell numbers in the SDN of Alzheimer's disease patients were found to be within the normal range for age and sex so that this condition does not seem to affect the SDN in any specific way (Swaab and Hofman, 1988).

Nucleus Tuberalis Lateralis

The lateral tuberal nucleus (NTL) is present in man and higher primates. In adulthood it contains some 60000 neurons whereas in Huntington's disease this number may be reduced to about 10000 (Kremer et al., 1990)

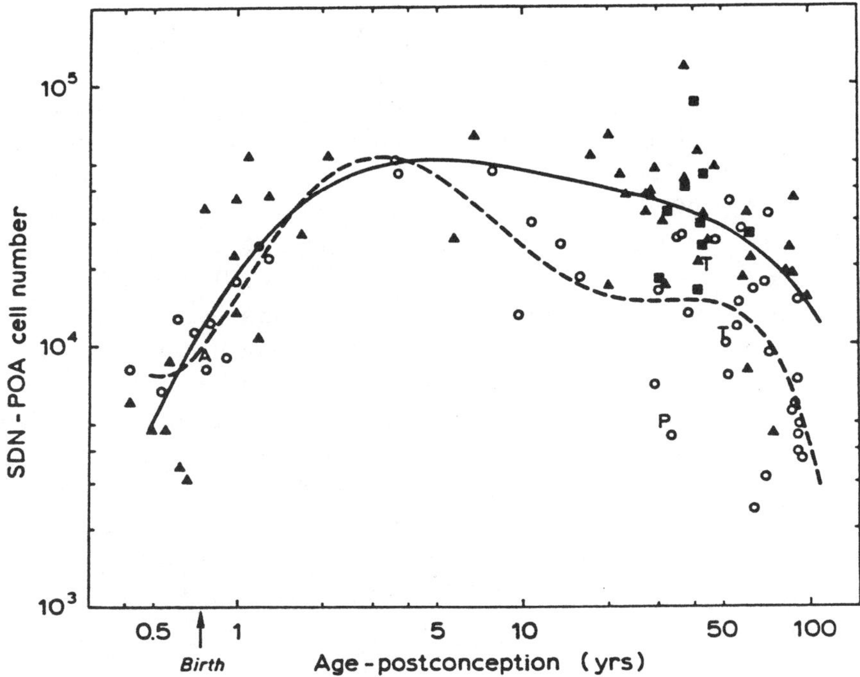

Figure 1. Development and sexual differentiation of the human sexually dimorphic nucleus (SDN) of the preoptic area of the hypothalamus. Log-log scale. Note that at the moment of birth the SCN is equally small in boys (△) and girls (○) and contains only about 22% of the cell number found at 2–4 years of age. The SDN cell number of a female neonate with a pituitary aplasia (A) is fully within the range of other neonates. Cell numbers reach a peak value around 2–4 years postnatally, after which a sexual differentiation occurs in the SDN due to a decrease in cell number in the SDN of women, whereas the cell number in men remains approximately unchanged up to the age of 50. Note that in men the reduction in cell number in senescence is less dramatic. The largest discrepancy in cell number between men and women is found around 30 years and in people older than 80, whereas the sexual dimorphism in the SDN cell number is least around the age of 60. The SDN cell number in homosexual men (□) does not differ from that in the male reference group. The cell number of the SCN of two male-to-female transsexuals (T) is within the female range, whereas the SDN of a woman with a Prader-Willi syndrome (P) is small. The curves are quintic polynomial functions fitted to the original data for males (solid line) and females (dashed line), with $F_s[5,49] = 10.05$, $P < 0.001$, and $F_s[5,39] = 7.32$, $P < 0.001$, respectively). From: Swaab and Hofman, 1988 (with permission).

depending on the age at onset of the disease and age at death. Pathological changes in the NTL have also been described in depression (Horn et al., 1988), Kallman's syndrome (Kovacs and Sheehan, 1982) and dementia with intracranial argyrophilic grains and silver-staining coiled bodies, containing straight filaments (Braak and Braak, 1989). In Alzheimer's disease, the number of NTL neurons did not differ from controls. The number of plaques in this nucleus was low, and they were exclusively of the amorphous type. Neurofibrillary tangles were rare in conventional silver stainings. Yet, immunocytochemical staining using the monoclonal antibody Alz-50 showed such an abundant reactivity of both perikarya and neurites that the NTL of Alzheimer's disease patients could even be recognized by the naked eye (Kremer et al., 1991). The NTL seems to represent a brain area in which Alzheimer's disease affects the neurons in a limited way, without further progress to the classical changes of silver staining of tangles and neuronal loss. Lesions in the lateral hypothalamus are known to be associated with a catabolic state. In Huntington's and in Alzheimer's diseases dementia is combined with severe weight loss associated with normal or even increased food intake, as is the case in the condition described by Braak and Braak (1989). Because NTL pathology is accompanied by cachexia also in this condition (H. Braak, personal communication) the NTL is hypothesized to play a role in feeding behavior and metabolism.

Suprachiasmatic Nucleus

The suprachiasmatic nucleus (SCN) is considered to be the major circadian pacemaker of the mammalian brain, coordinating hormonal and behavioral circadian rhythms (Rusak and Zucker, 1979). At birth the SCN contains some 13% of the vasopressin-expressing neurons and 20% of the total cell number found in adulthood (Fig. 2). Subsequently, cell numbers rise to maximum values around 1–2 years postnatally, after which they decrease gradually to some 50% in adulthood (Swaab et al., 1990). Age-related changes in circadian rhythms have been reported in man as well as in nonhuman species (Van Gool and Mirmiran, 1986). A fragmentation of sleep–wake patterns occurs in senescence, a phenomenon that is even more pronounced in Alzheimer's disease (Mirmiran et al., 1988; Witting et al., 1990). In Alzheimer's disease the disruptions of the circadian rhythms are often so severe that they are even thought to contribute to the mental decline (Fekete et al., 1985). In addition, sleep disturbances often lead to hospital admission of the elderly (Sanford, 1975). Total SCN cell numbers and numbers of vasopressin-expressing neurons were, therefore, determined during aging and in Alzheimer's disease. A marked decrease was found in SCN cell number in subjects of 80–100 years of age, while in Alzheimer's disease these changes were even more dramatic (Swaab et al., 1985; 1987). In this respect it is important to note that both the retina and the optic nerve, which provide direct and indirect input to the SCN, show degenerative changes in Alzheimer's disease (Hinton et al., 1986; Katz et al., 1989; Trick et al., 1989). In addition to degenerative changes, Alzheimer patients are generally exposed to less light than their age-matched controls (Campbell et al., 1988). As a result, both the input of the visual system to the SCN and the SCN itself seem to be seriously affected in Alzheimer's disease. The contribution of each of these components to circadian disturbances has yet to be investigated.

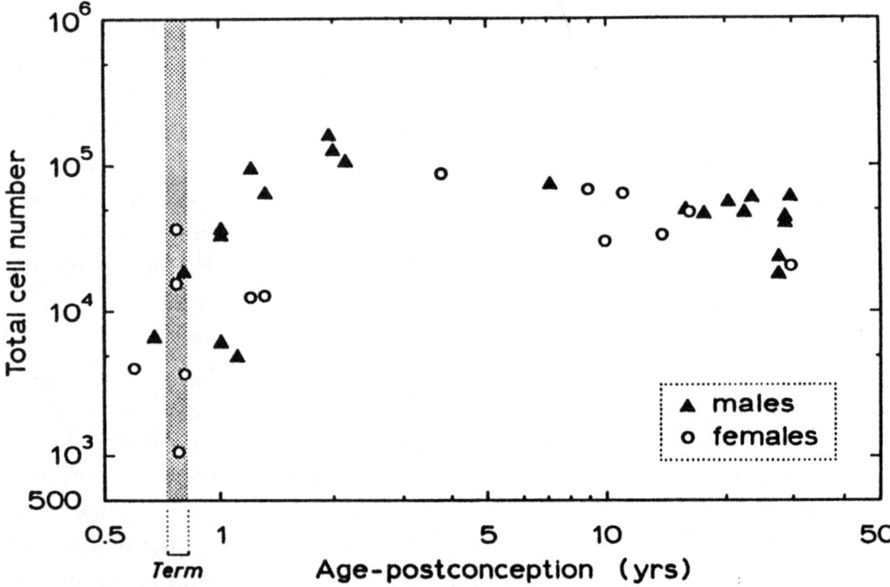

Figure 2. Development of the human suprachiasmatic nucleus (SCN) of the hypothalamus. Log-log scale. The period at term (38–42 weeks of gestation) is indicated by the vertical bar. Note that total cell number is low at the moment of birth (21% of the cell number found in adulthood). There is no difference in the developmental course of the SCN in boys and girls. Cell numbers around 1–1.5 years postnatally are more than twice the adult cell numbers. After these high levels a decrease to adult total cell number is found. From Swaab et al., 1990, with permission.

In the SCN of the aged rat, the total cell number remained unaltered, but the number of vasopressin- and vasopressin-intestinal-peptide (VIP)-expressing neurons in the SCN was significantly diminished (Chee et al., 1988; Roozendaal et al., 1987). These changes may be related to the diminished amplitude of circadian rhythms observed in aged rats (Van Gool and Mirmiran, 1983). Recent observations (Witting et al., in preparation) showed that by increasing the environmental light intensity, the circadian amplitude of sleep/wakefulness in old rats can be brought up to the level of young animals. This finding fits with the idea that activation might restore neuronal functions in aging (see below).

Supraoptic and Paraventricular Nucleus

The neurosecretory cells of the hypothalamic supraoptic and paraventricular nucleus (SON and PVN) produce the neuropeptides vasopressin and oxytocin, which are released into the blood circulation in the neurohypophysis. These neurons form a population of extremely stable cells. Neither in the course of normal aging nor in Alzheimer's disease patients were any significant loss in neurons or total cell number observed (Goudsmit et al., 1990). Various recent observations provide evidence for the hypothesis that activation of neurons may interfere with the process of aging, and thus prolong the life span of neurons or restore their function. This hypothesis is paraphrased as "use it or lose it." The SON and PVN neurons are not only metabolically highly active throughout life, but they are even additionally activated in senescence, as can be judged from the increase in the size of the vasopressin-containing perikarya (Fliers et al., 1985) and nucleoli (Hoogendijk et al., 1985), and the enhanced plasma levels of vasopressin (Frolkis et al., 1982) and neurophysins (Legros et al., 1980). Similar activation of vasopressin neurons was observed in the aged rat (Fliers and Swaab, 1983; Goudsmit et al., 1988) and is probably due to a loss of vasopressin receptors in the kidney during aging (Ravid et al., 1987). The hypothesis that the neurosecretory cell number is so stable because they are extra-activated needs to be tested experimentally, e.g., by long-term inhibition of these neurons during aging, in which case neuronal degeneration is expected to occur.

Activation Restores Innervation

An example of activation restoring a peptide system during aging was recently observed in the rat. Vasopressin innervation in the senescent male rat brain is particularly decreased in those regions where the fiber density in young adult males was shown to be dependent upon plasma levels of sex steroids (Fliers et al., 1985). Plasma testosterone levels and testicular weight decrease progressively with age in the rat (Ravid et al., 1987). However, when old rats were treated with testosterone for one month, the vasopressin innervation in the rat brain was restored. The vasopressin fiber system responding to the testosterone treatment

(Goudsmit et al., 1988) originates from the bed nucleus of the stria terminales (BST) and medial amygdala, where testosterone seems to stimulate vasopressin synthesis. In castrated male rats, in which axonal transport was blocked by means of colchicine, testosterone increased the number of BST neurons expressing vasopressin (Van Leeuwen et al., 1985). Moreover, vasopressin-producing neurons in the BST and amygdala were shown to contain steroid hormone receptors (Axelson and Van Leeuwen, 1990). Thus, the increased vasopressin staining in aged animals following testosterone treatment demonstrates how activation of the synthetic activity of peptidergic neurons (i.e., BST neurons) may reverse the age-related changes in their termination.

Summary and Conclusions

The human hypothalamus contains several nuclei manifesting a wide variety of changes in aging and Alzheimer's disease that might be related to a particular symptomatology.

In the first place, the SDN shows a decreased cell number during prepubertal development leading to sexual dimorphism and, subsequently, a decrease in cell number in both sexes during normal aging. The latter change is possibly related to a decrease in sexual activity and changes in hormone levels. In Alzheimer's disease cell numbers decrease at a similar rate as in normal aging (Swaab and Hofman, 1988).

The NTL might be involved in feeding behavior and metabolism. It does not show any decrease in neuronal numbers in Alzheimer's disease. Yet, a very strong Alz-50 staining is present in the NTL of these patients that is based upon a dense network of dystrophic neurites and numerous stained perikarya. The NTL in Alzheimer's disease patients seems, therefore, to be in an early phase of the disease process. In addition, we may conclude that Alz-50 is not simply a marker for cell death.

The SCN coordinates circadian rhythms. Its cell number decreases 50% from 1–2 years postnatally to adulthood and, subsequently, during normal aging. In Alzheimer's disease, the decrease in cell number is even more pronounced. This pathology might be the neural basis for the nightly restlessness observed in Alzheimer's disease, whereas alterations in the visual system (previously discussed) might contribute to these functional disturbances. It seems worthwhile, therefore, to try and see whether stimulation of the visual pathways by light therapy might increase the amplitude of the circadian rhythms in Alzheimer patients in a similar way as it does in the aged rat.

The cells of the SON and PVN are examples of neuron populations that seem to stay perfectly intact in aging and Alzheimer's disease. We hypothesize that this might be due to the activation of these endocrine cells during the aging process, a hypothesis that will be tested in animal experiments. We have already shown that neurons also may be activated in very old rats, if the right stimulus (e.g., testosterone for the BST) is offered.

For several reasons, the hypothalamus offers a unique brain structure for the study of basic questions on the changes in aging and Alzheimer's disease, since (1) its nuclei can be delineated, which enables determination of changes in cell numbers, (2) the various hypothalamic nuclei show a wide range of variations in aging and Alzheimer's disease, and (3) it becomes increasingly possible to relate the cell number and neuropathological changes in the different nuclei to functional changes, either in terms of behavior or in endocrine changes. Such studies might establish the beginning of a neglected topic in neuropathology, namely, the neuropathology of the human hypothalamus.

Acknowledgments. Brain material was obtained from the Netherlands Brain Bank, Amsterdam. The authors would like to express their thanks to Mr. B. Schillings for his secretarial help.

References

Axelson JF, Van Leeuwen FW (1990): Differential localization of estrogen receptors in various vasopressin synthesizing nuclei of the rat brain. *J Neuroendocr* 2:209–216

Braak H, Braak E (1987): The hypothalamus of the human adult: chiasmatic region. *Anat Embryol* 176:315–330

Braak H, Braak E (1989): Cortical and subcortical argyrophylic grains characterize a disease associated with adult onset dementia. *Neuropathol Appl Neurobiol* 15:13–26

Campbell SS, Kripke DF, Gillin JC, Hrubovcak JC (1988): Exposure to light in healthy elderly subjects and Alzheimer patients. *Physiol Behav* 42:141–144

Chee CA, Roozendaal B, Swaab DF, Goudsmit E, Mirmiran M (1988): Vasoactive intestinal polypeptide neuron changes in the senile rat suprachiasmatic nucleus. *Neurobiol Aging* 9:307–312

De Jonge FH, Louwerse AL, Ooms MP, Evers P, Endert E, Van de Poll NE (1989): Lesions of the SDN-POA inhibit sexual behavior of male Wistar rats. *Brain Res Bull* 23:483–492

Fekete M, Van Ree JM, Niesink RJM, De Wied D (1985): Disruption of circadian rhythms induces retrograde amnesia. *Physiol Behav* 34:883–887

Fliers E, Swaab DF (1983): Activation of vasopressinergic and oxytocinergic neurons during aging in the Wistar rat. *Peptides* 4:165–170

Fliers E, De Vries GJ, Swaab DF (1985): Changes with aging in the vasopressin and oxytocin innervation of the rat brain. *Brain Res* 348:1–8

Frolkis VV, Golovchenko SF, Medved VI, Frolkis RA (1982): Vasopressin and cardiovascular system in aging. *Gerontology* 28:290–302

Gorski RA, Gordon JH, Shryne JE, Southam AM (1978): Evidence for a morphological sex difference within the medial preoptic area of the rat brain. *Brain Res* 148:333–346

Goudsmit E, Fliers E, Swaab DF (1988): Testosterone supplementation restores vasopressin innervation in the senescent rat brain. *Brain Res* 473:306–313

Goudsmit E, Swaab DF (1989): Neuroendocrine aspects of normal and pathological aging. In: *Recent Advances in Basic and Clinical Neuroendocrinology*, Casanueva FF, Dieguez C, eds. Amsterdam: Elsevier Science Publishers, pp. 369–382

Goudsmit E, Hofman MA, Fliers E, Swaab DF (1990): The supraoptic and paraventricular nuclei of the human hypothalamus in relation to sex, age and Alzheimer's disease. *Neurobiol Aging*, in press

Hinton DR, Sadun AA, Blanks JC, Miller CA (1986): Optic nerve degeneration in Alzheimer's disease. *N Engl J Med* 315:485–487

Hofman MA, Swaab DF (1989): The sexually dimorphic nucleus of the preoptic area in the human brain: a comparative morphometric study. *J Anat* 164:55–72

Hoogendijk JE, Fliers E, Swaab DF, Verwer RWH (1985): Activation of vasopressin neurons in the human supraoptic and paraventricular nucleus in senescence and senile dementia. *J Neurol Sci* 69:291–299

Horn E, Lach B, Lapierre Y, Hrdina P (1988): Hypothalamic pathology in the neuroleptic malignant syndrome. *Am J Psychiatry* 145:617–620

Jacobson CD, Shryne JE, Shapiro F, Gorski RA (1980): Ontogeny of the sexually dimorphic nucleus of the preoptic area. *J Comp Neurol* 193:541–548

Katz B, Rimmer S, Iragui V, Katzman R (1989): Abnormal pattern electroretinogram in Alzheimer's disease: evidence for retinal ganglion cell degeneration? *Ann Neurol* 26:221–225

Kovacs K, Sheehan HL (1982): Pituitary changes in Kallman's syndrome. A histologic, immunocytologic, ultrastructural, and immunoelectron microscopic study. *Fert Steril* 37:83–89

Kremer HPH, Roos RAC, Dingjan G et al. (1901): Atrophy of the hypothalamic lateral tuberal nucleus in Huntington's disease. *J Neuropathol Exp Neurol* 49:371–382

Kremer HPH, Swaab DF, Bots G, Fisser B, Ravid R, Roos RAC (1991): The hypothalamic lateral tuberal nucleus in Alzheimer's disease *Ann Neurol* 29:274–284

Legros JJ, Gilot P, Schmitz S, Bruwier M, Mantanus H, Timsit-Berthier M (1980): Neurohypophyseal peptides and cognitive function: a clinical approach. In:

Progress in Psychoneuroendocrinology, Brambilla F, Racagni G, de Wied D, eds. Amsterdam: Elsevier Science Publishers, 325–337

Mirmiran M, Overdijk J, Witting W, Klop A, Swaab DF (1988): A simple method for recording and analysing circadian rhythms in man. *J Neurosci Meth* 25:209–214

Plum F, Van Uitert R (1978): Nonendocrine diseases and disorders of the hypothalamus. In: *The Hypothalamus*, Reichlin S, Baldessarini RJ, Martin JB, eds. New York: Raven Press, pp. 415–473

Ravid R, Fliers E, Swaab DF, Zurcher C (1987): Changes in vasopressin and testosterone in the senescent Brown-Norway (BN/BiRij) rat. *Gerontol* 33:87–98

Roozendaal B, Van Gool WA, Swaab DF, Hoogendijk JE, Mirmiran M (1987): Changes in vasopressin cells of the rat suprachiasmatic nucleus with aging. *Brain Res* 409:259–264

Rusak B, Zucker I (1979): Neural regulation of circadian rhythms. *Physiol Rev* 59:449–526

Sanford JRA (1975): Tolerance of debility in elderly dependents by supporters at home: its significance for hospital practice. *Br Med J* 3:471–473

Swaab DF, Fliers E (1985): A sexually dimorphic nucleus in the human brain. *Science* 224:314–318

Swaab DF, Fliers E, Partiman TS (1985): The suprachiasmatic nucleus of the human brain in relation to sex, age and senile dementia. *Brain Res* 342:37–44

Swaab DF, Roozendaal B, Ravid R, Velis DN, Gooren L, Williams RS (1987): Suprachiasmatic nucleus in aging, Alzheimer's disease, transsexuality and Prader-Willi syndrome. In: *Progress in Brain Research*, Vol. 72, Amsterdam: Elsevier Science Publishers, pp. 301–310

Swaab DF, Hofman MA (1988): Sexual differentiation of the human hypothalamus: ontogeny of the sexually dimorphic nucleus of the preoptic area. *Dev Brain Res* 44:314–318

Swaab DF, Hofman MA, Honnebier MBOM (1990): Development of vasopressin neurons in the human suprachiasmatic nucleus in relation to birth. *Dev Brain Res* 52:289–293

Trick GL, Barris MC, Bickler-Bluth M (1989): Abnormal pattern electroretinograms in patients with senile dementia of the Alzheimer type. *Ann Neurol* 26:226–231

Turkenburg JL, Swaab DF, Endert E, Louwerse AL, Van de Poll NE (1988): Effects of lesions of the sexually dimorphic nucleus on sexual behavior of testosterone-treated female Wistar rats. *Brain Res Bull* 21:215–224

Van Gool WA, Mirmiran M (1983): Age-related changes in sleep pattern of male adult rats. *Brain Res* 279:394–398

Van Gool WA, Mirmiran M (1986): Aging and circadian rhythms. In: *Progress in Brain Research*, Vol. 70, *Aging of the Brain and Alzheimer's Disease*, Swaab DF et al., eds. Amsterdam: Elsevier Science Publishers, pp. 255–279

Van Leeuwen FW, Caffe AR, De Vries GJ (1985): Vasopressin cells in the bed nucleus of the stria terminalis of the rat: sex differences and the influence of androgens. *Brain Res* 325:391–394

Witting W, Kwa IH, Eikelenboom P, Mirmiran M, Swaab DF (1990): Alterations in the circadian rest-activity rhythm in aging and Alzheimer's disease. *Biol Psychiatry* 27:563–572

PART III
Cognitive Science

Developing a Memory "Thermometer": A Theoretical Framework and Initial Results

G. J. Spilich and Stevo Knežević

It has been known for some time that memory disorders can accompany and even help define disease states; only relatively recently has medical science turned its attention to understanding the processes behind such disorders. The factors behind a growth of medical interest in memory disorders are many: aging of the general population, advances in our understanding of the neurochemistry of memory, and increased interest in diseases not easily diagnosed via computerized tomography scans. Furthermore, memory disorders are of interest to a wide variety of professional disciplines outside of neurology: family medicine, industrial medicine, clinical neuropsychology, and experimental psychology.

In order to deal effectively with a wide spectrum of memory disorders, we must have assessment tools that can not only detect and quantify such disorders but then differentiate one from another. At the present time, there is no biological test that accurately diagnoses and measures memory disorders ("Memory Testing," 1987); therefore we must rely upon those instruments that directly measure human cognitive performance to provide the background for making such assessments. It is commonly accepted that no psychological test that is currently and widely available is adequate to the task. In this article, we wish to suggest a framework for the development of clinical memory assessment instruments, and to describe the results of our efforts to improve the assessment of memory disorders.

Levels of Memory Testing

There are three levels of memory testing: rough bedside screening for the presence of a deficit, then more complex assessment for the differential diagnosis of a deficit, and lastly, in-depth explorations of the mechanisms of a deficit in order to advance both clinical practice and our understanding of memory phenomena. First level demands are satisfied by tests that meet the requirement of being short, simple and easily administered at bedside to patients able to communicate; we believe such tests already exist (Knopman and Ryberg, 1989). Our interest was in the devising of a test that would meet the demands of levels two and three.

Criteria for Memory Testing

There are at least four criteria that such a test should satisfy. First, a test should detect the presence of a memory deficit that is clinically relevant. To do so, a test must be sensitive enough to detect a disturbance in memory as early as possible in the course of the disease. After showing that it can separate pathological from normal cognitive processes, a test must then be able to separate patients into different groups. That is, it must be possible to administer the test to more than one clinical group and differentiate them with some reasonable level of success. Third, our focus changes from intergroup to intragroup differences; a test should discriminate among a wide variety of individual differences in function. Some examples are: distinguishing placebo from treatment effects, detecting drug responders from nonresponders, differentiating acute from chronic effects, and so on. Last, the ideal test should be linguistically portable.

The starting point for devising such an instrument is that a successful memory assessment tool would be based upon current conceptualizations of memory as a biological/cognitive phenomenon. Memory assessment is likely to be as complex a procedure as biochemical assay or photomicroscopy (Lancet, 1987), and so without a linkage between memory theory and test construction, a memory test is not likely to be successful. In this chapter we describe an instrument that is based upon a specific theoretical position in memory research that weighs the role that a carry-forward buffer or "working memory" plays in cognitive operations (Cohen, 1979, 1981 and her chapter in the present volume; Spilich, 1983, 1985; Spilich and Voss, 1983). Working memory is a hypothetical link between short-term memory (STM) and long-term memory (LTM); it is a limited capacity buffer whose contents are selected in response to the external demands of the task and the internal knowledge base of the individual, and is normally deployed during comprehension. This buffer operates to build contextual relations among ideas, and thereby builds coherence. An example of how differences in conceptual complexity in

LTM affect contextual processes during comprehension may clarify this point. Figure 1 represents a simple illustration of how the representation of a single concept would differ between an expert and a novice.

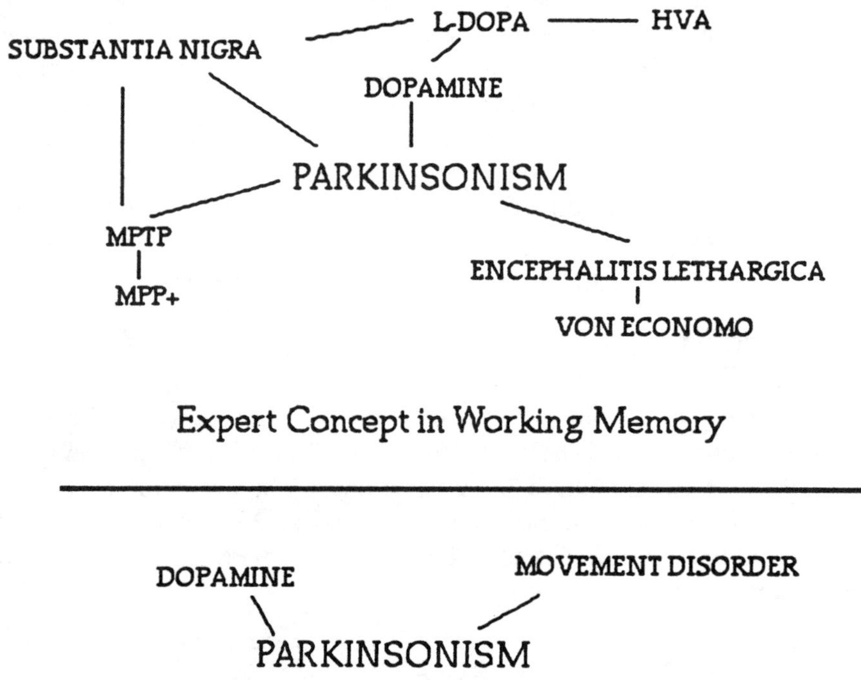

Figure 1. Example of Expert and Novice concepts in Long Term Memory.

In a discussion of the treatment of Parkinson's disease, several concepts would be automatically loaded from LTM into the working memory system of the knowledgeable diagnostician or scientist when the concept of Parkinsonism is activated. The availability of those concepts would both speed processing and deepen the resultant memory trace, making it more resistant to forgetting (Spilich et al., 1979). No such elaborate activation would automatically occur in the working memory of the beginning student, and so the relevant concepts would have to be either generated or retrieved and the linkage among these concepts would have to be explicitly drawn, if a full understanding is to occur. Explicitly drawing such links requires additional processing time and resources. It is exactly these changes in contextual operations that we believe to be of great importance in human cognition.

Working Memory and Dementia

It is known that breakdowns in the working memory system play a significant role in memory deficits such as Alzheimer's. We have found that a test that was designed to assess the operations of working memory can be useful not only in distinguishing normal from abnormal memory processes (Spilich, 1983, 1985) but can also distinguish various clinical and subclinical memory disorders (Milanovic et al., 1990). Furthermore, since this test is based upon general psycholinguistic processes rather than the meaning of specific concepts that might differ substantially from language to language, the test is relatively translatable into other languages. This test is known as the CETM, for contextual effects in text memory.

The CETM consists of a series of 24 units, presented serially to each subject, one unit per page. Each unit is comprised of a context consisting of three simple declarative sentences, followed by a target sentence presented underneath the context sentences and clearly separated from the context sentences by a dark line. There are three types of contexts: strong, weak, or no context. In a strong contextual relationship, the context enjoys a predictive and easily encodable relationship with the target sentence.

IT WAS FIVE O'CLOCK
JOHN WAS DOWNTOWN
HE WAS HUNGRY

JOHN WENT TO A RESTAURANT

Figure 2. Example of context-target unit.

Figure 2 illustrates the target sentence "John went to a restaurant" paired in conjunction with its strong context, as a subject would see it. Weak contexts are not illogical but require additional processing on the part of the processor for comprehension to occur. The corresponding three sentences for the weak context for the previously illustrated target sentence are:

"John was downtown."

"He looked in his wallet."

"There was some money there."

Last, one third of all targets for each subject are presented in the no context condition. Here, the target sentence was presented alone on the page, without a context.

There are four phases to the task: study, free recall, cued recall, and recognition. In the study phase, subjects are instructed to read the units aloud one at a time, and were told that while the purpose of the context was to "help them remember" the target, they were responsible only for remembering the target sentence. After the initial presentation phase, a free recall task follows. Then, the 16 contexts originally presented to that subject are re-presented for cued recall of the targets; finally, a forced choice recognition task paired one original sentence with several distractor sentences; subjects indicate which was the sentence they had seen at the study phase. Group differences in global performance measures such as overall performance can then be compared, along with qualitative group differences in performance in response to the three levels of contextual challenge posed by the CETM.

In this chapter, we briefly summarize one avenue in the development and subsequent testing of a modern memory assessment tool. Some portions of this work have already been reported elsewhere while others are drawn from work still in progress; we report only that which is relevant to this discussion. We hope to demonstrate the existence of a memory assessment tool that meets our four proposed characteristics of detection, discrimination, individual difference differentiation, and portability.

Detection of Dementia

In order to demonstrate sensitivity, a memory test must be able to detect the presence of a memory disorder and differentiate it from normalcy. Spilich and Voss (1983) have reported that a task whose focus is on the operation of contextual mechanisms can reliably distinguish the performance of young from aged, and memory-impaired from nonimpaired elderly.

Subjects. There were three subject groups in this study: college-aged, elderly normal, and elderly memory-impaired. The two aged groups were matched on WAIS vocabulary performance and equated on educational level; memory impairment was diagnosed by the WMS and DSM III criteria applied by the medical staff. All subjects were administered the CETM.

Results. In a comparison of college-aged (mean age = 19), normal elderly (mean age = 76) and memory-impaired elderly (mean age = 81), it was found that our task clearly differentiated each group from the others. These results are presented in Fig. 3.

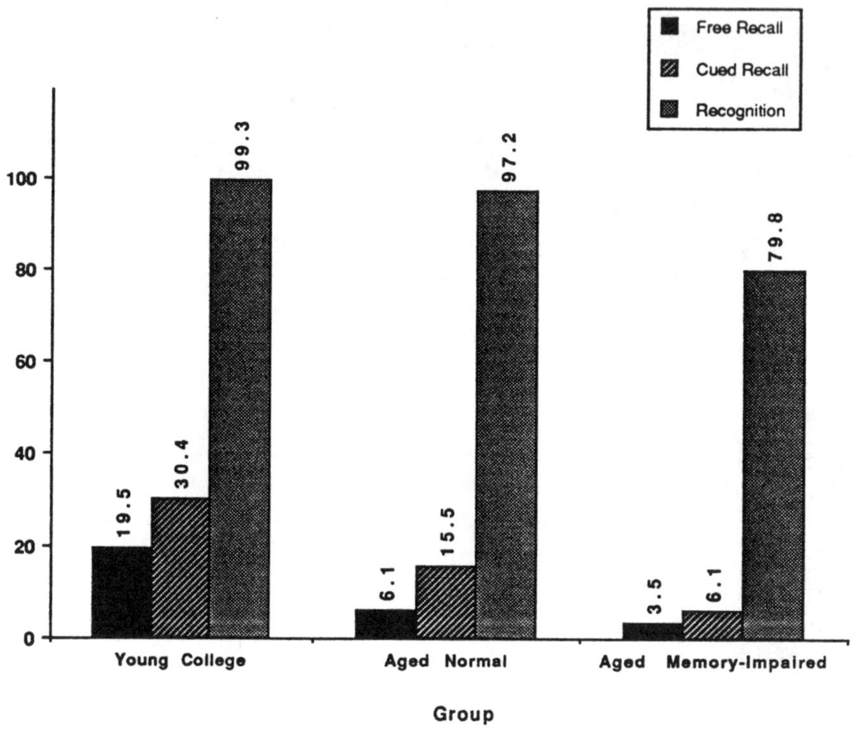

Figure 3. CETM performance by three groups on three memory tasks.

In a similar vein, Milanovic et al. (1990) have reported an investigation of the effects of solvent exposure upon cognitive performance. In this study, cognitive performance of workers exposed to a mixture of industrial solvents such as toluene and n-hexane were contrasted with an educationally matched group of nonexposed control workers at the same factory.

Subjects. There were 40 individuals in the solvent exposed group; 24 nonexposed individuals were in the control group. The impetus for this study was the self-reported complaints of the exposed workers regarding perceived loss of cognitive function. The exposed and control groups did not differ reliably in age, length of employment, or level of educational attainment. A team of specialists in occupational medicine, neurology, and psychology examined each patient; individuals were excluded who demonstrated signs of a degenerative nervous disorder, alcohol or drug abuse, or who had recently suffered from head trauma leading to loss of consciousness. Additional detail on subject characteristics and screening procedures is available elsewhere (Milanovic et al., 1990). Subjects were tested both before and after their work shift.

Results. The CETM was able to detect the effects of solvent exposure upon intellectual operations; a statistically reliable drop in cognitive performance was observed among exposed workers, $F(1,46) = 4.31$, $p. = .04$. These results are summarized in Figure 4.

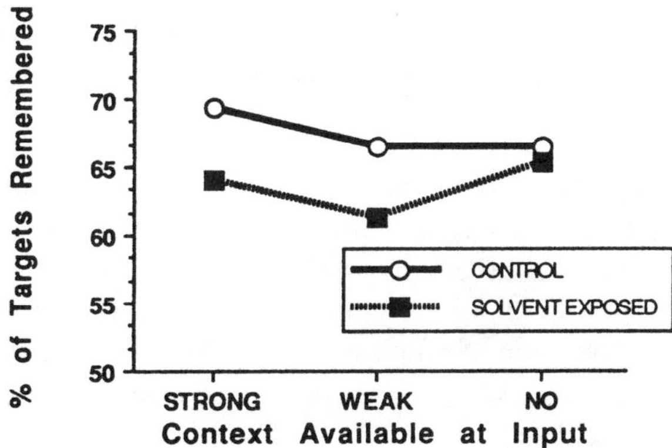

Figure 4. CETM performance by solvent-exposed and control groups for three levels of contextual challenge.

Summary. In the first example, the CETM detected the presence both of normal age effects and also of pathological aging effects upon memory. In the second example, the CETM was able to detect the presence of a subclinical deficit resulting from prolonged exposure to industrial solvents. These two studies, taken as a whole, support our claim that the test has demonstrated the ability to detect clinical and subclinical cognitive disorders and differentiate the performance of abnormal from normal individuals.

Discrimination Among Dementias

To meet the criterion of discrimination, a tool must be able to discriminate changes in information processing that are a result of different pathological conditions. Knežević et al. (1989) report the results of a study investigating the effects of pyritinol upon regional cerebral blood flow (rCBF) and cognitive performance in those individuals diagnosed as suffering from senile dementia of the Alzheimer's type (SDAT); the cognitive assessment was performed with a number of tests, including the CETM. We report here both a summary of the original results and additional unpublished information; specifically the effects of pyritinol therapy upon cognitive performance in those individuals diagnosed as suffering from multi-infarct dementia (MID), using the CETM as a dependent variable.

Subjects. Screening procedures using DSM-III criteria were used to generate a pool of subjects who were classified as either SDAT or MID; further details on the screening procedure are provided elsewhere (Knežević et al., 1989). There were a total of 26 SDAT individuals; mean age was 76.3; SD = 7.3. A total of 26 individuals were in the MID group; mean age was 76.0, SD = 8.0.

Materials and Procedure. The effect of pyritinol upon cognitive performance was evaluated in a crossover randomized clinical trial; drug effects were contrasted with placebo effects. An experimental version of our task was employed to assess cognitive performance; it was essentially identical to the original version (Spilich and Voss, 1983) with the exception that the cued recall task was not performed in order to reduce testing demands upon the subjects.

Results. Pyritinol was found to "normalize" the CBF activation pattern in SDAT individuals; this normalization was accompanied by a statistically significant increase in performance on the CETM. Specifically, pyritinol therapy was associated with an increase in CBF that was precentral and temporal in nature; this specific increase in CBF was associated with an improvement in cognitive performance. Placebo therapy was associated with a more global increase in CBF and also a slight improvement in cognitive performance. The corresponding CBF results for the MID group were less clear, but once again pyritinol appeared to exert a normalizing trend. As the purpose of the present chapter is to discuss memory assessment tools, we will not discuss the analysis of the regional cerebral blood flow data, which is presented in greater detail in the original article (Knežević et al., 1989)

As Figure 5 indicates, the CETM was able to distinguish the performance of the SDAT from the MID groups, $F(1,96) = 4.30$, $p. = .041$.

Summary. In order to meet the criterion of discriminability, a test must be able to separate one clinically relevant group from another. Our results indicate that the CETM can be used to differentiate SDAT from MID groups.

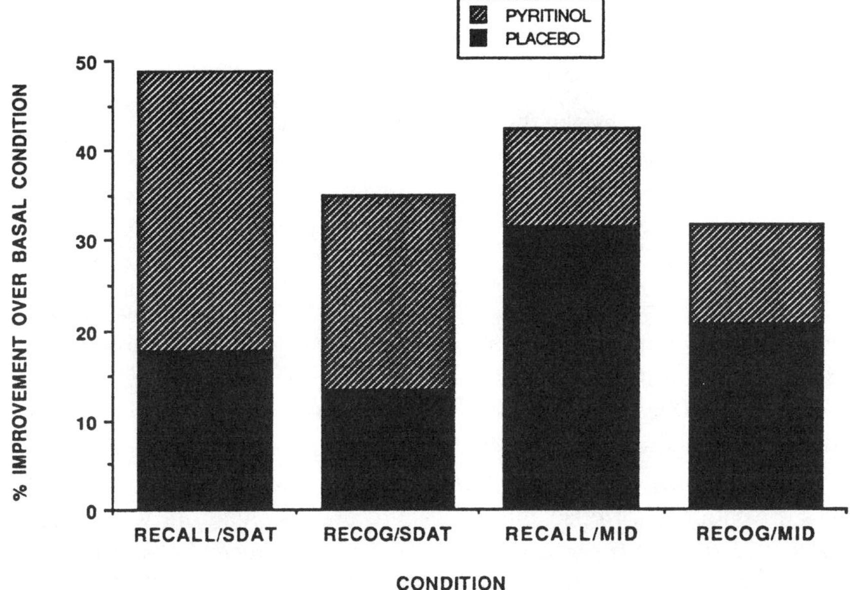

Figure 5. CETM improvement by Trial, Task, and Group.

Differentiation of Individual Differences

In order to meet this criterion, a test of cognitive performance must be able to refract the wide spectrum of response within a group of individuals into the separate components of individual response to a pathological state or a treatment.

For instance, one such component that might be expected within a global response is the differentiation between placebo and drug effects. While it is widely acknowledged that placebo effects exist, it is not as common to observe them in actual research. Table 1 illustrates both placebo effect and drug response by the SDAT and MID groups. These comparisons were on combined recall and recognition data; comparisons were Bonferroni-corrected for experiment-wide error rate and also for directionality of the comparison. Table 1 and Fig. 4 indicate that the CETM distinguished the effects of placebo from pyritinol for the SDAT group, and detected a reliable placebo effect for the MID group. The SDAT group also demonstrated a larger CBF response to pyritinol than did the MID group; this greater normalization of CBF by the SDAT individuals was accompanied by concomitant increases in performance on our task. It would appear then that the normalization of CBF brought about by pyritinol was in accord with changes in the patient's observed cognitive performance.

Table 1

Effect	SDAT	MID
Δ placebo	p. < .05	p. < .02
Δ pyritinol	p. < .002	p. < .19

Placebo and drug effects as a function of group.

We see other evidence of sensitivity to the components of group response when we return to the solvent exposure data previously introduced in work reported by Milanovic et al. (1990) but now focus on performance differences that distinguish pre- and postshift cognitive performance. Such a distinction is useful, as preshift performance is an index of the chronic effect of solvent exposure, whereas postshift performance provides a glimpse into the acute effects that follow a workday of exposure. Figure 6 portrays performance differences between controls and solvent-exposed workers for free recall portion of the CETM, which was the most sensitive portion of the test for this task.

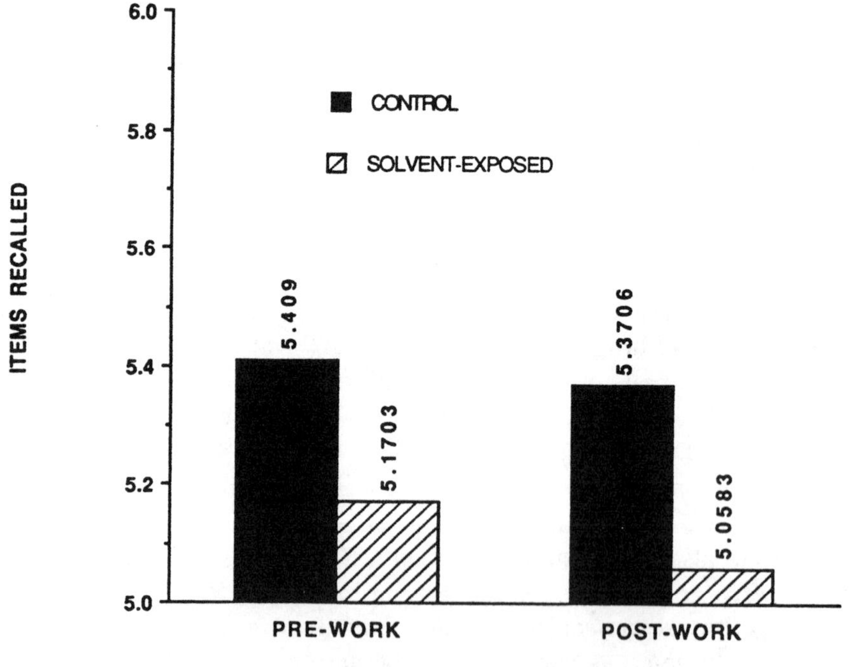

Figure 6. CETM free recall performance by time of measurement and group.

The drop in CETM performance measured postshift for the solvent-exposed workers was reliably greater than the corresponding change for the controls, F(1,74) = 3.74, p. = .05. Our interpretation of this data is that the CETM was able to separate the chronic and acute (that is, pre- and postshift) effects of solvent exposure upon cognitive status. We note that this performance difference, while statistically reliable, was relatively slight.

Summary. The differentiation of placebo and drug effects in both SDAT and MID groups along with the separation of chronic and acute effects of solvent exposure upon cognitive performance indicates that the CETM has met with some success in demonstrating differentiability.

Portability

A task that is portable would be translatable into several languages. As the test is based upon general psycholinguistic principles rather than specific items, it is less likely to be affected by the idiosyncratic characteristics of each language than is a test that relies heavily upon the meaning of specific items.

At present, there are versions of the CETM in four languages (Spilich, 1989): the original in English, a second version in Serbo-Croatian, the common language of Yugoslavia, and new versions in Spanish and also in German. We have found these versions to be useful and so suspect that other language versions are possible.

Conclusion

The goal of this chapter was to provide an outline for the development of memory assessment tests, and to relate our progress in the development of a test designed to detect memory disorders, differentiate one from another, discriminate individual differences, and to do so in two languages. We feel that we have obtained sufficient evidence to conclude that modest success in reaching our goals has been met.

References

Cohen G (1979): Language comprehension in old age. *Cognitive Psychology* 11:412–429

Cohen G (1981): Inferential reasoning in old age. *Cognition* 9:59–72

Knežević S, Mubrin Z, Risberg J, Vucinic G, Spilich G, Gubarev N, Wannenmacher W (1989): Pyritinol treatment of SDAT patients: Evaluation by psychiatric and neurological examination, psychometric testing and rCBF measurements. *International Clinical Psychopharmacology* 4:25–38

Knopman DS, Ryberg S (1989): A verbal memory test with high predictive accuracy for dementia of the Alzheimer type. *Archives of Neurology* 46:141–145

Memory testing: No thermometers available. *Lancet* 12 September 1987:604–606

Milanovic L, Spilich G, Vucinic G, Knežević S, Ribaric B, Mubrin Z (1990): Effects of occupational exposure to organic solvents upon cognitive performance. *Neurotoxicology and Teratology* 12:657–660

Spilich GJ (1983): Life-span components of text processing: Structural and procedural differences. *Journal of Verbal Learning and Verbal Behavior* 22:231–244

Spilich GJ (1985): Discourse comprehension across the span of life. *Aging and Human Performance*. New York: Wiley

Spilich GJ (1989): The CETM: A test for clinically relevant information processing deficits. Preliminary Manual, version 1.0, March, 1989. Available from the author, c/o Psychology Department, Dunning Science Center, Washington College, Chestertown, MD 21620, USA

Spilich GJ, Vesonder GT, Chiesi HL, Voss JF (1979): Text processing of domain-related information for individuals with high and low domain knowledge. *Journal of Verbal Learning and Verbal Behavior* 18:275–290

Spilich GJ, Voss JF (1983): Contextual effects upon text memory for young, aged-normal, and aged-impaired individuals. *Experimental Aging Research* 9: 45–49

Biological Bases of Specific Language Impairment (Developmental Aphasia)

Dorothy Bishop

When a child is brought to a pediatrician because language is not developing normally, there are several investigations that are typically carried out. The child's hearing is tested, and cognitive and social development are assessed. The medical history is scrutinized to see if there is any evidence of neurological disease or trauma early in life, and the child is examined for neurological signs and abnormalities of the speech apparatus. The parents are interviewed about the family circumstances. Although in many cases one might find some explanation for the child's language disorder in the course of these investigations, all too often this is not the case. The child appears to be physically normal with adequate hearing, good nonverbal ability and normal social relationships, and the medical history and home background are unremarkable. Traditionally the terms "developmental aphasia" or "developmental dysphasia" have been used to describe cases of abnormal language development of unknown origin. However, this terminology is not entirely satisfactory. Use of a neurological label implies similarities with adult acquired dysphasia, but this is usually misleading since, as we shall see, there is seldom any indication of brain damage in these children. Furthermore, the label suggests a clear-cut syndrome, whereas in fact a wide range of problems is included in this definition. For these reasons, more neutral terms such as "specific developmental language disorder" or "specific language impairment" are frequently preferred (Bishop and Rosenbloom, 1987).

A major question of interest concerns the biological basis of specific language impairment. There are three main classes of explanation to be considered: (1) focal brain damage; (2) abnormal development of neuronal connections as a consequence of inadequate sensory input; and (3) abnormal neurological maturation due to prenatal influences on the developing brain.

Hypothesis 1: Focal Brain Damage

We know that, in adults, aphasia is a consequence of lesions of certain areas of the left hemisphere, so we might expect specific language problems in children to have a similar underlying cause. Although few children with language impairment suffer from neurological diseases or head injury in the first few years of life, the possibility remains that some degree of brain damage might be sustained around the time of birth.

Perinatal hazards

Pasamanick (1952) popularized the notion that brain injury sustained around the time of birth might cause a range of developmental disorders, with his hypothesis of a continuum of reproductive casualty. According to this hypothesis, prematurity and complications of pregnancy and delivery can lead to varying degrees of brain injury insufficient to cause death but sufficient to impair neurological function later in life. The advent of ultrasound has made it possible to study the neurological sequelae of prematurity and other perinatal hazards much more directly, and has confirmed that neurological impairment is particularly likely in those neonates with ultrasound indications of periventricular hemorrhage or other lesions (cf. Stewart et al., 1987).

Although it might seem reasonable to attribute developmental language disorders to this type of early brain injury, there is remarkably little evidence to support this hypothesis. Pasamanick and his colleagues were successful in using birth records to demonstrate a high rate of perinatal complications in a range of neurodevelopmental disorders, but when they turned their attention to speech impairments, they found no differences in perinatal histories between 290 speech-impaired children and those of a control group (Pasamanick et al., 1956). It should be noted, however, that they studied a heterogenous sample of children, including stutterers and those with voice disorders, and they did not distinguish specific language impairment from more general developmental delay.

Numerous studies have approached this issue from the other direction, by following the progress of infants who are subject to major perinatal hazards, e.g., children of very low birth weight. Results vary from study to study, not least because of changes in neonatal medicine that affect the management and survival of high risk infants. Although some studies have reported specific language defects in premature infants (e.g., Fitzhardinge and Steven, 1972), it is much more common to find reports of more general developmental delays affecting nonverbal as well as verbal skills. For instance, in a recent large-scale Canadian study of children of low birth weight a wide range of psychological tests was used (Dunn, 1986). The low birth weight group was significantly impaired relative to controls on language tests, but this was not a selective effect: nonverbal psychological tests showed a similar pattern, suggesting that when brain injury is sustained in the perinatal period it depresses all aspects of mental development to a similar extent.

Brain imaging techniques

Perinatal hazard is a very indirect index of neurological damage. A handful of studies have used brain imaging techniques such as CT scan to identify brain lesions in older children known to have language disorders. Practical and ethical considerations make such studies rare, and where they have been conducted they tend to involve children with particularly severe language difficulties, often with accompanying behavioral disorders. Caparulo et al. (1981) found markedly abnormal CT scans in 6 out of 16 language-disordered children. Three of the children with abnormal scans were described as having comprehension problems of some severity, and others had associated motor or behavioral problems. Dalby (1977a, b) described abnormal dilation of the temporal horns, especially on the left side, in 46 out of 87 language-disordered children, but this sample included many who were untestable or below normal limits on nonverbal IQ tests, as well as a substantial minority who were epileptic. When purer samples of children with more selective language disorders are studied, results are less striking. Rosenberger and Hier (1980) reported no brain lesions on CT scans in a sample of 53 learning-disabled children, which included many with selective verbal impairments. Harcherik et al (1985) studied only nine children with developmental language disorder, but their investigation is of particular value because they used blind analysis of CT scans and excluded children with additional neurological handicaps. They found no evidence of specific abnormalities in this group. Overall, it seems that developmental language disorder is seldom associated with CT scan abnormalities unless the child has additional neurological problems.

Electroencephalography

Results from electroencephalography depend on the type of language disorder studied. EEG abnormalities, typically in the form of spike-waves in the temporal lobes, are frequently found in those rare children who develop severe comprehension problems after a period of normal development, the so-called Landau-Kleffner syndrome of acquired epileptic aphasia (Landau and Kleffner, 1957). EEG abnormalities are not common in other types of congenital language disorder where there is no discontinuity in developmental progress (Waldo et al., 1978).

Outcome of children with demonstrable brain lesions

Another way of evaluating the brain-damage hypothesis is to consider the language development of children known to have sustained a clear-cut focal brain lesion early in life. The most extreme case one can consider is that of children who undergo hemidecortication. This operation involves surgical removal of the cortex on one side to control intractable seizures. Although surgery may not be performed until middle childhood or adulthood, the pathology is usually present from the first year of life. Early reports of cases undergoing hemidecortication for infantile hemiplegia noted the surprising absence of aphasia after left hemidecortication (Basser, 1962). Although intelligence was often depressed, verbal skills did not appear to be selectively affected. More recent claims that left hemidecortication is associated with a distinct pattern of subtle syntactic deficits (Dennis and Kohn, 1975; Dennis and Whitaker, 1976), have been questioned on methodological grounds (Bishop, 1983, 1988a).

If removal of the entire cortex of the left hemisphere does not lead to specific language deficits, it might seem reasonable to argue against small left hemisphere lesions as a cause of developmental language disorder. However, one needs to be cautious here. If there is a lesion of the left hemisphere that destroys the traditional language areas, then language develops in the right hemisphere (Rasmussen and Milner, 1977). It is possible, however, that a child might sustain a lesion that encroached on the language areas but was insufficient to bring about reorganization of cerebral lateralization. In that case, language would continue to be mediated by a damaged left hemisphere. If this were so, then one would expect to see signs of aphasia in children with early focal injuries of the left hemisphere.

Bishop (1988a) reviewed the literature on the outcome of hemiplegic children. Where the brain damage occurs before birth or during the first year of life, there are few differences in the language development of those with left and right hemiplegia. As in hemidecortication, the usual outcome with lesions of either hemisphere is that general intelligence is somewhat depressed, but selective verbal deficits are rare, and if and when language difficulties are found, these do not resemble the types of disorder seen in children with specific language impairment.

Can the hypothesis explain associated characteristics of specific language impairment?

There is one more line of evidence that can be used when evaluating causal hypotheses, and that is to consider how far a postulated causal factor can account for associated characteristics of specific language impairment. One characteristic of relevance to the brain-damage hypothesis is the motor impairment frequently described in language-impaired children (Bishop,1990a; Bishop and Edmundson, 1987; Johnston et al., 1981; Robinson, 1987). At first glance, the association of motor impairment with language difficulties might seem to support an explanation in terms of underlying brain damage. However, Johnston et al. (1981) and Bishop (1990a) noted that the motor deficits that are found are typically not lateralized and they involve a slow rate of motor movements, rather than abnormalities of reflexes or tone. They may therefore be regarded more as indications of neurological immaturity than of focal damage.

Another characteristic of specific language impairment is that males are far more likely to be affected than females. To reconcile this fact with an etiological explanation in terms of brain damage one would have to argue either that early brain damage was more common in males than females, or that its consequences were different for the two sexes. There is indeed evidence that boys are more likely than girls to suffer from cerebral palsy (Rutter et al., 1970) and they are at greater risk of having very low birth weight (Stewart et al., 1987). However, the sex difference is not substantial enough to explain the large preponderance of boys with specific language impairment. There is some support for the notion that the outcome of brain damage differs for males and females. Taylor (1969) used data on sex differences in inception rate of epilepsy to argue that the male brain is more immature than the female brain and hence more vulnerable to certain types of insult early in life.

Hypothesis 2: Inadequate Sensory/Perceptual Input

There is ample evidence from animal experiments that in order to develop normally the brain needs appropriate sensory input. The best-known example of this comes from the study of development of the visual system of the cat (see Atkinson et al., 1982, for a review). Animals who are deprived of visual experience or given distorted visual input early in life develop aberrant patterns of neuronal interconnectivity that can be reversed only if normal stimulation is restored while the animal is still young. This work has led to the conclusion that unless adequate stimulation is received during a certain critical period of development, permanent perceptual deficits can result, even if normal stimulation is subsequently provided.

Fluctuating conductive hearing loss associated with secretory otitis media

Ruben and Rapin (1980) reviewed the literature on plasticity of development in the auditory system. They noted that animal experiments provided evidence for a critical period during which the immature organism must receive adequate auditory input in order for the auditory system to develop normally. If there is an analagous critical period in humans, then even mild conductive hearing losses early in life might have permanent detrimental effects on auditory processing. This would have immense practical significance, because secretory otitis media is a common treatable disease, affecting as many as 50% of preschool children in some populations that have been screened (Klein and Rapin, 1988). In this disease, fluid collects in the middle ear, often causing a conductive hearing loss of 20 to 30 dB. Traditionally, conductive losses of this magnitude have been regarded as unimportant, because in adults such a loss has little or no impact on functional comprehension. However, the suggestion has been raised that a mild loss of this kind may be important during a critical period while the child is still learning language.

One might imagine that it should be straightforward to investigate the role of otitis media in causing specific language impairment by comparing the prevalence of this disease in language-impaired and control children. However, there are immense methodological problems. The first problem is that we may fail to detect a difference between language-impaired and control children because the otitis media has resolved by the time the study is conducted. Otitis media is often asymptomatic: studies in which whole populations have been screened frequently find many children with previously undetected middle ear disease. This means that parents may have been unaware that their child was suffering from middle ear disease, so retrospective information about past medical history is bound to be unreliable.

The second methodological problem is that we may get spurious associations between middle ear disease and language impairment because the language-impaired child is more likely than other children to be fully investigated by an otolaryngologist. Bishop and Edmundson (1986) found that while parents of language-impaired children did not report any more episodes of otitis media than those of control children, the language-impaired children were significantly more likely to have attended a hospital outpatient department for investigation of otitis media, and to have been treated by insertion of grommets or tympanostomy tubes (see Fig. 1).

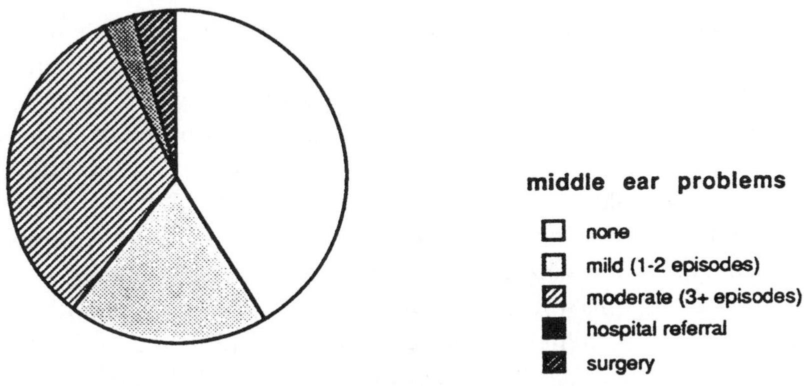

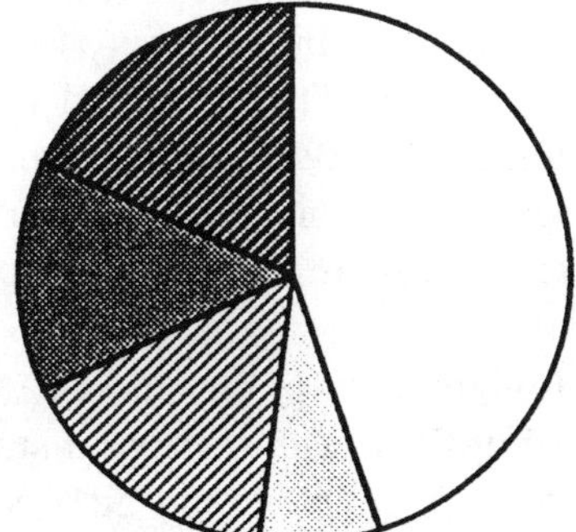

Figure 1. Data from Bishop and Edmundson (1986) based on retrospective parental report of history of middle ear problems in language-impaired and control samples. If children were referred to a hospital department and/or treated surgically, they are included in that category, irrespective of the number of episodes of otitis media.

One possible conclusion to draw from this finding is that language-impaired children are indeed more likely to have prolonged episodes of otitis media that interfere significantly with hearing. However, Bishop and Edmundson noted that we could not ignore an alternative explanation, namely that presence of a language disorder in a child is one factor that will prompt the physician to look for evidence of otitis media. Furthermore, if a child is found to have otitis media but is progressing normally, the physician is more likely to adopt a conservative treatment policy than if a child with otitis media is failing to develop language.

These methodological difficulties make it very difficult to come to firm conclusions on the basis of studies that compare the frequency of otitis media in language-impaired and control groups. They also throw into question the validity of numerous studies that have adopted the contrasting approach of comparing language attainments in children recruited from otolaryngology clinics with those of control children, e.g., the seminal study of Holm and Kunze (1969). Language disorder does not cause otitis media, but it will influence the detection and treatment of the disease, meaning that the clinic samples in such studies are likely to be biased toward including language-impaired children.

A much more decisive test of the hypothesis that otitis media causes specific language impairment is provided by studies in which a whole population is screened for otitis media at regular intervals so that children who have had several episodes of otitis media can be compared with a control group with no history of ear disease. Results of such studies have been conflicting, with some reporting verbal deficits in children with frequent episodes of otitis media (Silva et al., 1986; Teele et al., 1984), and others finding no evidence of any language impairment (Lous and Fiellau-Nikolajsen, 1984; Roberts et al., 1986). The important point to note, however, is that even where language deficits have been reported, these are very small in magnitude (see Bishop, 1988b). There is no evidence that otitis media is a sufficient cause for clinically-important language difficulties in children.

Bishop (1988b) argued that it is difficult to explain the sex ratio in specific language impairment in terms of otitis media, because a study by Biles et al. (1980) found that this disease affects boys and girls equally. However, a review of studies on risk factors by Chalmers et al. (1989) concluded that, although the data were very variable from study to study, there was some evidence that episodes of otitis media were more prolonged in boys than girls.

Another characteristic associated with specific language impairment that is hard to explain in terms of otitis media is motor impairment. However, it has been suggested that otitis media could affect motor development by interfering with vestibular function (Silva et al., 1982).

The evidence for otitis media as the main cause of specific language impairment is weak but the possibility remains that it may interact with other causal factors. Bishop and Edmundson (1986) found that, although perinatal hazards and neurological disease were not unduly common in a sample of language-impaired children, there was an unusually high frequency of cases in which a history of otitis media co-occurred with a high perinatal/neurological risk score (see Fig. 2).

Factors that were unrelated in control children were significantly associated in language-impaired children, suggesting that the combination of recurrent otitis media with hazardous perinatal history may be sufficient to cause language impairment, while either risk factor alone has little effect.

Inadequate language stimulation from parents

It is self-evident that in order for a child's brain to develop the capacity to process language, the child must be exposed to language. Mogford and Bishop (1988) noted that lay persons and professionals often suggest that developmental language problems are the result of inadequate language stimulation at home.

Despite the widespread belief in this explanation, research studies have provided little convincing evidence of differences in mother–child interaction between language-impaired and control children. Conti-Ramsden (1985) noted that such differences as have been described can be explained in terms of the mother adjusting her speech to the child's language level. Furthermore, in many families language-impaired children co-exist alongside highly articulate siblings. Perhaps the strongest evidence against a "language environment" explanation of specific language impairment comes from studies that assess language development in children who are subjected to an abnormal language input early in life. The clearest group to study are hearing children whose parents are congenitally deaf. In such cases, the mother may use very little intelligible spoken language to the child. Surprisingly, most children growing up in this situation show little evidence of any language delay, provided the child's own hearing is normal (see Schiff-Myers, 1988, for a review). Language appears to be remarkably robust, and able to develop despite severely restricted experience of normal speech.

The final piece of evidence against the "inadequate stimulation" hypothesis is that it cannot easily explain the sex ratio, nor can it account for the high rate of motor clumsiness in this population. While one would not want to deny that the quality and quantity of language heard by the child can affect language development, it does not seem realistic to attribute severe language disorders in children to inadequate stimulation, unless the verbal deprivation is very extreme indeed.

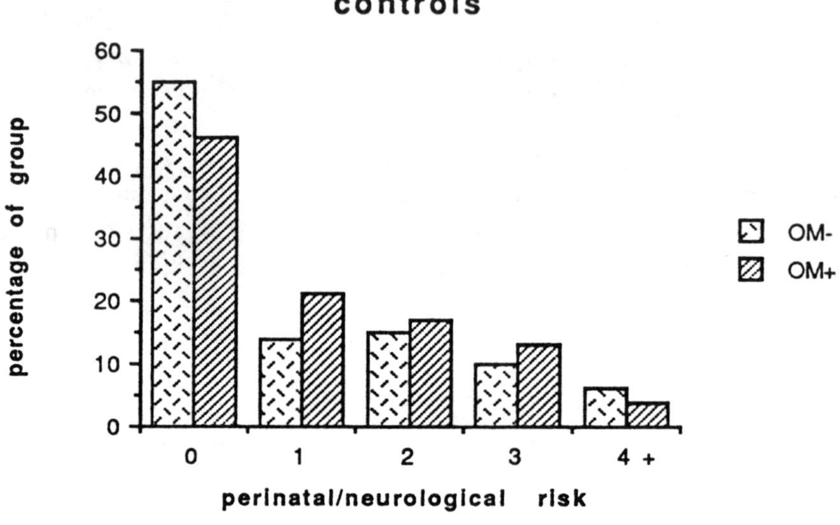

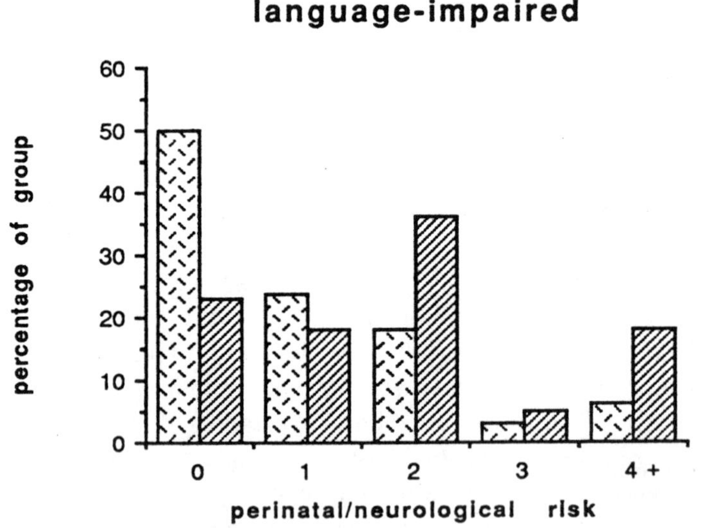

Figure 2. Associations between otitis media and perinatal/neurological risk in language-impaired and control samples. Children are coded as OM+ if they suffered from six or more episodes of otitis media and/or were referred to a hospital outpatient department for treatment of otitis media. The perinatal/neurological risk score is described by Bishop and Edmundson (1986).

Hypothesis 3: Abnormal Neurological Development

It has frequently been suggested that developmental disorders affecting spoken and written language may be the consequence of a delay or disturbance in the normal process of neurological maturation. Although this seems an attractive explanation, it has been difficult to test because of the lack of reliable procedures for investigating neurological development in living humans, and because the theory is too general and vague. However, with the advent of new techniques for analyzing the developing brain, there is some prospect of progress. Most of the studies that have been conducted have looked at individuals suffering from specific reading retardation ("developmental dyslexia"). Such studies may well be relevant to our understanding of children with specific language impairment, because at least some types of reading problems seem to be on a continuum with more severe disorders involving spoken language, and may represent the same condition at a later stage of development.

Rate of myelination

One promising approach involves the use of high-resolution magnetic resonance imagery (MRI) to study the development of myelination in the brains of living subjects (Levene et al., 1982). A study has been carried out using MRI scanning with individuals suffering from specific language impairment (Jernigan and Tallal, 1990; Tallal and Katz, 1989), but to date no data have been reported on patterns of myelination in affected children.

Morphological brain asymmetry

In the majority of people, the brain is physically asymmetrical. Studies of the parietal and temporal lobes have shown that areas of the left hemisphere known to be involved in language comprehension are typically larger than the corresponding regions on the right (Geschwind and Levitsky, 1968). Another type of morphological asymmetry is left occipital petalia, i.e., a more marked indentation of the skull that results from greater protrusion of the left occipital lobe, which can be identified from computerized tomography (CT scan) (LeMay, 1977). Several studies using CT scan or MRI to investigate the brains of individuals with developmental reading problems and specific language impairment have reported a reduction in morphological brain asymmetry (Haslam et al., 1981; Harcherik et al., 1985; Rosenberger and Hier, 1980; Rumsey et al., 1986; Tallal and Katz, 1989).

It is tempting to speculate that lack of morphological asymmetry in disordered groups reflects lack of functional asymmetry for language processing, especially since relationships between morphological and functional asymmetries have been demonstrated for both handedness and speech lateralization in samples without language disorders. However, a review by Bishop (1990a) concluded that such evidence as exists indicates that individuals with disorders of written or spoken language usually have left-hemisphere language representation. The problem is not that language is unlateralized, but that it is lateralized and mediated by an inadequate left hemisphere.

It was originally assumed that the larger size of the left planum temporale seen in the majority of people arose because the left hemisphere developed earlier and grew more rapidly than the corresponding area on the right. However, Galaburda et al. (1987) showed that in brains that lacked the usual asymmetry, the planum temporale was larger on both sides than in brains that were asymmetrical. They concluded that the normal asymmetry arose through more extensive cell loss in the right hemisphere, rather than from more rapid growth in the left.

Cytoarchitectonic studies

Galaburda et al. (1985) studied the cytoarchitectonic structure of brains of individuals who suffered from developmental dyslexia. In these brains they found evidence of abnormalities of neuronal migration, with disruption of the normal laminar structure of the cortex, and areas of focal dysplasia. Such results must be interpreted cautiously, given the lack of controls. The methodology needs to be improved by use of quantitative methods and blind analysis of control and experimental brains. Nevertheless, this is an exciting line of work, which one would like to see extended to the study of brains of individuals with specific language impairment.

Behavioral evidence

Rutter (1984) noted that if a developmental disorder is caused by a maturational lag, then one might expect it to disappear as the child grew older. Bishop and Edmundson (1987) set out to study the development of a group of language-impaired children over time to see whether some children did grow out of their difficulties. They found that many children who presented with a specific language impairment at four years of age did look entirely normal 18 months later, when assessed on a comprehensive battery of language tests. A further follow-up at eight years of age indicated that these children continued to make good progress at school (Bishop and Adams, 1990).

It had been anticipated that this subgroup of children might differ from the remainder of the sample in terms of associated etiological factors or pattern of language impairments, but the only differences that were found were quantitative ones: i.e., the children who had a good outcome had less severe language disorders when first assessed.

This study also provided insights into the nature of the motor clumsiness associated with specific language impairment. When first assessed, language-impaired children did poorly on a peg-moving task relative to a control group, but there was no evidence of lateralized impairment. However, the motor deficit changed over time, showing close links with language development. In those children who recovered from early language delay, the motor deficit also disappeared. Bishop and Edmundson argued that this result was consistent with the idea that the motor impairment reflected underlying immaturity of the nervous system, rather than the consequences of neuronal damage.

What Factors Influence Neurological Development?

We know that the early development of the brain can be affected by a range of prenatal influences, including nutritional status of the mother and exposure to teratogens or infections (see Sparks, 1984, for a review). The role of such influences in causing language disorders in children has barely been investigated. However, two further factors are the focus of current research interest: hormonal influences and genes.

Geschwind and Galaburda (1985) proposed a detailed explanation of the biological basis of developmental disorders of written and spoken language in which testosterone is seen as having a vital role in controlling the rate of neurological maturation. According to their theory, high rates of testosterone *in utero* delay neurological development and can lead to disturbances of neuronal migration, resulting in the focal dysplasias described above. Another postulated consequence is a reduction in the usual morphological and functional asymmetry of the cerebral hemispheres, which is thought to lead to developmental language disorders and left-handedness. This theory is consistent with the male preponderance in specific language impairment. However, individuals with adrenogenital syndrome, which leads to elevated levels of testosterone *in utero*, have normal verbal abilities (Pennington et al., 1987), whereas those with congenital androgen deficiency do not differ from normal controls in handedness or functional cerebral lateralization (Cappa et al., 1988). Furthermore, Bishop (1990a, b) noted that although language-impaired children tend to be clumsy, they do not have unusually high rates of left-handedness.

Genes are known to be important in determining the timing of developmental events, so it is plausible to suppose that any disruption of neurological maturation might have a genetic basis. It is rare to encounter instances of language disorders that follow a Mendelian pattern of inheritance, although an interesting family has recently been documented by Hurst et al. (1990) (see Fig. 3).

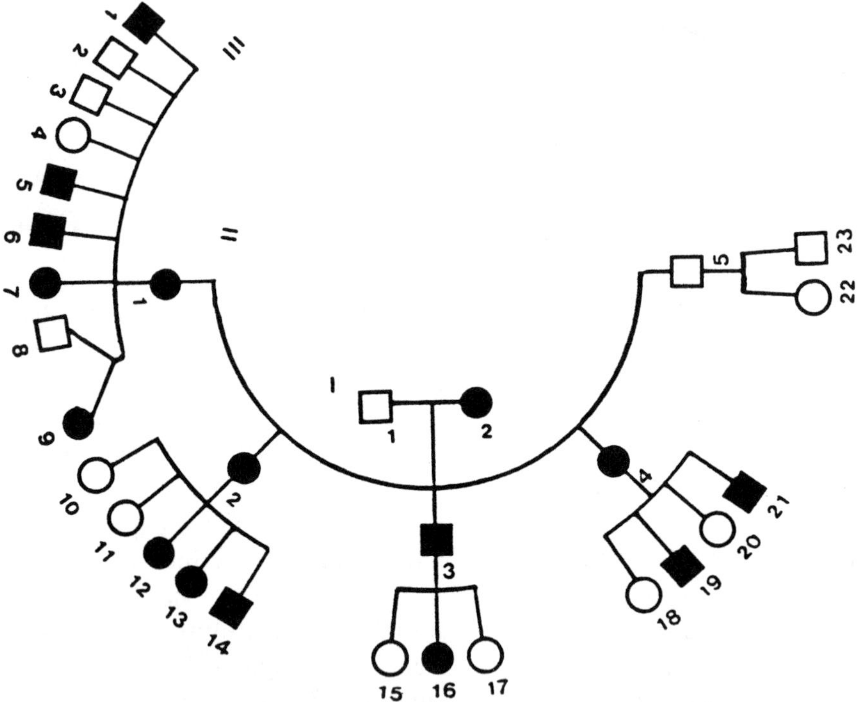

Figure 3. Pedigree of family affected by a severe specific language impairment. Affected members are shown in black. Reprinted from Hurst et al. (1990).

More commonly, one finds familial aggregation of disorders not fitting any obvious Mendelian pattern (Bishop and Edmundson, 1987; Lewis, 1990; Robinson, 1987; Tallal et al., 1989). I am currently carrying out a twin study with the aim of estimating the importance of a genetic contribution to specific language impairment by comparing concordance rates in identical and nonidentical same-sex twin pairs. Preliminary data on 30 twin pairs indicate that whereas dizygotic twins are sometimes discordant for specific language impairment, monozygotic twins are nearly always concordant.

Of course, a genetic explanation is not incompatible with the hormonal hypothesis; we do not know the mode of action of the genes involved, and it could be that they have an effect by influencing the sensitivity of target tissues to circulating androgens.

Geschwind (1983) noted that genetic explanations were for many years unfashionable, not least because they appeared to carry unpalatable eugenic overtones and were thought to imply therapeutic nihilism. However, as we gain a better understanding of the mode of action of different genes a more optimistic approach is warranted.

In conclusion, it seems likely that prenatal factors that affect the process of neurological maturation are more important in causing specific language impairment than either brain injury damage or inadequate stimulation. Genes appear to be important determinants of specific language impairment, although other factors, such as perinatal hazard and conductive hearing loss early in life, may assume importance if they affect a child who is already at risk.

References

Atkinson J, Barlow HB, Braddick O (1982): The development of sensory systems and their modification by experience. In: *The Senses*, Barlow HB, Mollon J, eds. Cambridge: Cambridge University Press

Basser LS (1962): Hemiplegia of early onset and the faculty of speech with special reference to the effects of hemispherectomy. *Brain* 85:427–460

Biles RW, Buffler PA, O'Donnell AA (1980): Epidemiology of otitis media: a community study. *American Journal of Public Health* 70:593–598

Bishop DVM (1983): Linguistic impairment after left hemidecortication for infantile hemiplegia? A reappraisal. *Quarterly Journal of Experimental Psychology* 35A:199–207

Bishop DVM (1988a): Language development after focal brain damage. In: *Language Development in Exceptional Circumstances*, Bishop DVM, Mogford K, eds. Edinburgh: Churchill Livingstone

Bishop DVM (1988b): Otitis media and developmental language disorder: a reply to Gordon. *Journal of Child Psychology and Psychiatry* 29:365–368

Bishop DVM (1990a): Handedness and developmental disorder. *Clinics in Developmental Medicine 110*: London: MacKeith Press/Blackwell Scientific

Bishop DVM (1990b): Handedness, clumsiness and developmental language disorders. *Neuropsychologia*, 28:681–690

Bishop DVM, Adams C (1990): A prospective study of the relationship between specific language impairment, phonological disorders and reading retardation. *Journal of Child Psychology and Psychiatry* 31:1027–1050

Bishop DVM, Edmundson A (1986): Is otitis media a major cause of specific developmental language disorders? *British Journal of Disorders of Communication* 21: 321-338

Bishop DVM, Edmundson A (1987): Specific language impairment as a maturational lag: evidence from longitudinal data on language and motor development. *Developmental Medicine and Child Neurology* 29:442–459

Bishop DVM, Rosenbloom L (1987): Classification of childhood language disorders. In: *Language Development and Disorders. Clinics in Developmental Medicine* (double issue), Nos. 101–2, Yule W, Rutter M, eds. London: MacKeith Press/Blackwell

Caparulo BK, Cohen DJ, Rothmann SL, Young JG, Katz JD, Shaywitz SE, Shaywitz BA (1981): Computed tomographic brain scanning in children with developmental neuropsychiatric disorders. *Journal of the American Academy of Child Psychiatry* 20:338–357

Cappa SF, Guariglia C, Papagno C, Pizzamiglio L, Vallar G, Zoccolotti P, Ambrosi B, Santiemma V (1988): Patterns of lateralization and performance levels for verbal and spatial tasks in congenital androgen deficiency. *Behavioural Brain Research* 31:177–183

Chalmers D, Stewart I, Silva P, Mulvena A (1989): *Otitis media with effusion in children—the Dunedin study.* Clinics in Developmental Medicine 108. Oxford: Blackwell Scientific; Philadelphia: Lippincott

Conti-Ramsden G (1985): Mothers in dialogue with language-impaired children. *Topics in Language Disorders* 5:58–68

Dalby MA (1977a): Aetiological studies in language retarded children. *Neuropaediatrie*, supplement to vol 8:499–500

Dalby MA (1977b): Diagnose und Aetiologie der Dysphasie. In *Dysphasie im Kindesalter*, Elstner W, Karlstad H, eds. Oslo: Universitetsforlatet

Dennis M, Kohn B (1975): Comprehension of syntax in infantile hemiplegics after cerebral hemidecortication: left-hemisphere superiority. *Brain and Language* 2:472–482

Dennis M, Whitaker H (1976): Language acquisition following hemidecortication: linguistic superiority of the left over the right hemisphere. *Brain and Language* 3:404–433

Dunn HG, ed. (1986): *Sequelae of low birthweight: the Vancouver study.* Clinics in Developmental Medicine 95–96. Oxford: Blackwell Scientific/ Philadelphia: Lippincott

Fitzhardinge PM, Steven EM (1972): The small-for-date infant: II. Neurological and intellectual sequelae. *Pediatrics* 50:50–57

Galaburda AM, Sherman GF, Rosen GD, Aboitiz F, Geschwind N (1985): Developmental dyslexia: four consecutive patients with cortical anomalies. *Annals of Neurology* 18:222–233

Galaburda AM, Corsignlia J, Rosen GD, Sherman GF (1987): Planum temporale asymmetry, reappraisal since Geschwind and Levitsky. *Neuropsychologia* 25:853–868

Geschwind N (1983): Genetics: fate, change, and environmental control. In: *Genetic Aspects of Speech and Language Disorders.* Ludlow CL, Cooper JA, eds. New York: Academic Press

Geschwind N, Galaburda A (1985): Cerebral internalization. Biological mechanisms, associations and pathology. I, II, and III. *Archives of Neurology* 42:428–459; 521–552; 634–654

Geschwind N, Levitsky W (1968): Left-right asymmetries in temporal speech region. *Science* 161:186–187

Harcherik DF, Cohen DJ, Ort S, Paul R, Shaywitz BA, Volkmar F R, Rothman SLG, Leckman JF (1985): Computed tomographic brain scanning in four neuropsychiatric disorders of childhood. *American Journal of Psychiatry* 142:731–734

Haslam RHA, Dalby JT, Johns RD, Rademaker AW (1981): Cerebral asymmetry in developmental dyslexia. *Archives of Neurology* 38:679–682

Holm VA, Kunze LH (1969): Effect of chronic otitis media on language and speech development. *Pediatrics* 43:833–839

Hurst JA, Baraitser M, Auger E, Graham F, Norell S (1990): An extended family with a dominantly inherited speech disorder. *Developmental Medicine and Child Neurology* 32:352–355

Jernigan TL, Tallal P (1990): Late childhood changes in brain morphology observable with MRI. *Developmental Medicine and Child Neurology* 32:379–385

Johnston RB, Stark RE, Mellits ED, Tallal P (1981): Neurological status of language-impaired and normal children. *Annals of Neurology* 10:159–163

Klein S, Rapin I (1988): Intermittent conductive hearing loss and language development. In: *Language Development in Exceptional Circumstances*, Bishop DVM, Mogford K, eds. Edinburgh: Churchill Livingstone

Landau W, Kleffner F (1957): Syndrome of acquired aphasia with convulsive disorder in children. *Neurology* 7:523–530

LeMay M (1977): Asymmetries of the skull and handedness. *Journal of the Neurological Sciences* 32:243–253

Levene MI, Whitelaw A, Dubowitz V, Bydder GM, Steiner RE, Ranell CP, Young IR (1982): Nuclear magnetic resonance imaging of the brain in children. *British Medical Journal* 285:774–776

Lewis BA (1990): Familial phonological disorders: four pedigrees. *Journal of Speech and Hearing Disorders*: 55:160–170

Lous J, Fiellau-Nikolajsen M (1984): A 5-year prospective case-control study of the influence of early otitis media with effusion on reading achievement. *International Journal of Pediatric Otorhinolaryngology* 8:19–30

Mogford K, Bishop DVM (1988): Five questions about language acquisition considered in the light of exceptional circumstances. In: *Language Development in Exceptional Circumstances*, Bishop DVM, Mogford K, eds. Edinburgh: Churchill Livingstone

Pasamanick B (1952): Patterns of research in mental hygiene. *Psychiatric Quarterly* 26:577–589

Pasamanick B, Constantinou F, Lilienfeld AM (1956): Pregnancy experience and the development of childhood speech disorders. *American Journal of Diseases of Children* 91:113–118

Pennington BF, Smith SD, Kimberling WJ, Green PA, Haith MM (1987): Left-handedness and immune disorders in familial dyslexics. *Archives of Neurology* 44:634–639

Rasmussen T, Milner B (1977): The role of early left-brain injury in determining lateralization. *Annals of the New York Academy of Science* 299:255–269

Roberts JE, Sanyal MA, Burchinal MR, Collier AM, Ramey CT, Henderson FW (1986): Otitis media in early childhood and its relationship to later verbal and academic performance. *Pediatrics* 78:423–430

Robinson RJ (1987): Introduction and overview. In: *Proceedings of the First International Symposium on Specific Speech and Language Disorders in Children.* London: Association for All Speech Impaired Children

Rosenberger PB, Hier DB (1980): Cerebral asymmetry and verbal intellectual deficits. *Annals of Neurology* 8:300–304

Ruben RJ, Rapin I (1980): Plasticity of the developing auditory system. *Annals of Neurology* 89:303–311

Rumsey JM, Dorwart R, Vermess M, Denckla MB, Kruesi MJ, Rapoport JL (1986): Magnetic resonance imaging of brain anatomy in severe developmental dyslexia. *Archives of Neurology* 43:1045–1046

Rutter M (1984): Issues and Prospects in Developmental Neuropsychiatry. In: *Developmental Neuropsychiatry*, Rutter M, ed. Edinburgh: Churchill Livingstone

Rutter M, Graham P, Yule W (1970): *A neuropsychiatric study in childhood.* Clinics in Developmental Medicine 35–36. London: SIMP

Schiff-Myers N (1988): Hearing children of deaf parents. In: *Language Development in Exceptional Circumstances*, Bishop DVM, Mogford K, eds. Edinburgh: Churchill Livingstone

Silva P, Kirkland C, Simpson A, Stewart IA, Williams SM (1982): Some developmental and behavioral problems association with bilateral otitis media with effusion. *Journal of Learning Disabilities* 15:417–421

Silva PA, Chalmers D, Stewart I (1986): Some audiological, educational and behavioral characteristics of children with bilateral otitis media with effusion —a longitudinal study. *Journal of Learning Disabilities* 19:165–169

Sparks SN (1984): *Birth Defects and Speech-Language Disorders.* San Diego: College Hill Press

Stewart AL, Reynolds EOR, Hope PL, Hamilton PA, Baudin J, Costello AM, Bradford BC, Wyatt JS (1987): Probability of neurodevelopmental disorders estimated from ultrasound appearance of brains of very preterm infants. *Developmental Medicine and Child Neurology* 29:3–11

Tallal P, Katz W (1989): Neuropsychological and neuroanatomical studies of developmental language/reading disorders: recent advances. In: *Brain and Reading*, von Euler C, Lundberg I, Lennerstrand G, eds. New York: Stockton Press

Tallal P, Ross R, Curtiss S (1989): Familial aggregation in specific language impairment. *Journal of Speech and Hearing Disorders* 54:167–173

Taylor DC (1969): Differential rates of cerebral maturation between sexes and between hemispheres. *Lancet* 2:140–142

Teele DW, Klein JO, Rosner BA, and The Greater Boston Otitis Media Study Group (1984): Otitis media with effusion during the first three years of life and development of speech and language. *Pediatrics* 74:282–287

Waldo MC, Cohen DJ, Caparulo BK, Young JG, Prichard JW, Shaywitz BA (1978): EEG profiles of neuropsychiatrically disturbed children. *Journal of the American Academy of Child Psychiatry* 17:656–670

Brain/Cognition Relationships: Interpreting the Effects of Focal Brain Damage

Rita Sloan Berndt

The normal human brain carries out a large number of complicated cognitive activities. This volume discusses, from a number of different perspectives, 'exceptions' to the normal state. That is, it considers a variety of ways in which cognitive activity can be affected by interruption of the maturational unfolding of the brain and by the physiological changes that accompany aging and disease. An important, though often neglected, requirement of research addressing these 'exceptions' is the understanding of the normal, mature relationship between the brain and cognitive activity. There must exist a "paradigm case" toward which the system evolves during development and from which it departs in conditions of pathology—a normal baseline against which immature and pathological conditions can be compared. There are a variety of methodologies available that have provided information about this paradigm case, i.e., about normal brain/cognition relationships.

Some types of cognitive functions can be investigated in animal subjects, allowing the possibility of mapping very specific components of the cognitive system to well-defined neuroanatomical sites probed by means that cannot be applied to humans. This approach is demonstrated for the memory system described in the present volume by Bachevalier and Mishkin, and has also been fruitfully applied to higher visual processes (e.g., Ungerleiter and Mishkin, 1982) and to attentional mechanisms (e.g., Petersen et al., 1987). In all of these cases, the expectation is not only that the areas isolated in primate brain have analogues in human brain, but that the functional fractionation of the cognitive components at issue will describe the human system as well (see, for example, Posner and Petersen, 1990).

At least some of the findings that have emerged from these animal studies have been shown to provide a reasonable model of human brain/cognition correlations in research using Positron Emission Tomography (PET), in which regional cerebral metabolic activity can be identified while normal human subjects perform specific cognitive tasks (e.g., Corbetta et al., 1990). In fact, imaging studies of brain metabolic activity promise to provide useful and very specific information in the future about the functional role of specific brain regions in a variety of cognitive activities (Posner et al., 1988.) To date, however, PET scanning and other indices of metabolic activity in normal humans have not provided much more than intriguing glimpses of differential brain activity during simple tasks. Our assumptions about brain/cognition relationships come from other sources, primarily two quite different types of studies: experiments inducing lateral asymmetries in normal subjects, and investigations of the effects of focal brain injury on cognition.

Several experimental methodologies have been developed for investigating the extent to which each of the cerebral hemispheres is preferentially engaged in the normal performance of specific tasks. In visual hemifield experiments, a simple stimulus is presented very briefly to one of the visual fields, and it is assumed that information conveyed by the stimulus is processed by the contralateral hemisphere prior to any information cross-over to the other side. In dichotic listening studies, competing auditory stimuli are presented to each ear; the subjects' biases in reporting the information received by each ear is assumed to represent an index of the extent that the stimulus is preferentially processed in the hemisphere contralateral to the side of stimulation (see Springer and Deutsch, 1981, for review of these methods). A large number of studies have been conducted using these paradigms, and despite criticisms of the methods and identification of numerous problems in interpretation (Blumstein et al., 1975; Sergent, 1983, 1988), they continue to provide the basis for many of our assumptions about the role of the two hemispheres in cognitive processes.

Although these studies of hemispheric processing in normal subjects have yielded some interesting generalizations about cerebral asymmetries, an even more influential source of information about the relative contributions of the two hemispheres to cognition has come from studies of the effects of focal brain damage on the cognitive system. Observations of the types of symptoms that occur following unilateral cortical damage have led to the view that many cognitive functions are asymmetrically executed in the brains of most individuals. It has been argued on the basis of the effects of focal lesions, for example, that the right hemisphere is responsible for performing certain types of visuospatial functions (Benton, 1985), for interpreting and producing emotional "tone of voice" (Heilman et al., 1985), and for carrying out some of the operations involved in producing or appreciating music (Gates and Bradshaw, 1977), while the left hemisphere is specialized for language (Broca, 1861), skilled motor movements (Heilman et al., 1983), and the generation of visual images (Farah, 1984).

Many other studies of the effects of focal brain damage have attempted to provide more detailed maps of brain/function relations (Damasio and Damasio, 1989; Geschwind, 1965), which have required not only more detailed neuroanatomical information, but also better specification of the cognitive activity that the brain damage has compromised. This chapter will consider the assumptions underlying the interpretation of the cognitive effects of focal lesions, and will attempt to illustrate some of the difficulties inherent in this methodological approach.

Normal Brain Organization, Normal Cognitive Function, and the Brain-damaged Patient

The critical assumptions underlying the lesion approach to brain/cognition correlation have been discussed in some detail (Damasio and Damasio, 1989; Caramazza, 1986; Saffran, 1982). A few of the most crucial issues will be illustrated here. One overriding problem encountered when studying the effects of focal lesions on the cognitive system is the uncertain and often misinterpreted role of the symptom in the structure/function correlation. As stated most conservatively by Damasio and Damasio: "major risks exist in the intellectual exercise necessary to conceptualize normal structure and function from damaged structure and deviant function...." (1989, p. 17). The behavioral manifestations of brain damage, i.e., the abnormal functioning of previously normal adults, is assumed to reflect some aspects of the normal cognitive system, with other aspects removed or disrupted. Sorting out the impaired processes from the spared processes is a difficult task, and among other things, appears to require a model of how the normal system was organized prior to the brain insult. Such models are frequently absent, or are inadequate to the task of providing detailed predictions of how symptoms might reflect a disruption of the premorbid system. Moreover, most of the symptoms that we attempt to interpret—those evident on neurological or neuropsychological examination—involve complex behaviors that might exploit numerous cognitive operations in their normal exercise. Thus, quite different types of deficits to the normal system might produce an identical symptom.

The successful interpretation of the effects of a focal lesion on cognitive function appears to rely on the specification of two separate elements on each side of the structure/function equation. For each side of the basic equation

Equation 1:

$$\text{Brain region} \approx \text{Cognitive function}$$

one theoretical and one methodological element must be specified, i.e.,

Equation 2:

$$\text{Neuroanatomical model} + \text{localization technique} \approx \text{cognitive model} + \text{neuropsychological tests and controls.}$$

On the "brain" side of equation 2, a successful interpretation of the cognitive effects of focal brain lesions requires specification of the neuroanatomical model that is assumed, including details about the size of the unit that is under study (e.g., hemisphere, lobe, "functional subsystem", etc.), and the relationship that is assumed to obtain between and among various elements. In addition, some means must be available for localization of the lesion (autopsy, neuroimaging, etc.) that provides information at the level of description required by the target model.

On the "cognition" side of the equation, a description of the cognitive function under study must be articulated that considers, minimally, the separable components that normally participate in the execution of that function, and the ways in which each target component might interact with other cognitive processes and representations. Behavioral tests must be available, or must be developed, that probe patients' symptoms to disclose the relative involvement and sparing of each element within the cognitive model.

This kind of analysis makes clear that progress in one of the four components of the expanded equation, without concomitant progress in the other components, although it may be of interest in its own right, will not by itself result in better understanding of how cognitive processes are carried out by the normal brain. Thus, for example, recent technological improvements that allow more precise localization of brain lesions will not lead to better understanding of structure/function relationships without considerable attention to the theoretical and methodological requirements on the "cognitive function" side of the equation.

Hemispheric/Cognitive Function: Generating Mental Images (An Example)

When the neuroanatomical level of description adopted for study is the hemisphere, the lesion methodology can be supplemented by other types of techniques (mentioned above) in attempting to localize function to one side of the brain. Thus, focus on the cerebral hemisphere as the neuroanatomical unit of analysis simplifies the "brain" side of equation 2 relative to more precise attempts at localization, to be discussed below. Nonetheless, using lesion studies to infer lateralization of a cognitive function is not without pitfalls.

The example used to illustrate some of these problems is the cognitive function of "generating a mental image," and it is chosen for several reasons. At first glance, it appears to be a relatively encapsulated function that could, if impaired, give rise to measurable symptoms; in addition, its lateralization has been the topic of some debate and is not yet settled. For many years, the ability to generate and manipulate mental images was assumed to be a function of the right hemisphere (Springer and Deutsch, 1981), perhaps because it is a function that seems inextricably linked to ("right hemisphere") basic visual/spatial processes. Several recent reviews of the evidence in support of a right hemisphere privilege for image generation have raised questions about this assumption, however. Reviews of the lesion data have led to the arguments that both hemispheres participate in the generation of mental images (Ehrlichman and Barrett, 1983), and that image generation is actually a left hemisphere function (Farah, 1984). Thus, all possible positions appear to be supportable from the evidence available from brain-damaged patients.

It is informative to consider how well the available evidence meets the minimal requirements of the equation set out above. On the "cognitive function" side, the requirement for a model of the target function appears to be met, in fact, unusually well. Kosslyn (1987) and Farah (1984) have each proposed detailed models of the component processes involved in normal mental imagery, and have developed experimental tests that putatively dissociate the various components from one another during normal processing. But what about the second element on the functional side of the equation? Tests must be available which can be performed by brain-damaged subjects that allow unambiguous attribution of symptoms to a particular component of the model, in this case, generation of a mental image. Although such tests have been developed for use with normal subjects, and have been employed successfully in tests of patients with focal brain injury (Kosslyn et al., 1985) and with surgically severed corpus callosum (Kosslyn et al., 1985), most of the lesion studies on which the hypothesis of left hemisphere specificity for image generation is based do not report tests that are adequately controlled to provide clear evidence of the source of patients' symptoms.

The primary problem is that the cognitive function at issue—generating a mental image—clearly requires intact abilities in other cognitive domains (e.g., memory, visual perception) that are themselves subject to impairment in conditions of focal brain damage. The claim that a patient suffers from an image generation deficit requires the demonstration that impaired performance on tests of image generation does not result from impairments to these other processes (Farah, 1984). Few of the cases that motivate claims for either a left or a right hemisphere superiority for image generation allow the assumption that these other, related cognitive functions are intact (see Sergent, 1990b, for review.) The situation is equally unclear on the "brain" side of the equation, as much of the evidence for unilateral image generation is based on post-hoc consideration of patients whose lesions (as well as their symptoms) are poorly described.

Because the neuroanatomical level of description at issue here is the hemisphere, rather than more specific regions within a hemisphere, there are other methods available to address the question. These techniques have also yielded conflicting information with regard to the hemisphere that is specialized to generate mental images. As noted above, experimental techniques have been developed for use with normal subjects that suggest preferential participation of each of the cerebral hemispheres in specific cognitive activities. Farah (1986) reported results of a tachistoscopic lateralized visual discrimination task with normal subjects that supported left hemisphere specialization for image generation. Sergent (1989), using a different task but the same basic experimental paradigm, found evidence for superior image generation in the right hemisphere. Similarly contradictory results have been found in studies of the disconnected cerebral hemispheres of patients who have undergone surgical resection of the forebrain commissures ("split brain" patients). Kosslyn et al., (1985) and Farah et al. (1985) have reported left hemisphere superiority for image generation in two split brain patients, while Corballis and Sergent (1988) found some evidence for right hemisphere superiority in another such patient.

This brief review indicates that unambiguous evidence supporting strict lateralization of a cognitive function to one of the cerebral hemispheres is difficult to accumulate, even when one of the variables required by the brain/function equation is well specified. In the case of the localization of the ability to generate visual images, the component cognitive elements required for the function have been clearly defined. Nonetheless, attempts to localize those elements to one or the other of the hemispheres have led to contradictory results for all three of the methodological approaches that have been used to address the issue. These contradictions may arise because of failure to meet adequately the requirements set out above for equating brain regions to cognitive function. Alternatively, as argued by Sergent (1990a), the very idea that one of the hemispheres is uniquely adapted to carry out all of the activities required for a complex function may be incorrect. This last possibility is explored in the Discussion section, below.

Specific Regions within a Hemisphere: Language Processing (An Example)

The complexities that are evident in attempts to attribute even well-described cognitive processes to one of the cerebral hemispheres are exacerbated when the goal is to isolate more specific functions within a hemisphere. Localization of cognitive functions within a hemisphere cannot rely on a convergence of information from a variety of sources; studies of split brain patients and methods that probe hemispheric activity selectively in normal subjects are not relevant to the task at hand. When the cognitive function at issue is unique to humans, animal models cannot provide a framework upon which hypotheses can built, and against which they can be tested. In fact, for more than 100 years, the only source of information about the relationship of many cognitive functions to specific regions within a hemisphere has come from studies of patients with focal brain lesions. Identification of the appropriate neuroanatomical regions, as well as the componential description of the cognitive operations that are compromised after brain injury, are made more difficult because of the absence of converging data from other sources that might constrain the development of hypotheses. Despite these problems, fairly specific models have been offered that link circumscribed neural structures to particular cognitive abilities (Damasio and Damasio, 1989; Geschwind, 1965; Heilman et al., 1983; Roeltgen, 1985), based on the symptoms of patients with focal brain lesions.

The cognitive system that will be used to examine the methods employed for localizing function within a hemisphere is the auditory–verbal language system. Language is chosen for discussion here for several reasons. First, a well-accepted neuroanatomical model exists that attributes specific language components to circumscribed regions within the left hemisphere. Second, language has been studied from a sufficient number of perspectives (e.g., linguistics, psycholinguistics) that an explicit cognitive model of the language system should be within reach. Third, the symptoms that can be used to link brain regions and language components (i.e., the symptoms of aphasia) are readily apparent (for the most part) and have been subjected to analysis for more than a century. Fourth, language is a uniquely human function that cannot be approached in experiments with animal subjects. Thus, the language system would seem to provide optimal possibilities for the mapping of cognitive functions to specific brain areas using the lesion method.

Contemporary views of the organization of the language system in the brain assume that the left hemisphere of most individuals is the site of virtually all activity. The first evidence that the left cerebral hemisphere plays a privileged role in the processing of language is usually credited to Pierre Paul Broca (1861), who reported the occurrence of aphasia in a series of patients following focal left hemisphere damage. Although recent claims have been made that the right hemisphere also plays a role in some aspects of language processing (e.g., Chiarello, 1988; Ross, 1981), the specialization of the left hemisphere for most aspects of language is supported by the higher probability of aphasia following left than right hemisphere injury.

Recent studies of brain-damaged patients have confirmed the importance of an intact left hemisphere in language processing, regardless of the characteristics of the language involved or the modality in which it is expressed. Thus it appears that left hemisphere damage leads to aphasia with roughly the same probability regardless of the structural properties of the language (Menn and Obler, 1990). Even more compelling is evidence that congenitally deaf adults whose native language is expressed in the visual/spatial modality (American Sign Language) become aphasic after focal left hemisphere injury in the same proportions as do speakers of languages expressed in the auditory–verbal modality (Poizner et al., 1987). Thus it would appear that the left cerebral hemisphere of most individuals is indeed specialized for the execution of language.

More specific claims about the functions of selected regions within the left hemisphere in the processing of language are also long-standing, dating from Wernicke's (1874) description of the different types of aphasia that occur following damage to different parts of the left hemisphere. This traditional neuroanatomic model of language representation was updated and elaborated by Geschwind (1965) to provide an account of several different aphasic syndromes (as well as other cognitive deficits following focal brain injury). The Geschwind "disconnection" model has survived relatively unchanged during a twenty-five year period of unprecedented advancement in neural imaging technology, and remains the standard model of language organization that is taught to neurologists and neuroscientists. For example, Fig. 1 is taken from the second edition of a popular American neuroscience textbook (Kandel and Schwartz, 1985). The cortical regions involved in a simple language task such as naming an object are described as follows:

> ...naming an object is thought to involve transfer of visual information from the retina to the lateral geniculate nucleus, and from there to the primary visual cortex, Brodmann's area 17; the information then travels to higher visual cortex, area 18, and then to a specific region (the angular gyrus—Brodmann's area 39) of the parietal-temporal-occipital association cortex. This region is thought to be concerned with the association of different sensory modalities—visual, auditory, and tactile. From here the information is projected to Wernicke's area in the superior posterior temporal lobe (Brodmann's area 22), where the auditory perception of the word is formed and where memory images are stored that represent the word as seen or heard. Once the pattern is formed, it is conveyed, by means of the arcuate fasciculus, to Broca's area, where the perception of language is translated into the grammatical structure of a phrase and where the memory for word articulation is stored. This information about the sound pattern of the phrase is then conveyed to the facial area of the motor cortex, which controls articulation, so that the words that name the visual object can be spoken. (Mayeux and Kandel, 1985, pp. 693–694.)

The evidence upon which this detailed model is based comes entirely from studies of the effects of focal brain injury on the language performance of previously-normal adult speakers. The assertion that Wernicke's area (area 22), for example, is the area "where the auditory perception of the word is stored" follows from descriptions of the primary symptoms of patients with focal damage to this area, which appear to involve the processing of the sounds of the language. The idea that Broca's area is involved in translating words into grammatical structures, and in storing memories about how words are articulated, follows from observations that patients with damage to Broca's area show symptoms that involve articulation and apparent failure of grammatical operations. Thus, the symptoms found in different varieties of aphasic disorders have been rather directly interpreted as descriptors of the complex cognitive operations that are localized in circumscribed areas of the cerebral cortex.

Let us analyze these claims in light of the requirements set out above (equation 2) for mapping brain regions to cognitive function. In order to simplify the task, we will focus on the putative functions of Broca's area. An equation might be written as follows to express the data base on which the relationship between Broca's area and its related language functions are predicated:

Equation 3:

$$\text{Broca's area lesion} \approx \text{Broca's aphasia symptoms.}$$

On the "brain" side of the equation, the neuroanatomical area is the frontal operculum, the third convolution in the frontal lobe. On the "cognitive function" side of the equation we have only the symptoms that have been described as Broca's aphasia; it remains to be seen whether or not a cognitive model, along with tests and controls, can be offered that allow interpretation of the symptoms of Broca's aphasia as indicating impairments to specific cognitive functions. The first step is to define the symptoms that constitute Broca's aphasia, and the second is to assure that a lesion to Broca's area is necessary and sufficient to produce that set of symptoms.

The description of Broca's aphasia offered by Mayeux and Kandel (1985) includes the following:

> ...Comprehension is usually preserved, but language output is impaired and nonfluent (verbal output is decreased). This ranges from almost complete muteness to a slowed, deliberate speech that uses very simple grammatical structures. Patients with Broca's aphasia use only key words. They express nouns only in the singular, verbs in the infinitive or participle, and they often eliminate articles, adjectives, and adverbs altogether. For example, a patient with Broca's aphasia when asked to say 'the large gray cat' may say 'gray cat'.
>
> Psycholinguistic studies of patients with Broca's aphasia have shown that these omissions are even more dramatic in more complex sentences. Here we can see the second characteristic of this defect: a breakdown in the construction and coordination of constituent phrases.... (Mayeux and Kandel, 1985, pp. 695.)

Despite the complexity of this description, and the fact that the symptoms appear to involve aspects of language processing (speech motor control, syntactic processing) that are not intuitively related to one another, the assumption has been that the typical co-occurrence of these symptoms indicates that all of the components whose impairment gives rise to these symptoms are localized in the same brain region.

Empirical support for this traditional view of the functions of Broca's area is scanty. It has been known for many years that the full syndrome of Broca's aphasia (described above) is caused, not by cortical damage limited to Broca's area, but by a much more extensive lesion involving large portions of the perisylvian language areas with considerable subcortical extension (Mohr, 1976). When lesions are limited to the classical Broca's area, symptoms are mild and transient (Mohr, 1976; Tonkonogy and Goodglass, 1981). Alexander and colleagues have reported detailed CT-scan correlates of aspects of Broca's aphasia symptoms in nine patients with lesions in the region of left frontal operculum (Alexander et al., 1990). These authors have identified a variety of symptom patterns (including "classic" Broca's aphasia) that appear to be related to specific lesion extensions. These patterns, summarized in Table 1, are interpreted as an indication that the syndrome of Broca's aphasia is made up of identifiable, and dissociable, constituent elements, each one of which may relate to a specific neural structure or set of structures.

Table 1

Lesioned Area(s) on CT-Scan	Symptom(s) produced
frontal operculum only	reduced output; failure to initiate; intact repetition; disturbed articulation
lower motor cortex only	articulation disturbance; dysprosody;
subcortical frontal areas only (periventricular white matter)	phonemic paraphasias; articulation disturbance
frontal operculum, lower motor cortex, and subcortical areas	*acute* severe nonfluent aphasia affecting speech, repetition, writing
above three areas, *plus* limbic-frontal periventricular pathways (medial subcollosal fasciculus)	classic, chronic Broca's aphasia

Summary of lesion/symptom correlations found in CT-scan study of nine cases of aphasia following lesions in the vicinity of "Broca's area" (adapted from Alexander et al., 1990).

In this study, the lesion responsible for classic, chronic Broca's aphasia was again shown to involve much more than Broca's area and included substantial involvement of subcortical structures. It would appear, therefore, that the "brain" side of the equation mapping Broca's area to speech production has inappropriately delimited the neuroanatomical regions that normally participate in this activity.

The analysis of Alexander and co-workers clearly indicates problems on the "cognitive function" side of the equation as well. The syndrome that has been described as Broca's aphasia appears to involve dissociable symptoms, each of which may reflect one or more deficits to components of the language system. Even if we focus only on the ability to produce speech sounds (and ignore, for the moment, the "grammatical elements") at least three separate symptoms can be distinguished: failure to initiate speech; disturbance of motor execution (dysarthria); and substitutions among speech sounds (phonemic paraphasias). Each of these appears to be related to different lesion sites and extents, and undoubtedly represents a functional impairment to a different component of the language production system. To date no unified cognitive model of speech production has been offered that allows unambiguous attribution of these symptoms to independent components that could be linked to the neuroanatomical sites identified.

Most of the effort to develop the "cognitive function" side of equation 3 has focused on the grammatical symptoms described in patients with Broca's aphasia. As noted in the description of this syndrome (quoted above), such symptoms have been said to involve omission of some kinds of words (articles, adjectives and adverbs) and the grammatical endings that allow nouns to be expressed in other than the singular, or verbs in finite form, as well as structural simplification of sentences. As recently as ten years ago, these symptoms were interpreted to be a manifestation of a single impairment of syntactic processing abilities, and, therefore, potentially localization in the brain (e.g., Berndt and Caramazza, 1980, 1981). More recent analyses have indicated that these symptoms can dissociate from one another—for example, the structural simplification of sentences can occur without the omission of grammatical elements (Berndt, 1987). Further, quantitative analysis of the narrative speech of these patients demonstrates that individual patients show quite idiosyncratic patterns of grammatical symptoms, with one patient tending toward frequent omission of one type of grammatical element (e.g., verb endings) and retention of another (e.g., articles), while a second, superficially similar patient shows an exactly opposite pattern of impairment (Saffran et al., 1989).

This variability in the precise form that grammatical symptoms can take in Broca's aphasia, along with other dissociations among symptoms that have been assumed to occur together necessarily (see Berndt, 1991, for review), have led to the view that the classical aphasia categories describe groups of patients who differ from one another in important ways that indicate very different underlying impairments to the language system (Caramazza, 1986). Within this view, attempts to localize cognitive functions by correlating the lesion sites and symptoms of groups of patients is doomed to failure. To the extent that selective functional deficits can be identified within an individual patient, a related lesion site may be able to be identified. As the specificity of the cognitive analysis becomes better defined, however, it becomes increasingly unlikely that the level of detail possible from the analysis of naturally-occurring lesions will allow the linking of specific functions with circumscribed brain regions.

This somewhat pessimistic outlook that has emerged from the study of Broca's aphasia can be extended to attempts to localize the language functions that have been implicated in other aphasic syndromes. Although it has been considerably less well studied, Wernicke's aphasia has also been shown to arise from lesions that extend considerably beyond "Wernicke's area" (Kertesz, 1983), and appears to describe a heterogeneous set of symptoms that occur to greater or lesser extent in different patients. Careful analysis of the speech production (Ryalls, 1986) and the sentence production (Butterworth and Howard, 1987) of Wernicke-type patients show symptoms (e.g., phonetic errors; omissions of grammatical elements) that are difficult to distinguish from symptoms found among Broca's aphasics. The situation for "conduction aphasia" is similarly complicated. The lesion site responsible for this syndrome has been widely debated (Damasio and Damasio, 1980), and at least two very different underlying functional deficits have been identified that produce the selective repetition impairment that is the hallmark of conduction aphasia (Berndt, 1989).

The primary conclusion that can be drawn from this review is that the relatively straightforward neuroanatomical atlas for language shown in Fig. 1, which was formulated almost entirely from the performance of brain-damaged patients, greatly simplifies (and even distorts) what must be a much more complex relationship. A secondary conclusion is that the lesion method of cognitive localization in humans, although it is clearly useful in identifying some gross correlations between anatomy and function, may be incapable of generating the database necessary to map well-defined cognitive functions to specific brain areas.

Summary

This selective review has indicated that the lesion method of localizing cognitive functions in the brain has a long and productive history. Much of what we now assume to be true about how the brain allocates its activities comes from observations of the effects of focal brain lesions on

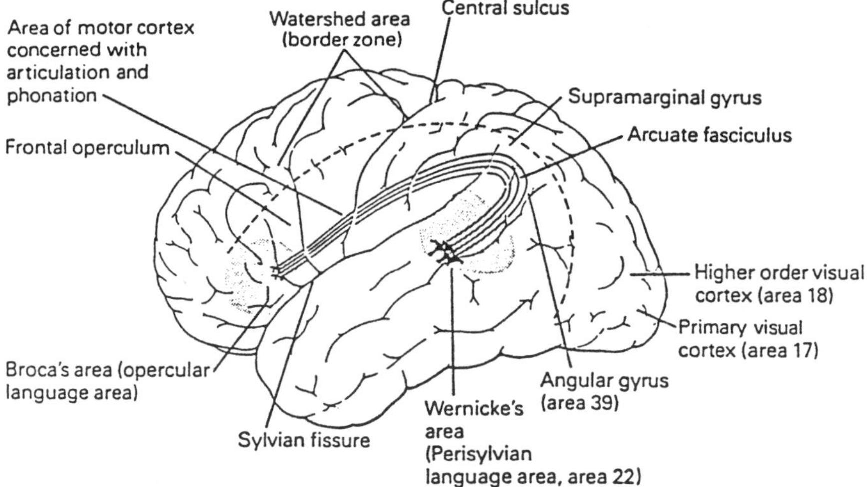

Figure 1. Primary language areas of the brain (reprinted, with permission, from Kandel and Schwartz, 1985).

cognitive activity. Some of these assumptions have withstood scrutiny using newly-emerging experimental methodologies and cognitive analyses. Other assumptions, although they have not withstood such scrutiny, have been very slow to pass from our repertoire of unquestioned beliefs. Further, there are indications that the lesion method itself is extremely limited in its potential for further elucidation of brain/function relationships.

Three factors can be identified that undermine confidence in the lesion method as a means of localizing cognitive activities. One very important problem that has not been raised earlier is the strong likelihood that the identification of a structural lesion does not effectively indicate the extent of the actual disruption to brain activity that results from that lesion. PET-scan studies of metabolic aberrations associated with focal cerebrovascular lesions have shown widespread effects quite remote from the lesion itself (Kushner et al., 1987). As PET and other metabolic measures are improved and applied to a wider range of patients, a much clearer picture should emerge of the neurophysiological effects of structural brain lesions.

A second general issue that appears to undermine the simple logic of the lesion method is the suggestion that isolated brain regions—perhaps even regions within different hemispheres—interact with one another in complicated ways in the execution of even a simple cognitive activity. Recent investigations of "split brain" patients, for example, have demonstrated remarkable integration of information between the two hemispheres, presumably accomplished in subcortical brain structures (Sergent, 1987, 1990a). These studies demonstrated that different information, presented tachistoscopically to each disconnected hemisphere, could be integrated in some tasks (e.g., for making semantic judgments or magnitude comparisons), even when one of the hemispheres showed no conscious awareness of the identity of the information it received. Such data suggest that the execution of a cognitive activity can exploit information processed in more than one brain region, and that these regions may be some distance from one another. Furthermore, subcortical brain areas, which as noted above have been inappropriately ignored in neuroanatomical models of higher function, may play an important role in the integration and coordination of information from diverse brain regions.

Although these possibilities are necessarily speculative at this time, the point is that new information from human studies, as well as from animal lesion studies in which brain areas can be precisely defined (Goldman-Rakic, 1988), suggest that even simple cognitive activities may involve numerous spatially distributed brain areas. As this kind of information accumulates, it becomes increasingly unlikely that "centers" exist in the brain that are uniquely and preferentially specialized to carry out a single cognitive function.

The third problem inherent in interpreting the effects of brain lesions on cognitive functions is the possibility that individual differences exist among normal subjects as to how cognitive activities are carried out in their brains. An important issue that can be identified in this regard is the expertise of the subject in performing the task at hand. There are some indications, for example, that training and experience with musical notation alters the hemisphere involved when listening to music from the right (untrained) to the left (trained) (Gates and Bradshaw, 1977). Sergent (1990b) has suggested that a similar dynamic may underlie the contradictory evidence for a left or right hemisphere specialization for generating a mental image. Many of the case reports of imagery deficits were carried out because the impairment occurred in a subject with clear premorbid skills in visual imagery (artists, engineers, etc.); these cases typically showed a left hemisphere locus of lesion. The suggestion that extensive experience with a particular cognitive function alters the way the brain carries out that function supports the speculations set out above that multiple brain regions may be potentially involved in a given cognitive activity. Moreover, such a suggestion also indicates that attempts to link any cognitive function to a single specialized brain region may be difficult, if not impossible.

A final issue relevant to the question of individual differences in brain/cognition relationships is a methodological one involving the interpretation of group vs. individual subject data. As noted in the discussion of the localization of the aphasias, heterogeneity among classically-defined groups of patients has led to the suggestion that studies of such groups of patients cannot be expected to yield data interpretable as representing specific deficits to the cognitive system (Caramazza, 1986). Although this point is debatable (see Saffran et al., 1989), it is difficult to counter the claim that detailed study of individual cases yields much clearer information than do group studies as to the nature of the deficits that give rise to particular symptoms. When the goal is to perform a cognitive functional analysis of the deficit, individual differences in cerebral organization are irrelevant, and may even be helpful. Abnormal brain organization (such as found in conditions of "crossed aphasia") can be very useful in producing dissociations among symptoms that tend to co-occur when the "typical" brain is damaged (see Berndt et al., 1991, for discussion).

The situation is quite different, however, when the goal is the localization of function in the brain. Any single case, no matter how well studied, could prove to be idiosyncratic in the cerebral representation of the function studied. Even the split-brain patients, whose "lesions" are likely to be at least comparable to one another, show heterogeneity of performance on similar tasks that suggests important differences in how cognitive activities are carried out by their disconnected hemispheres. It would appear, then, that the localization of function through interpretation of pathological conditions will require systematic accumulation of information about large numbers of similar patients if such localization is to be interpreted as generally applicable. For all of the reasons that have been listed here, the designation of patients as "similar" to one another is not an easy matter but must be taken very seriously in any attempt to link symptoms to lesion sites.

The conclusion that must be drawn from these suggestions is that the localization of cognitive functions in the human brain will prove to be a difficult enterprise, and will require converging information from a variety of methodologies. Ultimately, some method that has not yet been discovered may be required to move beyond the impasse that appears to have been reached. In the meantime, however, related research that focuses on goals other than localization can be expected to yield important information. For example, neurophysiological investigations of the distribution and integration of neural activity in animals will continue to define the minimum complexity that can be expected with respect to human functions. Cognitive analyses of the effects of brain lesions, which can be carried out without regard to the localization of those lesions, will continue to help define the atomic functional components that are required for the execution of complex cognitive activities

Acknowledgments. The preparation of this manuscript was supported by grant No. R01-DC00262-06 to the University of Maryland School of Medicine. The author is grateful to Michael P. Alexander and Michael Weinrich for comments on an earlier draft of this chapter.

References

Alexander MP, Naeser MA, and Palumbo C (1990): Broca's area aphasias: Aphasia after lesions including the frontal operculum. *Neurology* 40:353–362

Benton A (1985): Visuoperceptual, visuospatial and visuoconstructive disorders. In: *Clinical Neuropsychology*, Vol 2., Heilman K, Valenstein E, eds. New York: Oxford

Berndt RS (1987): Symptom co-occurrence and dissociation in the interpretation of agrammatism. In: *The Cognitive Neuropsychology of Language*, Coltheart M, Sartori G, Job R, eds. New Jersey: Lawrence Erlbaum Associates

Berndt RS (1989): Repetition in aphasia: Implications for models of language processing. In: *Handbook of Neuropsychology*, Boller R, Grafman J, eds. Amsterdam: Elsevier Science Publishers

Berndt RS (1991): Sentence processing in aphasia. In: *Acquired Aphasia*, 2nd Edition, Sarno MT, ed. New York: Academic Press

Berndt RS, and Caramazza A (1980): A redefinition of the syndrome of Broca's aphasia: Implications for a neuropsychological model of language. *Applied Psycholinguistics* 1:225–278

Berndt RS, and Caramazza A (1981): Syntactic deficits in aphasia. In: *Acquired Aphasia*, 1st Edition, Sarno MT, ed. New York: Academic Press

Berndt RS, Mitchum CC, and Price TR (1991): Short-term memory and sentence comprehension: An investigation of a patient with crossed aphasia. *Brain* 114:263–280

Blumstein SE, Goodglass H and Tartter V (1975): The reliability of ear advantage in dichotic listening. *Brain and Language* 2:226–236

Broca PP (1861): Remarques sur le siege de la faculte du langage articule, suives d'une observation d'aphemie. *Bull Soc Anat* 2:330–357

Butterworth B, Howard D (1987): Paragrammatisms. *Cognition* 26:1–37

Caramazza A (1986): On drawing inferences about the structure of normal cognitive systems from the analysis of impaired performance: The case for single-patient studies. *Brain and Cognition* 5:41–66

Chiarello C (1988): *Right Hemisphere Contributions to Lexical Semantics*. Berlin: Springer-Verlag

Corballis MC, Sergent J (1988): Imagery in a commissurotomized patient. *Neuropsychologia* 26:13–26

Corbetta M, Miezin FM, Dobmeyer S, Shulman GL, Petersen S E (1990): Attentional modulation of neural processing of shape, color, and velocity in humans. *Science* 248:1556–1559

Damasio H, Damasio A (1980): The anatomical basis of conduction aphasia. *Brain* 103:337–350

Damasio H, Damasio AR (1989): *Lesion Analysis in Neuropsychology.* New York: Oxford University Press

Ehrlichman H, Barrett J (1983): Right hemisphere specialization for mental imagery: A review of the evidence. *Brain and Cognition* 2:55–76

Farah MJ (1984): The neurological basis of mental imagery: a componential analysis. *Cognition* 18:245–272

Farah MJ (1986): The laterality of mental image generation: A test with normal subjects. *Neuropsychologia* 24:541–551

Farah MJ, Gazzaniga MS, Holtzman J, Kosslyn SM (1985): A left hemisphere basis for mental imagery? *Neuropsychologia* 23:115–118

Gates A, Bradshaw J (1977): The role of the cerebral hemispheres in music. *Brain and Language* 4:403–431

Geschwind N (1965): Disconnexion syndromes in animals and man. *Brain* 88: 237–294, 585–644

Goldman-Rakic PS (1988): Topography of cognition: Parallel distributed networks in primate association cortex. *Annual Review of Neuroscience* 11: 137–156

Heilman KM, Rothi L, Kertesz A (1983): Localization of apraxia-producing lesions. In: *Localization in Neuropsychology*, Kertesz A, ed. New York: Academic Press

Heilman KM, Bowers D, Valenstein E (1985): Emotional disorders associated with neurological disease. In: *Clinical Neuropsychology*, 2nd Edition, Heilman KM, Valenstein E, eds. New York: Oxford University Press

Kandel ER, Schwartz JH eds. (1985): *Principles of Neural Science*, 2nd Edition. New York: Elsevier North Holland

Kertesz A (1983): Localization of lesions in Wernicke's Aphasia. In: *Localization in Neuropsychology*, Kertesz A, ed. Orlando, FL: Academic Press

Kosslyn SM (1987): Seeing and imagining in the two cerebral hemispheres: a computational approach. *Psychological Review* 94:148–175

Kosslyn SM, Berndt RS, Doyle TS (1985): Imagery and language processing. In: *Attention and Performance XI*, Posner MI, Marin OSM, eds. Hillsdale, NJ: Lawrence Erlbaum

Kosslyn SM, Holtzman J, Farah MJ, Gazzaniga MS (1985): A computational analysis of mental image generation: Evidence from functional dissociations in split brain patients. *Journal of Experimental Psychology: General* 114:311–341

Kushner M, Reivich M, Fieschi C, Silver F, Chawluk J, Rosen M, Greenberg J, Burke A, Alavi A (1987): Metabolic and clinical correlates of acute ischemic infarction. *Neurology* 37:1103–1110

Mayeux R, Kandel ER (1985): Natural language, disorders of language and other localizable disorders of cognitive function. In: *Principles of Neural Science*, 2nd Edition, Kandel ER, Schwartz JH, eds. New York: Elsevier North Holland

Menn L, Obler LK (1990): *Agrammatic Aphasia: A Cross Language Narrative Sourcebook.* Amsterdam: John Benjamins

Mohr JP (1976): Broca's area and Broca's aphasia. In: *Studies in Neurolinguistics*, Vol. 1, Whitaker H, Whitaker HA, eds. New York: Academic Press

Petersen SE, Robinson DL, Morris JD (1987): Contributions of the pulvinar to visual spatial attention. *Neuropsychologia* 25:97–105

Poizner H, Klima ES, Bellugi U (1987): *What the Hands Reveal about the Brain.* Boston: Bradford Books

Posner MI, and Petersen SE (1990): The attentional system of the human brain. *Annual Review of Neurosciences* 13:25–42

Posner MI, Petersen SE, Fox PT, Raichle ME (1988): Localization of cognitive operations in the human brain. *Science* 240:1627–1631

Roeltgen D (1985): Agraphia. In: *Clinical Neuropsychology*, Heilman KM, Valenstein E, eds. New York: Oxford University Press

Ross ED (1981): The aprosodias. *Archives of Neurology* 38:561–569

Ryalls JH (1986): A study of vowel production in aphasia. *Brain and Language* 29:48–67

Saffran EM (1982): Neuropsychological approaches to the study of language. *British Journal of Psychology* 73:317–337

Saffran EM, Berndt RS, and Schwartz MF (1989): The quantitative analysis of agrammatic production: Procedures and data. *Brain and Language* 37:440–479

Sergent J (1983): Role of the input in visual hemispheric asymmetries. *Psychological Bulletin* 93:481–512

Sergent J (1987): A new look at the human split brain. *Brain* 110:1375–1392

Sergent J (1988): Some theoretical and methodological issues in neuropsychological research. In: *Handbook of Neuropsychology*, Vol. 1, Boller F, Grafman J, eds. Amsterdam: Elsevier

Sergent J (1989): Image generation and processing of generated images in the cerebral hemispheres. *Journal of Experimental Psychology: General* 15:170–178

Sergent J (1990a): Furtive incursions into bicameral minds. *Brain* 113:537–568

Sergent J (1990b): The neuropsychology of visual image generation: data, method and theory. *Brain and Cognition* 13:98–129

Springer SP, Deutsch G (1981): *Left Brain, Right Brain.* San Francisco: Freeman

Tonkonogy V, Goodglass H (1981): Language function, foot of the third frontal gyrus, and rolandic operculum. *Archives of Neurology* 38:486–490

Ungerleiter LC, Mishkin M (1982): Two cortical visual systems. In: *Analysis of Visual Behavior*, Ingle DT, Goodale M A, Mansfield RJW, eds. Cambridge, MA: MIT Press

Wernicke C (1874): *Der Aphasiche Symptomencomplex*. Breslaer: Cohn and Weigart

A Spreading Activation Model of Age-related Deficits in Cognition

Gillian Cohen

Research on the effects of aging on cognition often results in a list or catalogue of those aspects of cognitive performance that are impaired by aging and a similar catalogue of other aspects of performance that are unaffected by aging. To advance beyond this stage we need to find a general framework, a cognitive theory that reveals how these various effects are related, and that will enable us to make predictions about what other kinds of tasks or processes are likely to be affected by age.

It is apparent that the effects of aging on cognition are both very complex and very selective, and this suggests that we need to look for multiple causes rather than a single, general factor affecting the whole cognitive system uniformly. Researchers have therefore attempted to identify the particular component parts or component processes in the cognitive system that become defective with age. The experimental evidence indicates that old age brings a reduction in the capacity (the number of items) that can be held in working memory; and a slowing down of the rate of processing. Between them, these two factors can account for many of the observed age effects.

However, Salthouse (1988) has suggested that age-related deficits in cognition can be explained more powerfully in terms of network models and spreading activation. Network models of knowledge representation (e.g., Anderson, 1976) are composed of nodes, representing concepts, which are inter-related by associative links or pathways. These links may vary in strength according to the degree of association between the concepts. Retrieval of information from the network occurs by a spread of activation. When one node is activated, the activation spreads via the linking pathways to associated nodes. Such a network is built up as increasing knowledge and experience produce a greater density of nodes and links.

There are several factors that can limit the efficiency of this mechanism:

(1) The number of nodes that can be simultaneously active in a single related pattern at any one time. This corresponds to working memory capacity.

(2) The total amount of activation in the entire network. Salthouse (1988) has suggested that this factor is equivalent to attentional capacity or mental energy.

(3) The speed or rate of spread of activation from node to node.

(4) The duration of a given pattern of activation as determined by the rate at which activation decays.

(5) The susceptibility of an existing pattern of activation to interference from new patterns of activation causing crosstalk within the network.

(6) The efficiency of an inhibitory gating mechanism limiting the spread of activation so that it is not diffused indefinitely throughout the network.

(7) The structure of the network; the existence and organization of nodes and links between them.

Because most of these seven factors are known to be affected by aging, a spreading activation model allows the possibility of providing a comprehensive account of age effects within a unified theoretical framework. Previous research has provided ample evidence for the first three factors, a reduction in working memory capacity; a slowing down of the rate of processing and a decrease in attentional capacity. (Salthouse, 1988). Susceptibility to interference is shown by the age-related increase in intrusion rates in free recall (Stine and Wingfield, 1987), and Hasher and Zacks (1988) have recently reviewed evidence that aging reduces the efficiency of inhibitory processes, allowing irrelevant information to receive activation and to enter working memory. However, there is no evidence for age-related structural differences in the knowledge network. Studies of the effects of semantic priming and category effects indicate that the same semantic relations remain intact in the memory networks of older people (Burke and Harrold, 1988). The priming studies do suggest, though, that new links take longer to establish (Howard, 1988).

To illustrate how the spreading activation model can be used to explain and interpret age differences, it is applied here to two different areas where age effects are relatively marked: memory for actions and memory for names.

Memory for Actions

One aspect of performance in everyday life that is affected by age-related cognitive deficits is the on-line monitoring of actions. Reason (1979) has developed a system of classifying absentminded 'slips of action', or action errors. The main types of errors are

(1) Repetitions occur when the actor forgets that an action has already been performed and does it again. These errors occur because the activation of the intention has not been purged after the action has been performed.

(2) Switches occur when the actor starts to perform an action sequence and involuntarily switches to another. Switches are caused by cross-talk between competing patterns of activation.

(3) Omissions occur when the actor omits a component action from a sequence. This type of error is caused when the level of activation is too low to trigger all the component actions.

(4) Reversals occur when the actor transposes the order of the component actions in a sequence. A possible cause of reversals is failure to inhibit activation of late component actions until earlier components have been performed.

(5) Blends occur when the actor mixes the component actions from different sequences and are also caused by crosstalk.

Examples of these different kinds of action errors can be drawn from a particular action schema. Norman (1981) developed a model in which an action sequence is represented as a hierarchical schema. The highest level in the hierarchy represents the overall goal, the general intention. At subordinate levels there are subschemas that correspond to the actions necessary to achieve this goal. Activation flows downwards from the intention level to trigger the subordinate actions. The subordinate actions can also be triggered by currently perceived stimuli.

Figure 1 shows an action schema for drinking tea. A repetition error would involve putting tea in the pot twice over. An example of a switch would be getting out the teapot and then washing it instead of putting in the tea. Forgetting to switch on the kettle would be an omission error. A reversal would be switching on the kettle before filling it with water, and a blend would be putting coffee in the teapot instead of tea. In this hierarchy the subschema of filling the kettle may be activated by the higher level intention to make tea and also by the stimulus of seeing the kettle.

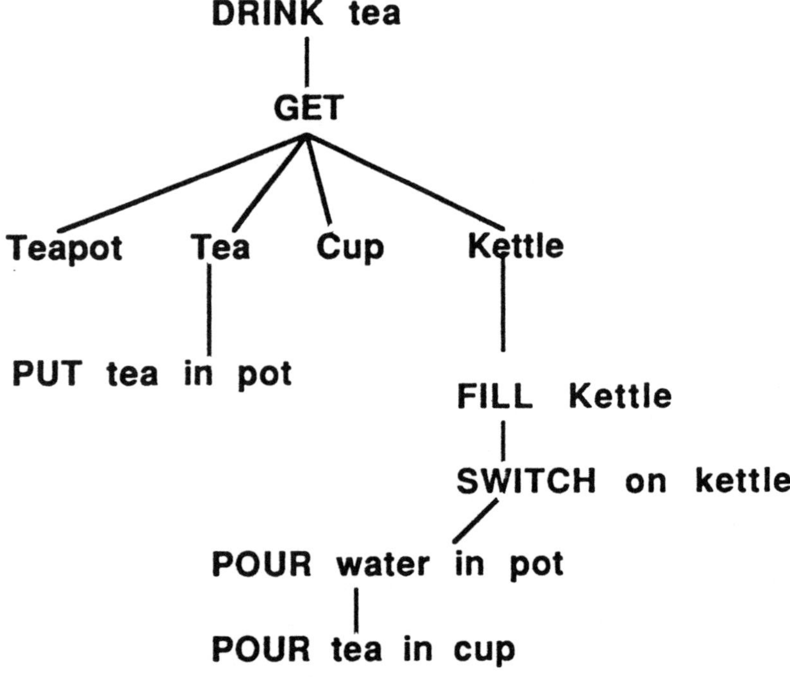

Figure 1. A hierarchical action schema for drinking tea.

Some of the absentminded action errors made by older people in everyday life can be fitted to this model. Cohen (1986) administered a Memory Lapse Questionnaire to young, middle-aged and elderly people. This questionnaire listed twenty different everyday action errors and asked subjects to rate how frequently they made each type of error on a scale of 1–5 from Never to Very Often.

There were three action errors that occurred more frequently in the elderly.

(1) The most commonly reported error occurs in conversation, when the elderly person forgets to say something he or she intended to say. This is an omission error, where there is a failure to implement an intention. Instead of triggering the action, the activation fizzles out or is disrupted by the ongoing conversation.

(2) Another common action error consists in going to a particular location, into a room, for example, and forgetting what he or she went there to do. In this case the action is only partially implemented: activation from the intention level has taken the actor to the correct destination but then failed to trigger any further subordinate actions. Sometimes a different action may be triggered by stimuli in the loca-

tion and there is a switch to a different action schema. The actor may go upstairs to fetch a sweater and come down with a book instead. In this case there has been crosstalk between the competing schemas for book fetching and sweater fetching.

(3) A third type of action error that is more common in the elderly is forgetting some object that is needed to carry out an action sequence, for example, going out to post a letter and forgetting to take the letter, going shopping and forgetting to take a purse, or making tea and forgetting to put any tea in the pot.

All these examples are omission errors. The results of the questionnaire showed that omission errors are the most frequent type of error made by elderly people, although this was not true of younger people for whom switches and blends are more frequent. Omission errors are caused because activation originating from the top of the schema hierarchy is too weak or too poorly maintained to trigger the lower levels.

It makes more sense to interpret these errors in terms of lowered levels of activation, or poorly maintained activation than it does to simply attribute them to reduced capacity of working memory. It is not the case that there are too many items to be held in working memory but rather that the activation originating at the intention level is not strong enough to trigger the lower level actions. If the spreading activation model is applied in conjunction with the actions hierarchy model, it has the advantage of being able to predict the kind of action errors that are most likely to occur.

Memory for Names

A second area in which a spreading activation model can be applied is memory for names. Difficulty in retrieving proper names is a common problem that increases in severity as people get older. Name blocks occur when a name that is known is temporarily inaccessible and cannot be retrieved. Name blocking occurs more frequently in old age although the incidence of name blocks is also increased in younger people if they are ill, stressed, or very tired.

The main characteristics of name blocks have been revealed by diary studies in which people are asked to record details of naturally occurring name blocks. Several studies by different researchers (e.g., Burke et al., 1988; Çohen and Faulkner, 1986) have yielded converging results and these findings can be interpreted within a spreading activation model.

(1) Blocks occur more frequently for proper names than for other words such as object names. According to the network model, proper names receive less activation than other words because they are relatively meaningless and lack semantic associations.

(2) Name blocking is less frequent in confrontation naming (that is, when the to-be-named person is present) than in naming from description (when the to-be-named person is described or referred to but is not present). This is perhaps because more cues are available in the confrontation situation and would therefore produce more activation.

(3) The fact that name blocks are occasional, so that failures occur on some occasions for names that are retrieved rapidly and easily on other occasions, indicates that a dynamic mechanism is involved.

(4) The majority of name blocks are temporary and are resolved without external help, that is, without asking someone else or looking it up. This is consistent with activation recovering from inhibition or being augmented by further cues.

(5) The majority of name blocks are partial rather than complete. Partial information (e.g., the first letter, length, number of syllables, or pattern of stress) is often available about the target name. This suggests that the target name is weakly activated but this activation is insufficient to produce retrieval. Another finding is that other names, nontarget candidates, which are related to the target phonologically, semantically, or contextually may come to mind. Subjectively these other names seem to block access to the target. This effect would be produced if, when the target name is cued, activation spreads to other related names. If one of these related names is more frequently or recently used it may have a lower threshold or a higher resting level of activation and so be triggered instead of the target.

(6) A reverse familiarity effect occurs for proper names. This refers to the paradoxical finding that whereas blocks for ordinary words, not proper names, are almost always for rare unfamiliar words, retrieval blocks for proper names are more often for names that are rated as well-known familiar ones.

(7) There are age differences in both the quantity and quality of name blocks. Blocks are more frequent in the elderly and are more complete. That is, there is less likely to be any partial information and related nontarget names are less likely to come to mind. These effects would be produced if elderly people have a lower level of activation.

(8) Retrieval processes to resolve the blocks involve several different types of strategy. One of the most successful is enriching the target

description by thinking about the person and the context in which the person is known. This generates more cues to augment the level of activation for the target name.

An alternative method is to abandon conscious search and wait for a pop-up. The exact nature of pop-up resolutions is disputed. They may occur because unconscious search processes are continuing. Or they may occur because some serendipitous cuing augments the activation of the target. In one example a subject had failed to recall the name 'Bell', but a pop-up resolution occurred when she heard a church bell.

(9) Retrieval of names and descriptions is asymmetrical. Failure to retrieve a name when presented with a face or a description is relatively common, but the opposite kind of failure, being unable to retrieve the description when presented with a known name does not occur. This asymmetry can be understood if we think of names and descriptions as hierarchically arranged with elements of a given description being shared by several names as shown in Fig. 2.

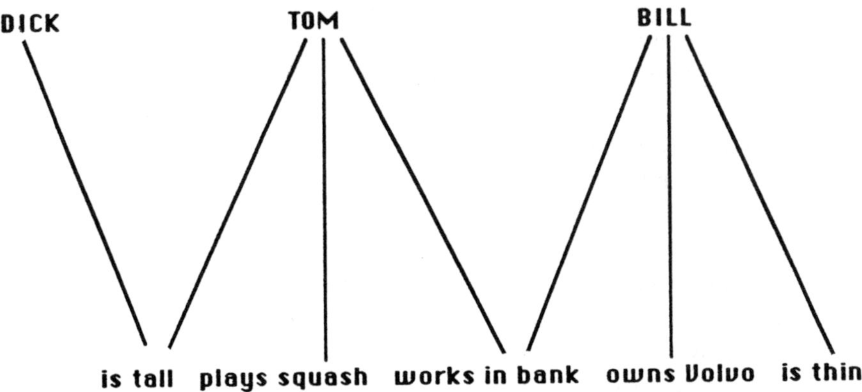

Figure 2. An overlapping organization of names and descriptions.

Cuing the description of Tom will partially activate several different names making it difficult to retrieve the target name. Cuing Tom's name will uniquely activate the relevant description so that it is easier to retrieve.

(10) Name confusions are not retrieval blocks, but are occasions when the wrong name is used to address or to refer to someone. These confusions occur most often within the family. Substituted names preserve sex and generation. Outside the family substituted names still tend to preserve context, so that, for example, the confused names are both names of colleagues, or both names of politicians. The

substituted name tends to be one that is more frequently or more recently recalled than the target name, again suggesting that such names require less activation.

All these findings can be interpreted *post hoc* within the framework provided by the spreading activation model. In addition, Cohen (1990) designed an experiment specifically to test whether this model can explain the reverse familiarity effect noted above.

It is puzzling and counterintuitive that the frequency effect for ordinary words is reversed in the case of proper names. Cohen and Faulkner (1986) found that 74% of name blocks occur for names of people who are rated as well known and similar results have been obtained in other diary studies. Names of people who are well known are used frequently, so, according to the normal frequency effect, these well-known names should be easier to retrieve. Furthermore, when someone is known well, a larger amount of knowledge about them is stored in memory, so there should be more potential cues for recall of the name.

The experiment that explores this paradox relies on Anderson's concept of the fan effect (King and Anderson, 1976). According to this, the more information is accumulated about a concept, the greater the number of pathways radiating from it, creating a *propositional fan.* In retrieval, activation spreads from the concept node, and the greater the amount of propositional fanning the longer the retrieval time. This is either because the paths are searched serially, or because activation is diffused along the multiple paths. The fan effect therefore creates the paradox that the more that is known about something the harder it is to retrieve this knowledge. Exceptions to this rule have been demonstrated. When the facts are unified in some way and can be integrated, the fan effect can be counteracted (McCloskey and Bigler, 1980; Smith et al., 1978). The spreading activation model therefore suggests that the reason why the names of well-known people are harder to retrieve than the names of people who are not known well is because more facts are known about them so the fan effect will be greater. It is also plausible to suppose that these facts may not always be well integrated. Facts about an individual person are often acquired in separate encounters at different times and places so that integration is likely to be imperfect. Elderly people would be more susceptible to the fan effect if their spread of activation was slower, or if they have an initially lower level of activation, so that when it is diffused down multiple paths it is insufficient to activate any name.

In this experiment, young, middle-aged, and elderly subjects were tested. They first learned 18 statements printed on a study sheet. The statements were presented in three conditions. Each consisted of a name and a personal attribute as shown in Table 1.

Table 1

No Fan	Fan	Crossed Fan
George likes cats	Jeff is a farmer	Tim is a doctor
Mark is tall	Jeff likes music	Tim likes beer
Tom has a boat	Jeff has a beard	Alan is a doctor
John likes films	Peter is a waiter	Alan has a bicycle
Bill has glasses	Peter is thin	David likes beer
Steve is a painter	Peter has a car	David has a bicycle

Examples of name–attribute statements in each condition.

In the No Fan condition each name is linked uniquely to one attribute. In the Fan condition each name is linked to three attributes and in the Crossed Fan condition each name is linked to two attributes each of which is also linked with one other name. Figure 3 shows how the statements from each condition would be represented in the network.

When the list had been learned to criterion there was a recognition test. The 18 target statements and 18 distractors (created by re-pairing names and attributes) were presented one at a time on the screen of an Apple Macintosh microcomputer. Subjects responded 'yes' to targets and 'no' to distractors by pressing designated keys. Table 2 shows the mean reaction times for each age group in each condition.

Table 2

Age	No Fan	Fan	Crossed Fan
Young	1698	1654	2107
Middle	1845	1904	2076
Elderly	2121	2474	2360

Mean reaction times for correct recognition responses, in msec.

The Elderly were slower than the Young and Middle-aged who did not differ significantly. For the Young and Middle-aged groups, Crossed Fan was slower than No Fan or Fan, which did not differ. For the Elderly, Fan and Crossed Fan were both slower than No Fan.

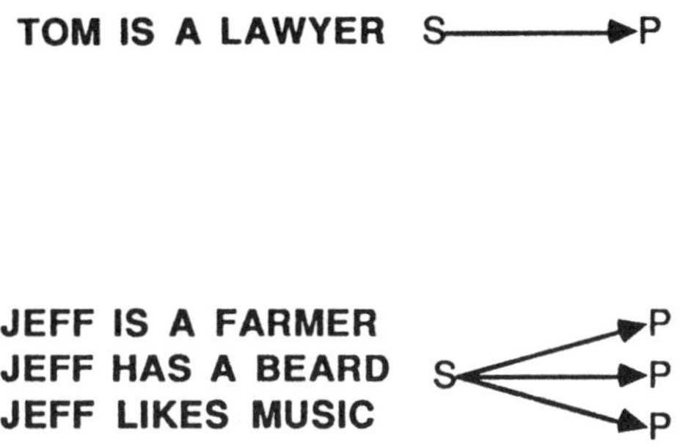

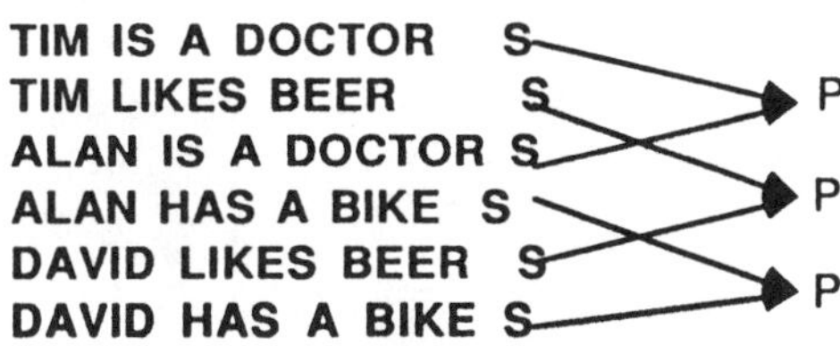

Figure 3. Representations of statements from the No Fan, Fan and Crossed Fan conditions (S = subject and P = predicate).

The main point to emerge is that the elderly are more susceptible to the fan effect. Because they have a slower rate of spread of activation or a lower level of activation they are particularly disadvantaged in the Fan conditions. Younger people are not slowed by fanning but only by crossed fanning. Young and middle-aged subjects reported that they integrated the three facts in the Fan condition by imagery (for example, imaging Jeff as a bearded farmer playing a violin in the fields). This strategy appears to have been effective in counteracting the fan effect, so that there was no increase in reaction time in this condition. In the crossed fan condition integration cannot be achieved in this way. It is more difficult to create distinctive images for a bicycling doctor, a bicycling beer drinker and a beer-drinking doctor, so the crossed fan effect cannot be counteracted by integration. Only two of the elderly group reported using any imagery in the task and the elderly group as a whole do not seem able to counteract the fan effect by using integrative imagery even in the uncrossed fan condition.

The results of this experiment shed some light on memory for names in everyday life. Some degree of cross fanning probably occurs in the memory representations of most of the people we know. Most of our acquaintances will have some things in common with some others as shown in Fig. 2. Activation of a personal description will be liable to partially activate several names. If none are activated above threshold level, the name block will be complete, but the wrong name may be triggered if its threshold is lowered by recent use or by context. If the activated description contains information unique to the target person, other names are less likely to receive activation, so the target name is more likely to be triggered.

Spreading activation models, together with the additional assumption that name nodes have thresholds that vary with frequency and recency of activation, can provide a comprehensive explanation for the observed characteristics of name retrieval and for the age deficits that are evident in memory for names.

Conclusions

Age-related deficits are evident in memory for actions and in memory for names. These deficits can be explained by a spreading activation model that provides both a unifying explanation and a source of predictions. This approach has a number of advantages over explanations based on information processing models, which attribute age effects to a reduction in processing resources. The spreading activation model distinguishes between different types of limitation on the cognitive system, and this makes it broad enough to encompass a wide range of phenomena and yet specific enough to yield clear and testable predictions. Another advantage of this approach is that it is easier to map onto a neurological level of explanation.

References

Anderson JR (1976): *Language, Memory and Thought.* Hillsdale, NJ: Lawrence Erlbaum Associates

Burke DM, Harrold RM (1988): Automatic and effortful process in old age: Experimental and naturalistic approaches. In: *Language, Memory and Aging,* Light LL, Burke DM, eds. New York: Cambridge University Press

Burke D, Worthley J, Martin J (1988): I'll never forget what's-her-name: Aging and tip-of-the-tongue experiences in everyday life. In: *Practical Aspects of*

Memory: Current Research and Issues, Volume 1, Gruneberg MM, Morris PE, Sykes RN, eds. Chichester: John Wiley and Sons

Cohen G (1986): Everyday memory in old age. Paper presented at the British Psychological Society meeting, London, December, 1986

Cohen G (1990): Recognition and retrieval of proper names: Age differences in the fan effect. *European Journal of Cognitive Psychology* 2(3):193–204

Cohen G, Faulkner D (1986): Memory for proper names: Age differences in retrieval. *British Journal of Developmental Psychology* 4:187–197

Hasher L, Zacks RT (1988): Working memory, comprehension and aging: A review and a new view. In: *The Psychology of Learning and Motivation*, Bower GH, ed. 22:193–225, New York: Academic Press

Howard DV (1988): Aging and memory activation: The priming of semantic and episodic memories. In: *Language, Memory and Aging*, Light LL, Burke DM, eds. New York: Cambridge University Press

King DRW, Anderson, JR (1976): Long term memory search: An intersecting activation process. *Journal of Verbal Learning and Verbal Behavior* 15:587–605

McCloskey M, Bigler K (1980): Focused memory search in fact retrieval. *Memory and Cognition* 8:253–264

Norman, DA (1981): Categorization of action slips. *Psychological Review* 88: 1–15

Reason, JT (1979): Actions not as planned: the price of automatization. In: *Aspects of consciousness*, Vol. 1, Underwood G, Stevens R, eds. London: Academic Press

Salthouse TA (1988): Effects of aging on verbal abilities:Examination of the psychometric literature. In: *Language, Memory and Aging,* Light LL, Burke DM, eds. New York: Cambridge University Press

Smith EE, Adams N, Schorr D (1978): Fact retrieval and the paradox of interference. *Cognitive Psychology* 10:438–463

Stine EAL, Wingfield A (1987): Process and strategy in memory for speech among younger and older adults. *Psychology and Aging* 2:272–279

Memory Changes in Aging

Fergus I. M. Craik

In this chapter I will discuss human memory from the standpoint of an experimental psychologist. Whereas neurophysiologists, neuroanatomists, and others describe the underlying machinery of memory and cognition, psychologists ask the question, "What is the *purpose* of the machinery, what does it do?" That is, the role of experimental psychology is to give an account of the functions of memory and other cognitive processes, in terms of both behavior and mental experience. The psychology of memory thus exists at a level of description and analysis that is different from those of biochemistry, anatomy, and physiology, although it relates to these other levels of description and, in a sense, depends on them. One analogy is with an oil painting, which could be analyzed in terms of color spectra or in terms of the chemistry of the paint pigments, although neither of these analyses would reveal *anything* about the subject matter of the painting—to understand that we must move to another level of description. Another analogy that has been drawn is to computers; by this analogy, the psychological level of description is like the software, whereas anatomy and physiology are represented by the hardware. Again, one domain depends on the other, although each is described in its own terms.

This chapter will focus on changes in memory in adults as a function of the normal aging process; some brief remarks will also be made concerning the similarity of memory changes in normal aging to those found in pathological conditions, and to those found during normal childhood development—does aging mirror development, and if so, do the similarities in behavior reflect similarities in the underlying neurophysiology?

Age-related Changes in Memory and Cognition

It is well established that, in general, cognitive processes decline in efficiency from early adulthood to old age (Kausler, 1982; Salthouse, 1982;

Welford, 1958). However, this general pattern becomes more complex as individual tasks are examined; some tasks show precipitous declines with age, whereas others show no age-related decrements, or even show improved performance levels in older people. For example, when the Wechsler Adult Intelligence Scale (WAIS) is examined, it is seen that the overall downward trend in performance across the adult years is composed of very sharp declines in some subtests (e.g., Digit-Symbol Substitution), but little if any decline in others (e.g., Vocabulary and Information) (Salthouse, 1982). One explanation of this differential pattern is that tasks that require the subject to make new mental connections and carry out novel computations—"Fluid Intelligence"—show age-related decrements, whereas tasks that draw on well-established mental routines—"Crystallized Intelligence"—hold up well into old age.

Memory tasks also show this mixed pattern of decline and immunity to change. Digit span (the longest series of unrelated numbers that a person can accurately reproduce) declines very little with age, for example, as does recognition memory (Craik, 1977, 1984). There has been much recent interest in a class of memory tasks referred to as "priming" or "implicit memory" tasks (Richardson-Klavehn and Bjork, 1988) and they also show very little change with age (Light et al., 1986). An example is word fragment completion in which the subject studies a list of words (containing such words as ELEMENT, ASSASSIN, ESTUARY) and is later given words with letters missing to complete. It is easier for subjects to complete words that were on the studied list (A____A____IN) than those that were not (____S____R____S____O). The ability to complete studied (or primed) words declines only slowly over time since study, and remains intact in older people and in amnesic patients (see Richardson-Klavehn and Bjork, 1988, for a review). Other tasks show very large and consistent decrements with age. Dichotic listening (the ability to reproduce two short strings of digits presented simultaneously, one string to each ear) declines in effectiveness in older adults, as does the free (unordered) recall of a list of unrelated words, and the ability to recall names of people and unusual objects (Craik, 1977; Poon, 1985).

What is the principle here? Why do some tasks show marked decrements and others show none? One possibility is that the different tasks described above tap different memory stores or different memory systems, and that these stores or systems are differentially affected by the processes of aging. I will argue that this approach is not fruitful, however, since tasks that show and do not show declines can be found associated with each proposed store or system. One influential division of memory is into short-term and long-term stores (e.g., Atkinson and Shiffrin, 1968).

On the basis of the findings that digit span and the size of the recency effect decline very little with age, yet free recall of words and retrieval of names does decline (e.g., Craik, 1977), it might be suggested that the short-term store is relatively invulnerable to aging, but that the long-term store is affected. Against this conclusion, however, are the findings that dichotic listening and performance on working memory tasks (see Hultsch and Dixon, 1990) *do* show large age-related decrements, and yet these tasks surely tap "short-term memory" also. In addition, recognition memory tasks and implicit memory tasks show rather small decrements, even though they must presumably reflect the operation of the long-term store.

Endel Tulving's (1983, 1987) suggested classification of memory into episodic, semantic, and procedural systems fares little better as a way of explaining the differential pattern of results with age. As a first approximation it seems reasonable to argue that episodic memory (for specific events) is particularly vulnerable to the aging process, whereas semantic memory (for general knowledge) is less so, and procedural memory (for well-learned motor and cognitive routines) perhaps not affected at all. The last statement regarding procedural memory does accord with the available evidence, but the other two face difficulties. Free recall is presumably an episodic memory task and does show age-related deficits, but recognition memory (also episodic) shows greatly reduced losses. In semantic memory, older people do retain general knowledge about the world (e.g., the Information subtest of the WAIS) but are notoriously poor at remembering names (also presumably a function of semantic memory).

My present conclusion is therefore that an account of age differences in memory couched in terms of proposed *structures* (stores or systems) is not very useful. Instead, I prefer an account that describes different types of processing, and ascribes age-related decrements to the particular difficulties that older people have with certain processing operations. Before discussing some empirical results, I will briefly outline this processing view of learning and remembering. More extensive accounts are given in previous publications (Craik, 1983, 1986).

A Processing Account of Remembering

When we reflect on our own experiences of memory it is natural to think that each remembered episode or piece of knowledge is stored discretely in the brain, rather like individual books in an extensive library. New learning is represented by adding further books to the appropriate section of the library, and successful retrieval is represented by some process (the librarian? consciousness?) finding and re-activating the wanted material. One problem with such accounts is that 100 years of intensive neurophysiological research has failed to identify the neural correlates of the "books"; the nature and location of memory traces (or "engrams") remain to be elucidated. A second problem is that damage to various cortical areas does not typically (although a few exceptions exist) result in loss of specific memories or classes of memories, as would be expected if areas of a library were destroyed. A third difficulty, at the psychological level, is that memories are not usually retrieved in a complete all-or-none fashion; more typically, some aspects of a remembered episode or of factual information are recovered, and further aspects may be recovered with extra effort or with additional cues. Such fragmentary and piecemeal retrieval would not be expected if memory records were stored as unified entities.

An alternative approach (Craik, 1983) is to view remembering as the psychological concomitant of neural *activity*, perhaps distributed rather widely in the cortex. This activity may occur primarily in areas whose principal function is perception; that is, the activity of remembering may be thought of as an aspect of the more general activity of perceiving. The major difference between perceiving and remembering is, of course, that the former takes place in the presence of the stimulus or event that is being perceived, whereas remembering takes place when the stimulus is no longer present. However, essentially the same cortical processes may be involved; in the first case largely driven by external stimulation, and in the second case initiated from within the brain itself.

At the psychological level we can say that perceiving depends both on the stimulus array and on previous perceptual learning; current perception is an interaction between current stimulation and the pre-existing perceptual machinery. Remembering may also be thought of as an interaction between incoming information (now referred to as "retrieval cues" or "context") and the same cortical machinery that has been modified by past experience in both general and specific ways. Thus "memory" is not a characteristic of the organism alone, but reflects the necessary *interaction* between the external environment and brain processes. We do not think of perceptual experiences lying dormant in the brain awaiting "activation" by the appropriate stimulus; rather, the perceptual machinery has the potential to construct specific percepts, given specific stimulation. Similarly, it may be suggested that remembering occurs as a function of the interaction between incoming stimulation and the learned modifications to perceptual and cognitive representations. It is clearly meaningless to ask "Where is the percept when we are not perceiving?" It does not exist, except in the form of the potential of the perceptual machinery to produce a given experience provided that the appropriate stimulus is present. But it may be equally meaningless to ask "Where is the memory trace when we are not remembering?" Again, it exists only in the form of a latent potential to produce a specific pattern of activity in the presence of appropriate stimulation (or "cuing"). No amount of careful dissection of cortical areas will ever disclose memory traces, for the simple reason that they do not exist as such (see also Rosenthal, 1988 for a similar argument).

By this view of remembering, learning or encoding consists simply of the modifications of general cognitive representations as a function of perception and comprehension of further events; that is, there are no special "learning" processes as such, normal perception and comprehension are sufficient (Craik and Lockhart, 1972). Retrieval, by this view, is the recapitulation of the original pattern of cortical activity that occurred when the event, idea, or fact was first experienced. Remembering is quite literally the re-cognition of the original event.

It follows that remembering will be most effective when as much as possible of the original event and its context are re-presented at the time of retrieval. In this case, present stimulation helps to reinstate the original configuration. In the absence of such stimulation, the system must "self-initiate" the appropriate cortical/mental operations; such self-initiated processing is likely to be effortful and to require substantial processing resources. The present suggestion with regard to the aging process is that older people have disproportionately greater difficulty with self-initiated processing (Craik, 1983, 1986). That is, when a memory task affords substantial environmental support—as in recognition memory—the appropriate processes are driven by the stimulus and context, and older people perform comparably to their younger counterparts; but when the task does *not* afford good environmental support (as in free recall, for example), and the subject must rely on self-initiated operations, older people will be at a disadvantage.

Arguably, then, when tasks require considerable mental manipulation (e.g., working memory tasks), effortful allocation of resources (e.g., dichotic listening), and recall of words or names in the absence of supportive cues, older people will exhibit substantially poorer performance levels than their younger counterparts. If the memory task is well supported by cues (e.g., recognition memory) or involves rather passive repetition of materials (e.g., digit span), or does not involve effortful, consciously controlled processing (e.g., "implicit memory" and procedural memory tasks) the older person's performance will be more closely comparable to that of young adults (see also Hasher and Zacks, 1979).

In summary, the present suggestion is that the differential pattern of memory loss exhibited in normal aging does not result from the differential vulnerability to the aging process of different memory stores or memory systems, but rather results from a decline in the effectiveness of effortful self-initiated processes, which in turn are required by some tasks much more than by others. Some experimental illustrations of these ideas follow in the succeeding sections.

Experimental Illustrations

Short-Term Memory

By the argument presented above, age-related memory losses should not appear and disappear as a function of the memory system tapped, but as a function of the types of processes required by each particular task. We might therefore expect that some short-term memory tasks would show losses and some would not, depending on the degree of effortful manipulation, and the amount of self-initiated processing required.

In a previous review (Craik 1977) I pointed out that several traditional short-term memory tasks show little or no age-related decline. These tasks included digit span, the recency effect in free recall (i.e., recall of the last three or four words from a 20-word list), and the Brown-Peterson task, in which three or four letters or words are recalled after a retention interval of say 20 sec filled with rehearsal-preventing activity. These tasks share the characteristic that they are relatively passive; they do not require the material to be manipulated or reorganized, nor must the subject divide attention between retention and other activities.

Such relatively passive short-term (or primary) memory tasks may be contrasted with tasks in which attention is divided, or in which the material must be actively reorganized. Working memory tasks have this characteristic. In working memory tasks, subjects must retain some material while also attending to and processing other material. By the present argument, age-related decrements would be expected in such situations.

The evidence supports this claim. Wright (1981) studied adult age differences in the working memory paradigm introduced by Baddeley and Hitch (1974); the task consists of verifying the truth of a sentence while simultaneously holding a string of words in mind. Wright found substantial age differences in this situation. A second paradigm that has been used extensively was formulated by Daneman and Carpenter (1980);

in their task, subjects are given a short series of sentences to verify (e.g., 2–6), and after all sentences have been presented, the subject must recall the last word of each of the presented sentences. Thus again the subject must both hold material in mind while simultaneously carrying out a simple problem-solving task. Light and Anderson (1985) have shown that older people do less well on this task.

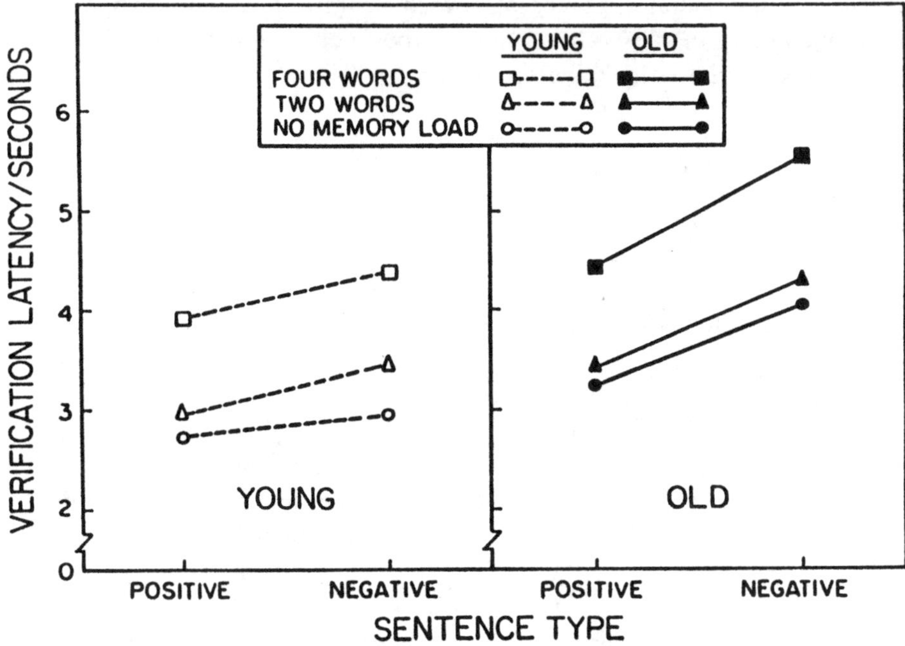

Figure 1. Sentence verification latencies for young and older subjects as a function of sentence complexity and concurrent memory load.

We have also found age-related decrements in working memory in my own laboratory. Using the Baddeley and Hitch paradigm, Morris et al. (1988) asked subjects in their early 20s and subjects in their 60s and 70s to verify a single sentence while holding 0, 2, or 4 unrelated words in mind. The single sentences were of the type "An elephant is usually bigger than a mouse" or "Children never like to play at the beach." That is, sentence complexity was varied by presenting sentences as positive or negative statements. Morris et al. (1988) had predicted that older subjects would be differentially penalized by increases in both types of complexity (or difficulty) in the working memory situation—both by increasing the concurrent memory load from 0 to 4 words, and by increasing the complexity of the sentence to be verified. Figure 1 shows that the time taken to verify the sentence does in fact increase with increases in both types of difficulty, but whereas the older subjects were relatively more penalized when dealing with negative as opposed to positive sentences, the older and younger subjects were penalized to the same degree by increases in the memory load. That is, one type of complexity interacted with age but the other did not.

In a second study, Gick et al. (1988) presented a set of 2, 4, or 5 sentences to be verified in the Daneman and Carpenter paradigm. The measure here is the proportion of the final words (e.g., "mouse" and "beach" in the preceding examples) of the presented sentences that subjects can recall after verifying all the sentences. In one condition the sentences were grammatically simple, in a second they were complex, and in a third condition no sentences were presented—simply the list of final words for the subject to recall. Again the authors predicted that older subjects would be differentially penalized by increases in both types of difficulty—the number of words or sentences presented, and the complexity of the sentences. However, Fig. 2 shows again that age interacted with sentence complexity, but did not interact with increasing set size.

These studies thus confirm the results of earlier experiments that older people have problems with working memory tasks, and suggest in addition that certain types of complexity present special difficulties to the elderly. One possibility is that the passive holding aspects of working memory are less affected by aging than are the active computational aspects, although this conclusion requires further experimental confirmation. In any event, the experiments support the general conclusion that age differences in working memory are substantial, whereas age differences in more passive short-term memory tasks ("primary memory") are typically slight or non-existent (Craik, 1977). It is not the memory *system* that determines this difference, but the type of processes required by the particular task; with tasks that require mental manipulation and reorganization being associated with age-related losses.

This general result has also been reported recently by Wingfield et al. (1988) and by Dobbs and Rule (1989). Arguably, the working memory tasks that show age-related losses require greater amounts of "processing resources" to carry them out, and also require greater amounts of control from the "central executive" aspect of working memory (Baddeley, 1986). There are reasons to believe that aging affects both of these correlates of successful higher level processing.

Longer-Term Episodic Memory

The cognitive aging literature provides a number of studies whose results support the notion that age-related decrements in memory are associated

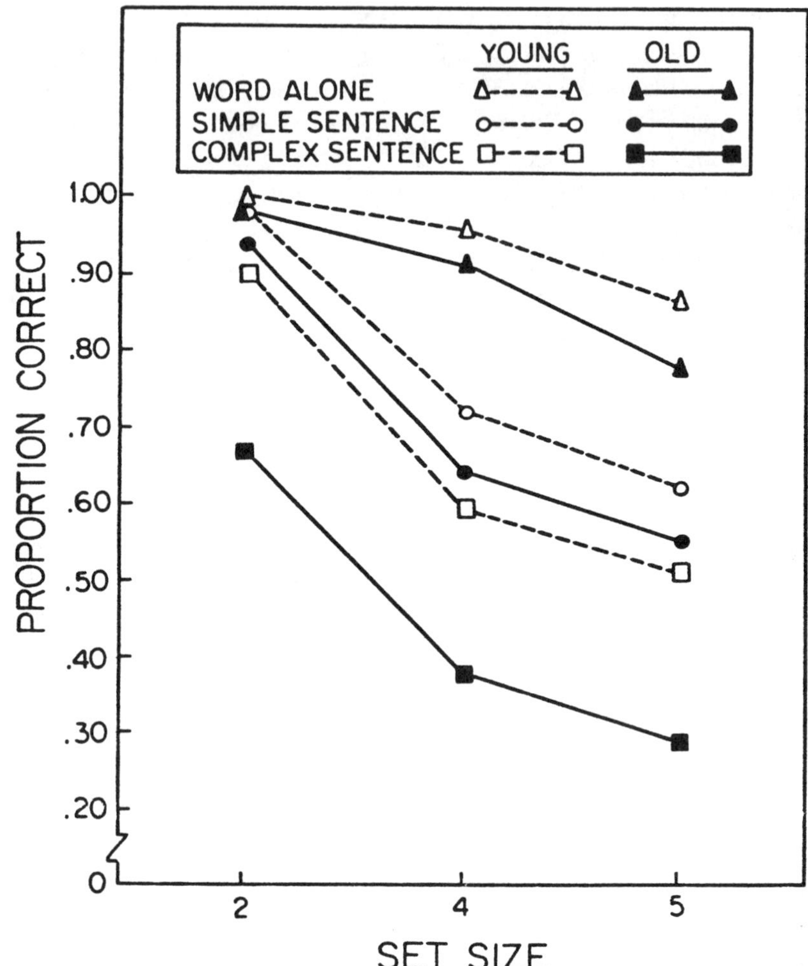

Figure 2. Proportions of words recalled by young and older subjects as a function of sentence complexity and memory set size.

with situations in which environmental support is weak and therefore much self-initiated processing is necessitated. Some of these studies are described in this section.

Waddell and Rogoff (1981) tested the spatial memory for 30 miniature items (model cows, animals, furniture, etc.) placed either in a set of featureless cubicles (like an array of office mailboxes) or in a model panorama composed of fields, woods, buildings, and trees. The subjects were middle-aged (mean = 46 years) and older (mean = 71 years) women. Waddell and Rogoff found that the middle-aged group performed better than did the older group at remembering the spatial position of each individual item in the cubicle condition (73% correct and 36% correct respectively) but that the two groups were essentially identical in the panorama condition (75% correct and 72% correct respectively). By the present account, the greater support provided by the meaningful panorama compensated for the older subjects' lack of spontaneous rich encoding demonstrated in the cubicle condition.

In a somewhat similar study, Sharps and Gollin (1988) presented 40 small objects to younger (mean = 20.5 years) and older (mean = 74.9 years) subjects for later free recall of the objects' names. In one experimental condition, the objects were simply shown to each subject, one at a time, for 5 sec each. In a second condition, the objects were placed in predetermined positions in a room; subjects were led on a tour round the room, viewing each object for 5 sec. In both cases they recalled as many objects as possible after a 2 min break. In the first condition the young and older groups recalled 18.0 and 9.3 objects respectively, and in the second condition the groups recalled 22.7 and 20.7 objects respectively. That is, although both groups benefited from the richer, more supportive context provided by the room, the older group benefited to a much greater extent.

Craik et al. (1987) tested the notion of environmental support under more traditional laboratory conditions. They presented young and elderly subjects with lists of 10 words to recall under various degrees of "support" at encoding and retrieval. In one condition the 10 nouns were both presented and recalled without cues of any kind. At the other extreme, each word was presented during the learning phase with a brief description (e.g., a type of bird—LARK; used in schools—BOOK), and this description was re-presented as a cue at the time of retrieval. Two intermediate conditions were also included, in which the description was provided during learning only or during retrieval only. In line with the environmental support notion, age differences were greatest under the free learning–free recall combination and least when the cues were provided during both encoding and retrieval.

The general conclusion from this work is that older people have greater difficulty in spontaneously producing rich elaborate encodings of events, and have corresponding difficulty at retrieval in self-initiating appropriate retrieval operations. Not all researchers share this view, however; see Burke and Light (1981) for a dissenting opinion.

As a final piece of empirical evidence, the topic of age differences in "source amnesia" will be described briefly. Source amnesia is observed when people remember a fact or a face, but cannot remember where or when they learned the fact or previously encountered the owner of the face. It is a common memory failure in everyone, but the phenomenon is greatly exaggerated in some amnesic patients—especially those with frontal lobe impairment (Schacter et al., 1984). If older people have greater difficulty in integrating experienced events with their contexts of occurrence (perhaps because of minimal frontal lobe dysfunction; Craik et al., 1990) it might be expected that normal aging would be associated with an increased incidence of source amnesia. Such a result was reported by McIntyre and Craik (1987). In outline, the paradigm consists of teaching people new facts, and testing both memory for the facts and memory for where the facts were learned. McIntyre and Craik showed that older people exhibited substantially greater amounts of source amnesia, and a later follow-up study by Craik et al. (1990) found that, within a group of older people, the incidence of source amnesia was related to the degree of subclinical frontal lobe dysfunction as indexed by neuropsychological tests.

In summary, then, the case being argued is that the differential pattern of memory loss found in older people is attributable to the types of operations required by various memory tasks. Memory is poor in older people to the extent that the task demands a great deal of consciously controlled, effortful, self-initiated activity. Speculatively, the difficulty that older people have with such mental operations may be related to a reduction in the processing resources ("mental energy") available to run off the required mental activities.

Three topics will be touched on very briefly in conclusion: The relationship (if any) between the memory "deficits" shown by older adults and young children, relative to the performance of young adults; the relationship between the memory deficits found as a function of the normal aging process and those resulting from pathological conditions; and the relations between age-associated brain changes and changes in memory functioning.

Does Aging Equal Development in Reverse?

It is tempting, if rather naive, to speculate that since many behavioral abilities improve during childhood to a peak in early adulthood and then decline during senescence, that the processes of aging somehow mirror those of child development. In the field of cognition there is certainly evidence that performance on many standard memory tasks increases until young adulthood and then tends to decline. There is also some evidence that the performance of children, like that of older adults, can be helped by the presence of a supportive context. In children as young as three months, Rovee-Collier and Fagen (1981) have shown that a learned kick to activate a toy depends on the same context being present at the test as was present during acquisition. In children aged 1.5 to 2 years, DeLoache and Brown (described by Brown et al., 1983) have shown that when a delay was imposed before the child was allowed to retrieve a hidden toy, children tested in an unfamiliar laboratory adopted various devices such as pointing at the hiding place, standing near it, or talking about it, at twice the level than children tested in their own home. Arguably, the "environmental support" provided by the familiar context had to be augmented by self-generated support in the unfamiliar setting. Finally, preliminary work in my own lab suggests that memory for source develops from younger children to young adults.

Despite these similarities, it is almost certainly unwarranted to conclude that the mechanisms are similar. Among other factors, episodic memory performance depends on the relevant knowledge base, use of strategies, and processing power (working memory). Whereas all three factors undoubtedly develop in children (Brown et al., 1983), it seems to me most unlikely that the same factors decline symmetrically during old age. In particular, there is good evidence that the knowledge base (semantic memory) holds up well (Perlmutter, 1978) and older adults also retain knowledge of memory strategies (Loewen et al., 1990). The effectiveness of processing at encoding and retrieval may well be the major factor in determining the older person's memory performance. Finally, even if the behavior is qualitatively similar, there is no guarantee that the underlying physiology is identical.

Is Pathological Aging an Exaggeration of Normal Aging?

Clearly the answer to this question is "no" at the level of the underlying neurology. The brain changes associated with Alzheimer's disease and other pathological conditions appear to be qualitatively different from those found in normal elderly people (Wisniewski, this volume). However, at the level of behavioral effects, the pattern of results found in normal aging often does appear to resemble the pattern found in, say, Alzheimer's disease, though in a much milder form. There is growing evidence that patients with frontal lobe lesions are deficient in encoding spatio-temporal context; they therefore exhibit source amnesia (Janowsky et al., 1989; Schacter, 1987) just as do normal elderly people (McIntyre and Craik, 1987).

As a second example, Nebes (1989) has recently reviewed the evidence on the ability of Alzheimer patients to retrieve words and other information from semantic memory. At least some of the evidence is in good agreement with the notion that Alzheimer patients, like normal older people, benefit substantially from a supportive semantic context. For example, Kirshner et al. (1984) showed that the ability of Alzheimer patients to name objects increased from masked drawings, to full drawings, to the objects themselves—that is, as the richness of the supplied information increased. Also, Nebes, Boller, and Holland (1986) found that patients were substantially slower (800 msec) than normal controls in completing non-constrained sentences (e.g., "They went to see the famous ______________"), but that the difference dropped to 200 msec with highly constrained sentences (e.g., "Father carved the turkey with a __________________"). It seems that at least some cognitive deficits of Alzheimer patients reflect very inefficient functioning of the underlying mechanisms, and that improvements in environmental support can improve the efficiency, to some degree at least. Again, however, we should be wary of concluding that similar behavior necessarily stems from similar neurophysiology.

Relations to Brain Changes?

Finally in this flurry of speculative remarks, can we say anything at this stage about the underlying correlates of memory dysfunction in the elderly? Given the complexity of brain mechanisms it is highly unlikely that age-related memory losses are associated with any one cause, but one related group of brain changes may at least prove to be a major component of the final story. At the behavioral level it is striking that essentially all the qualitative effects associated with memory decline in the elderly can be mimicked in young adults working under conditions of divided attention (Craik, 1982, 1983). There is also evidence that a very similar pattern of memory strengths and weaknesses is shown in connection with alcoholic intoxication and sleep deprivation (Nilsson et al., 1989). The fact that these varied conditions (intoxication, aging, sleep deprivation) resemble each other suggests the presence of some common component, and the fact that division of attention in normal young subjects is associated with the same pattern suggests that the common component may be a reduction in attentional resources. That is, the decline in memory efficiency may reflect the reduction (temporarily or permanently) of levels of cortical activation, since Petersen and others have shown in PET studies that behavioral attention is a correlate of enhanced blood flow (and presumably increased neural activity) in the relevant cortical regions.

Why should such activity levels be reduced in the elderly? One current suggestion has been made by Gold and his collaborators (Gold, 1987) in terms of reduced cortical glucose uptake by the older brain. Clearly, aging is accompanied by structural changes too (e.g., Morgan and May, 1990; Horvath and Davis, 1990) but from the standpoint of the present argument it is attractive to believe that one major correlate of memory dysfunction in the elderly is a reduction in the efficiency of neural processing following a reduction in the metabolic supply. Further, that this decreased efficiency can be compensated for (up to some limit at least) by an enhancement of appropriate guiding stimulation from the external environment.

However the story comes out eventually, it is clear that any final account of memory and its disorders must deal with the phenomena at several different levels of description (biochemical, anatomical, physiological, and behavioral). A fully developed science of memory must incorporate all these levels and also provide an account of how changes at one level reflect changes at the others.

Acknowledgments. This chapter was prepared with the help of a grant from the Natural Sciences and Engineering Council of Canada.

References

Atkinson RC, Shiffrin RM (1968): Human memory: A proposed system and its control processes. In: *The Psychology of Learning and Motivation*, Vol. 2, Spence KW, Spence JT, eds. New York: Academic Press, pp. 89—195

Baddeley AD (1986): *Working Memory.* London: Oxford University Press.

Baddeley AD, Hitch GJ (1974): Working memory. In: *The Psychology of Learning and Motivation*, Vol. 8, Bower GH, ed. New York: Academic Press, pp. 47–90

Brown AL, Bransford JD, Ferrara RA, Campione JC (1983): Learning, remembering, and understanding. In: *Handbook of Child Psychology*, Vol. III, Flavell JH, Markman EM, eds. New York: Wiley, pp. 77–166

Burke DM, Light LL (1981): Memory and aging: The role of retrieval processes. *Psychological Bulletin* 90:513–545

Craik FIM (1977): Age differences in human memory. In: *Handbook of the Psychology of Aging*, Birren JE, Schaie KW, eds. New York: Van Nostrand Reinhold, pp. 384–420

Craik FIM (1982): Selective changes in encoding as a function of reduced processing capacity. In: *Cognitive Research in Psychology*, Klix F, Hoffmann J, van der Meer E, eds. Berlin: Deutschen Verlag der Wissenschaften, pp. 152–161

Craik FIM (1983): On the transfer of information from temporary to permanent storage. *Philosophical Transactions of the Royal Society of London*, Series B 302:341–359

Craik FIM (1984): Age differences in remembering. In: *Neuropsychology of Memory*, Squire LR, Butters N, eds. New York: Guilford Press

Craik FIM (1986): A functional account of age differences in memory. In: *Human Memory and Cognitive Capabilities. Mechanisms and Performances*, Klix F, Hagendorf H, eds. Amsterdam: North-Holland, pp. 409–422

Craik FIM, Byrd M, Swanson JM (1987): Patterns of memory loss in three elderly samples. *Psychology and Aging* 2:79–86

Craik FIM, Lockhart RS (1972): Levels of processing: A framework for memory research. *Journal of Verbal Learning and Verbal Behavior* 11:671–684

Craik FIM, Morris LW, Morris RG, Loewen ER (1990): Relations between source amnesia and frontal lobe functioning in older adults. *Psychology and Aging* 5:148–151.

Daneman M, Carpenter PA (1980): Individual differences in working memory and reading. *Journal of Verbal Learning and Verbal Behavior* 19:450–466

Dobbs AR, Rule BG (1989): Adult age differences in working memory. *Psychology and Aging* 4:500–503

Gick ML, Craik FIM, Morris RG (1988): Task complexity and age differences in working memory. *Memory and Cognition* 16:353–361

Gold PE (1987): Sweet memories. *American Scientist* 75:151–155

Hasher L, Zacks RT (1979): Automatic and effortful processes in memory. *Journal of Experimental Psychology: General* 108:356–388

Horvath TB, Davis KL (1990): Central nervous system disorders in aging. In: *Handbook of the Biology of Aging*, 3rd edition, Schneider EL, Rowe JW, eds. San Diego: Academic Press, pp. 306–329

Hultsch DF, Dixon RA (1990): Learning and memory in aging. In: *Handbook of the Biology of Aging*, 3rd edition, Schneider EL, Rowe JW, eds. San Diego: Academic Press, pp. 258–274

Janowsky JS, Shimamura AP, Squire LR (1989): Source memory impairment in patients with frontal lobe lesions. *Neuropsychologia* 27:1043–1056

Kausler DH (1982): *Experimental Psychology and Human Aging*. New York: Wiley.

Kirshner HS, Webb WG, Kelly MP (1984): The naming disorder of dementia. *Neuropsychologia* 22:23–30

Light LL, Anderson PA (1985): Working-memory capacity, age, and memory for discourse. *Journal of Gerontology* 40:737–747

Light LL, Singh A, Capps JL (1986): Dissociation of memory and awareness in young and older adults. *Journal of Clinical and Experimental Neuropsychology* 8:62–74.

Loewen ER, Shaw RJ, Craik FIM (1991): Age differences in components of metamemory. *Experimental Aging Research* 16:43–48

McIntyre JS, Craik FIM (1987): Age differences in memory for item and source information. *Canadian Journal of Psychology* 41:175–192.

Morgan DG, May PC (1990): Age-related changes in synaptic neurochemistry. In: *Handbook of the Biology of Aging*, 3rd edition, Schneider EL, Rowe JW, eds. San Diego: Academic Press, pp. 219–254

Morris RG, Gick ML, Craik FIM (1988): Processing resources and age differences in working memory. *Memory and Cognition* 16:362–366

Nebes RD (1989): Semantic memory in Alzheimer's disease. *Psychological Bulletin* 106:377–394

Nebes RD, Boller F, Holland A (1986): Use of semantic context by patients with Alzheimer's disease. *Psychology and Aging* 1:261–269

Nilsson LG, Bäckman L, Karlsson T (1989): Priming and cued recall in elderly, alcohol intoxicated and sleep deprived subjects: A case of functionally similar memory deficits. *Psychological Medicine* 19:423–433

Perlmutter M (1978): What is memory aging the aging of? *Developmental Psychology* 14:330–345

Poon LW (1985): Differences in human memory with aging: Nature, causes, and clinical implications. In: *Handbook of the Psychology of Aging*, Birren JE, Schaie KW, eds. New York: Van Nostrand Reinhold, pp. 427–462

Richardson-Klavehn A, Bjork RA (1988): Measures of memory. *Annual Review of Psychology* 39:475–543

Rosenthal I (1988): *The Invention of Memory: A New View of the Brain*. New York: Basic Books

Rovee-Collier CK, Fagen JW (1981): The retrieval of memory in early infancy. In: *Advances in Infancy Research*, Vol. 1, Lipsitt LP, ed. Norwood, NJ: Ablex

Salthouse TA (1982): *Adult Cognition: An Experimental Psychology of Human Aging*. New York: Springer-Verlag

Schacter DL (1987): Implicit memory: History and current status. *Journal of Experimental Psychology: Learning, Memory, and Cognition* 13:501–518

Schacter DL, Harbluk JJ, McLachlan DR (1984): Retrieval without recollection: An experimental analysis of source amnesia. *Journal of Verbal Learning and Verbal Behavior* 23:593–611

Sharps MJ, Gollin ES (1988): Aging and free recall for objects located in space. *Journal of Gerontology* 43:8–11

Tulving E (1983): *Elements of Episodic Memory*. Oxford: Oxford University Press

Tulving E (1987): Multiple memory systems and consciousness. *Human Neurobiology* 6:67–80

Waddell KJ, Rogoff B (1981): Effect of contextual organization on spatial memory of middle-aged and older women. *Developmental Psychology* 17:878–885

Welford AT (1958): *Ageing and Human Skill.* London: Oxford University Press

Wingfield A, Stine EAL, Lahar CJ, Aberdeen JS (1988): Does the capacity of working memory change with age? *Experimental Aging Research* 14:103–107

Wright R (1981): Aging, divided attention, and processing capacity. *Journal of Gerontology* 36:605–614

PART IV
Clinical Applications

CT, MRI, and PET Studies of Hippocampal Pathology in Alzheimer's Disease

M. J. de Leon, G. Smith, A. E. George, T. McRae, J. Golomb, A. Convit, A. Kluger, W. Tsui, S. H. Ferris, and A. P. Wolf

Postmortem studies reveal numerous biochemical and neurotransmitter deficits in Alzheimer's disease (AD) patients relative to elderly controls. Many of these deficits are more pronounced in the hippocampus; than in the neocortex (for a review see Selkoe and Kosik, 1983). There are many structural hippocampal abnormalities documented in AD. They include neuronal loss, granulovacuolar degeneration, neurofibrillary tangles, and neuritic plaques (Kemper, 1984). In AD these neuropathological findings show a consistent topography with respect to the hippocampal formation and its projections. Pyramidal cells are the primary neuronal population involved. These affected cells are located within the entorhinal cortex (the neurons of origin of the perforant pathway), the subiculum, CA1 of the hippocampus, and the temporal association cortex. This pattern of degeneration results in the "isolation" of the hippocampal formation from its neocortical association areas (Hyman et al., 1984).

Structural hippocampal pathology, similar to that documented in AD, also occurs in aging Down patients and in nondemented elderly. There is data showing that in Down's syndrome the first site of AD-type degeneration occurs in the hippocampus (Mann and Esiri, 1988). These findings support the hypothesis of early hippocampal changes in AD. At autopsy, severe hippocampal degeneration has been documented in AD and is observed occasionally in the absence of neocortical deficits. These findings have led some investigators to characterize AD as "a hippocampal dementia" (Ball et al., 1985). De la Monte (1989) reported that hippocampi (Ammon's horn) both from AD patients and clinically normal elderly with neuropathological evidence of AD showed 10% volume reductions from "pure" elderly control values. Neocortical changes were found only in the demented AD group, suggesting the primacy of the hippocampal changes. With increasing frequency, the hippocampi of normals, from the seventh to tenth decade, manifest neuronal loss, neurofibrillary tangles, granulovacuolar degeneration, and neuritic plaques (for review see Kemper, 1984). It is largely unknown how prevalent these changes are and if these changes are related to the mild memory changes so often reported in clinical studies of nondemented normal elderly. Moreover, it is unknown if the long-lived individuals with AD-like hippocampal changes will develop neocortical changes and AD-like clinical symptoms.

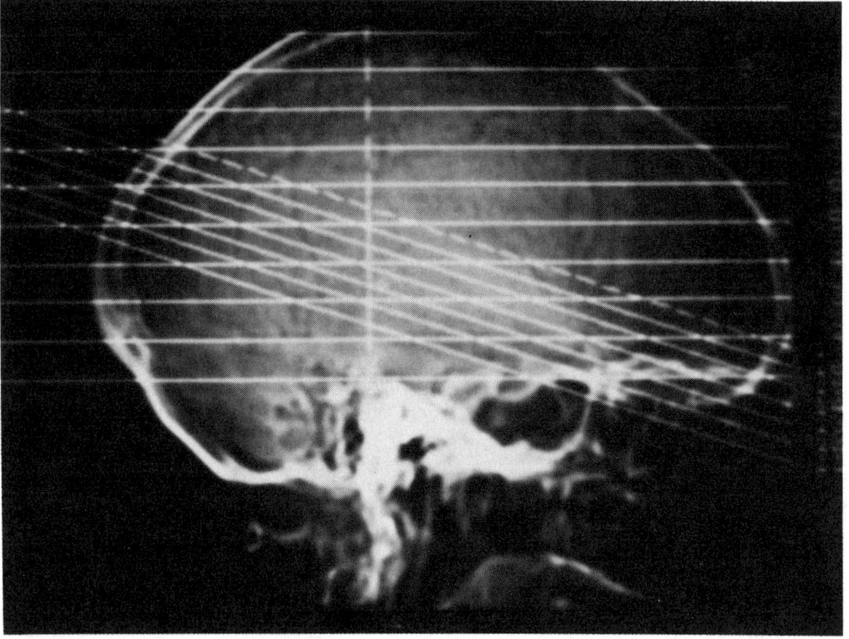

Figure 1. Lateral CT scout depicting the conventional canthomeatal plane (scans 1–10) and the negative angulation (scans 11–17).

The observations of hippocampal pathology in the absence of neocortical pathology in normal elderly and in nondemented Down's syndrome cases suggests that an early focus of pathology in AD is in the hippocampus. However, at present there is no direct neuropathological evidence to support this contention. Longitudinal imaging of hippocampal changes is now providing the first opportunities to directly confirm the primacy of hippocampal change in early AD patients and individuals at risk for AD.

CT and MRI Hippocampal Studies

Cross-sectional studies

Only a few *in vivo* studies of the hippocampus have been reported in the AD literature. Seab et al. (1988) found that hippocampal volumes on coronal MRI were reduced in AD from elderly normal values. Our cross-sectional negative angulation CT studies (see Fig. 1) of the hippocampal and choroidal fissures found hippocampal atrophy to be of diagnostic value (de Leon et al., 1988; de Leon et al., 1989; George et al., 1990). Moreover, our longitudinal follow-up of nondemented elderly with baseline hippocampal atrophy suggest this measure has value as a potential early diagnostic marker for AD (de Leon et al., 1989 and see Fig. 1).

The CT scans of 76 AD patients, 27 questionably demented patients, and 72 controls, were evaluated subjectively on the basis of hippocampal atrophy. We have found on CT and MRI, and later confirmed at autopsy, that these atrophic appearances are caused by CSF accumulations secondary to parenchymal volume loss (see Fig 2).

The hippocampi were separately rated for each hemisphere using a four-point scale (0 = none, 1 = questionable, 2 = mild–moderate, and 3 = severe). Using a cutoff rating of > 2, hippocampal atrophy was significantly more prevalent both in questionably demented and AD groups than in controls (overall 77% versus 22%). The atrophy occurred more frequently in the advanced stages of AD (as assessed by the Global Deterioration Scale; Reisberg et al., 1982). The prevalence of the hippocampal atrophy is depicted in Fig. 3.

Consistent with the neuropathological literature, our normal controls with hippocampal atrophy were significantly older than those without evidence of atrophy (75.2 versus 68.9 years). In the normal elderly, the atrophy was clearly age dependent (see Fig. 4). In AD patients, atrophy was equally distributed across all age groups.

Atrophy severity scores for both hemispheres were averaged to derive a composite score. The composite score was subjected to analysis of variance to determine the significance of the severity of atrophy. Relative to normal controls, hippocampal atrophy was significantly and increasingly more severe in each of the three levels of impairment: questionable dementia, mild AD, and moderate to severe AD. From the sixth to eighth decade, atrophic changes in AD patients were more severe than in normals. Hippocampal atrophy was equally severe in both AD and normal groups in the ninth decade. These data suggest that the diagnostic utility of the hippocampal atrophy measure is limited to patients under 75 years of age.

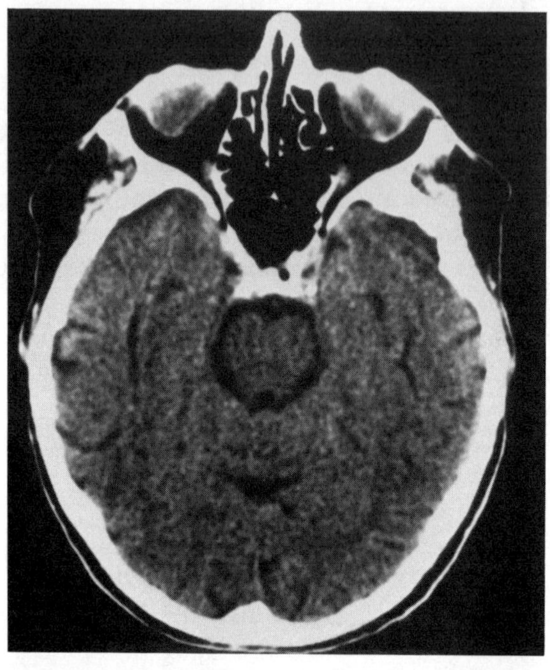

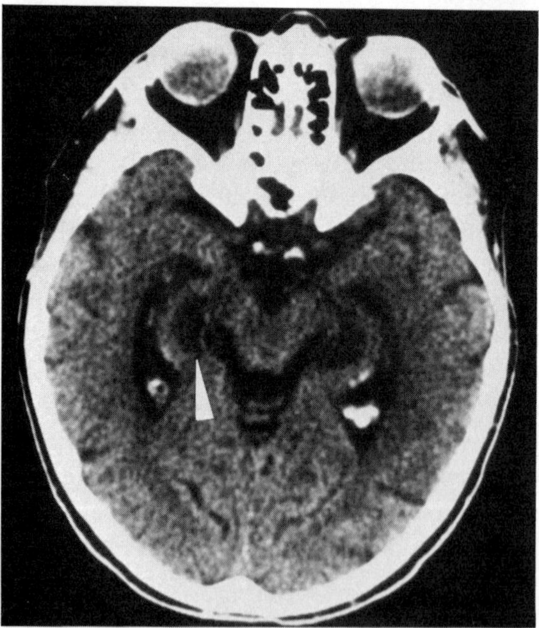

Figure 2. Transaxial CT scans depicting the hippocampal region. Top study is normal and bottom study shows severe bilateral CSF accumulation in the region of the choroidal and hippocampal fissures (arrow).

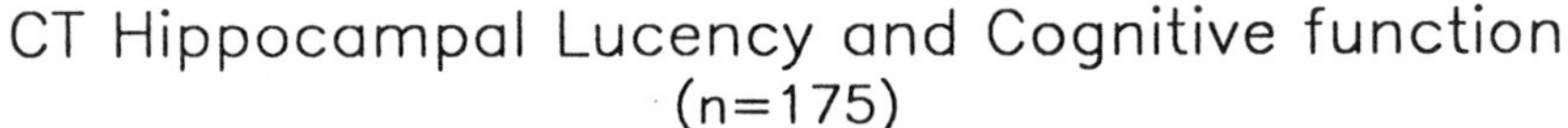

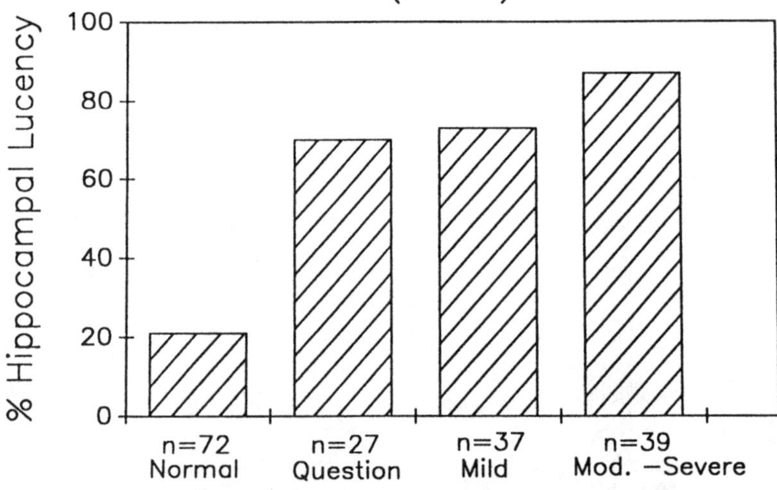

Figure 3. The prevalence of hippocampal atrophy in normal elderly and Alzheimer patients with different severities of cognitive impairment.

Recently, we successfully demonstrated an objective validation for the subjective estimates of dilated hippocampal and parahippocampal CSF spaces. The validation procedure used contiguous 4-mm-thick T_1-weighted coronal MRI to determine the volumes of CSF in the region of the choroidal and hippocampal fissures and the perimesencephalic cisterns. In all studies, the CSF volume was determined between the interpeduncular cistern and the cerebral aqueduct, spanning four slices perpendicular to the plane of the hippocampus. The results for eight AD patients and five age-matched elderly controls indicated that CSF volume, derived using an individualized threshold procedure, was significantly associated with the subjective ratings of dilated choroidal/hippocampal fissures derived from the 5 mm negative angulation transaxial T_1 studies ($r = .86$, $p < .001$). In the region of maximum CSF accumulation posterior to the pes hippocampus, the MRI determined volume of CSF was increased in AD by 87% ($p < .04$, one-tailed). At the same slice location, the hippocampal and parahippocampal gyrus parenchymal volumes were decreased by 35% in the AD patients ($p < .01$).

Longitudinal studies

The results of the cross sectional studies suggest that hippocampal changes occur early in the course of AD. In order to assess the early diagnostic value of hippocampal changes, a three-year follow-up was

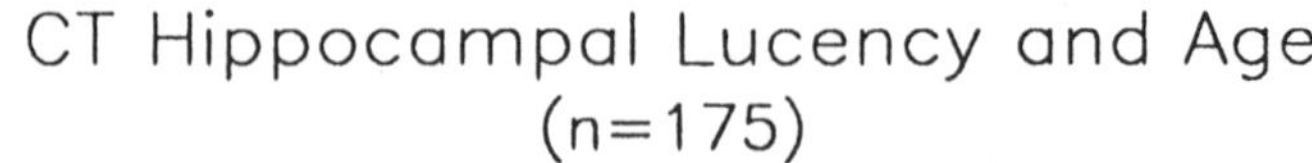

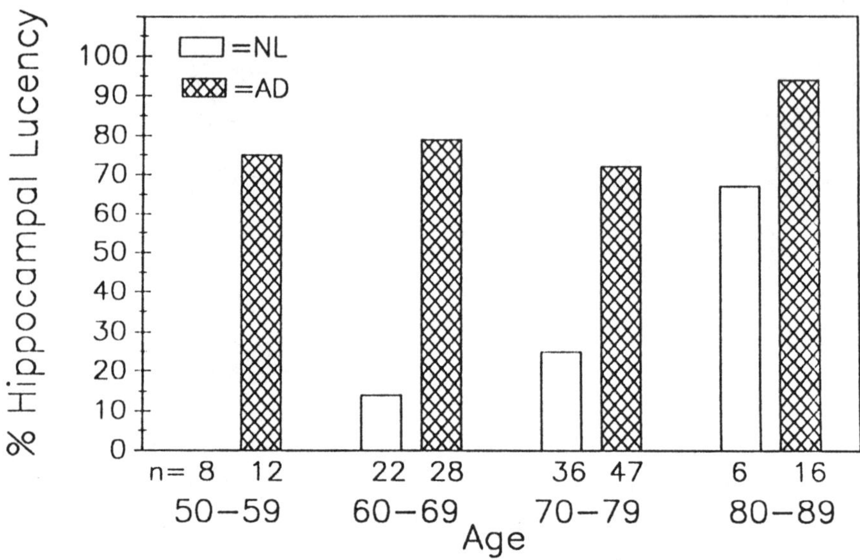

Figure 4. The prevalence of hippocampal atrophy for normal elderly and Alzheimer patients as a function of age.

carried out in the normal elderly group (N = 28) and the questionable dementia group (N = 20). Over the three year interval, 11 out of 20 of the questionable group and 1 out of 28 of the controls deteriorated to receive the diagnosis of AD (GDS $\geq$ 4). Baseline hippocampal ratings $\geq$ 2 were present in 91% of the decliners and absent in 81% of the non-decliners (chi square = 20.02, $p < .001$). These studies point to the need for the longitudinal study of the temporal relation between hippocampal change, neocortical pathology, and the development of intellectual dysfunction. Longitudinal study is also essential for an understanding of the early stages of AD, its progression, and how it differs from the normal aging process. Because only one of our normal elderly deteriorated, we feel that the period of observation must be extended beyond three years in order to evaluate the predictive risk of the hippocampal change for cognitive deterioration.

PET Studies

Positron Emission Tomography investigations using relatively low resolution equipment (i.e., > 1 cm in all planes) have not been able to directly sample the relatively small hippocampus. Consequently, these studies have focused on relatively large neocortical and subcortical samples. Of course, these studies have been limited in the anatomic definition of the underlying substrate sampled. Consequently, in the interest of sample purity, areas potentially containing partially volumed CSF and/or other large contaminating structures are typically avoided and at times corrections are estimated based on the corresponding CT or MRI scans.

Despite the limitations in spatial resolution, there is a large literature describing the diagnostic utility of PET in studies of AD (Benson et al., 1983; Cutler et al., 1985; de Leon et al., 1983; Duara et al., 1986; Ferris et al., 1980; Foster et al., 1984; Friedland et al., 1985; Haxby et al., 1986). There is a consensus that temporal lobe and temporo-parietal reductions in glucose utilization are typically the most salient changes among the global changes seen in AD. Unfortunately, there is very little longitudinal data published and consequently little is known about progressive patterns of brain change or early features or predictors of dementia. Moreover, there are no reports addressing hippocampal metabolic changes using high resolution PET (i.e., less than 8 mm resolution in all planes). With respect to longitudinal PET glucose metabolism studies of AD, the few published works (Duara et al., 1986; Frackowiak et al., 1981; Jagust et al., 1988) suggest overall that the longitudinal changes reflect what was observed cross-sectionally. Our own data also support this observation.

We conducted a cross-sectional study using low spatial resolution PET VI data on 44 AD patients and 21 normal elderly controls. The mean age for the control group was 69.4 ± 6.5 years and for the AD group 68.3 ± 8.8 years. The results indicated that the greatest glucose metabolism differences were found in the temporal, parietal, and frontal association areas. Lesser involvement was found in subcortical grey structures (e.g., thalamus and basal ganglia and in the primary sensory and motor areas). When comparing mild dementia cases (GDS 3 and 4) with severe dementia cases (GDS greater than 5) the greatest reductions were found in the temporal, parietal, and frontal association areas.

Recently we completed the three-year longitudinal PET VI follow-up of 14 of the AD patients and 15 of the controls. The results indicated that over the follow-up interval the AD patients showed the greatest declines in the parietal lobe (47%) and temporal lobe (43%) followed by the frontal lobe (36%). The longitudinal declines of about 10–15% in basal ganglia, thalamus, and cerebellum were not significant. The normal controls did not show any differences over the study period. Therefore, the regional pattern of metabolic deficit observed cross-sectionally is also observed longitudinally. The neocortical areas of greatest initial metabolic vulnerability to the AD process show the greatest decrements over time.

It is important to note that the results observed cross-sectionally and longitudinally were in excess of the structural atrophic change observed. The use of either measures of ventricular dilatation (a ventricle to brain ratio) or measures of regional sulcal CSF accumulation derived from CT as covariates in the analysis of the metabolic data did not alter the significance of the results.

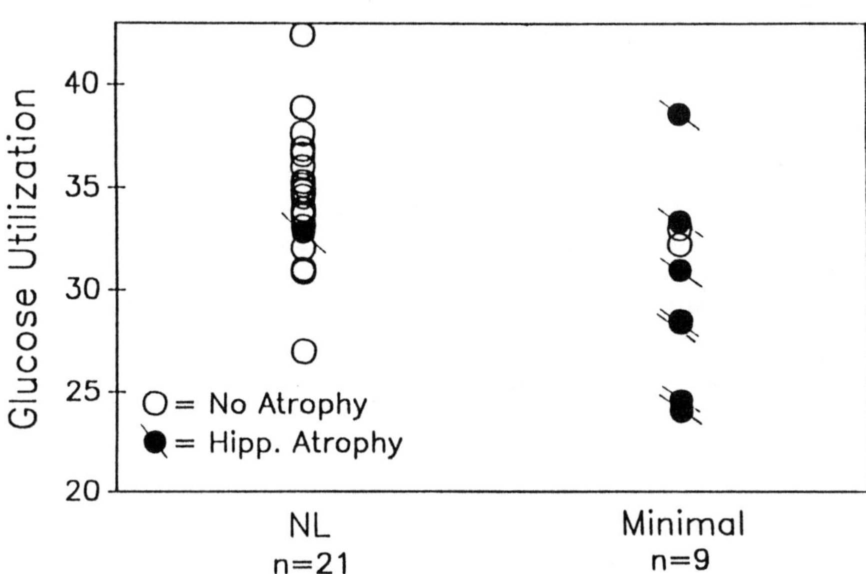

Figure 5. The comparison of the PET glucose utilization study of the temporal lobe and the CT study of the hippocampal region in the differentiation of normal elderly from minimally impaired (nondemented) elderly patients.

In order to assess whether structural hippocampal atrophy is observed before the neocortical metabolic reduction, the metabolic rates for the nondemented subjects with and without hippocampal atrophy were compared. Hippocampal atrophy was found in 1 out of 21 normal and 7 out of 9 questionable patients, whereas temporal lobe glucose metabolism and ventricular enlargement did not reveal any group differentiation (Fig. 5). Therefore, with respect to the predictors of change, our longitudinal data show that CT hippocampal atrophy precedes both CT ventricular enlargement and PET cortical temporoparietal reductions in the questionably affected patients. The observations that structural hippocampal change precedes neocortical involvement further suggests that hippocampal involvement occurs earlier in the course of AD. The use of the high resolution PET will enable future studies to examine whether metabolic change in the hippocampus precedes structural hippocampal change.

The baseline metabolic rate of the temporal lobe is of some predictive value in determining the extent of subsequent metabolic temporal lobe reductions. Specifically, the greater the baseline deficit the greater the absolute or the relative reduction at follow-up (see Fig. 6). Similarly, the more demented the patient is at baseline the greater the longitudinal reduction in metabolic rates. Such data suggest that the rate of patient brain change is not constant but rather accelerating. With respect to diagnosis, in mild to moderate AD patients, the progressive PET metabolic deficits and the cross sectional hippocampal atrophy are of equal value. Both these measures are superior to longitudinal and cross sectional determinations of ventricular enlargement in discriminating AD from normals.

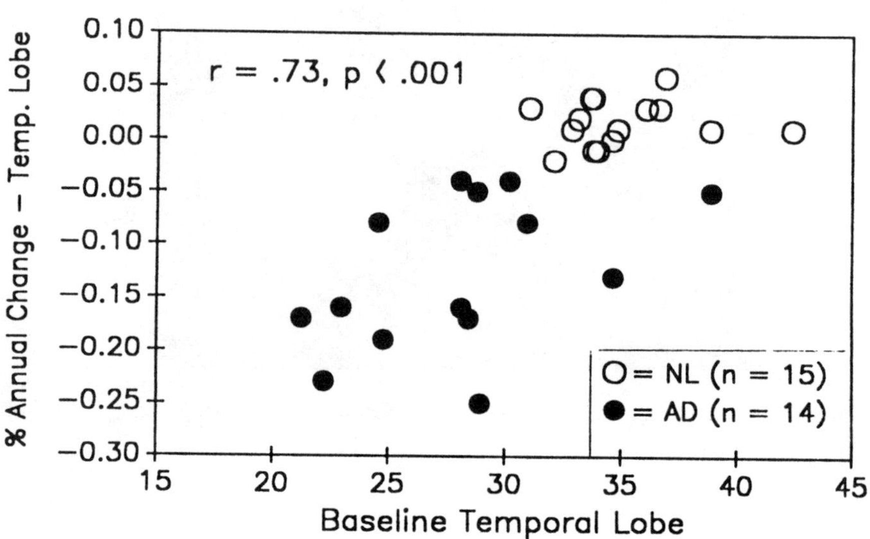

Figure 6. The annual rate of glucose utilization change in the temporal lobe as a function of baseline temporal lobe glucose utilization.

Using the high resolution CTI 931 PET camera (6.8 mm resolution in all planes) and a 25° negative angulation scan protocol, we can now evaluate the hippocampal region accurately. Using PET and MRI scans (3.4 mm thick) we studied 9 elderly controls and 14 AD patients. The mean ages were 70 ± 10 and 72 ± 7 and the mean GDS was 1.5 ± .5 and 4.7 ± .8, for the controls and patients, respectively. Region-of-interest analyses, including the hippocampus, were performed with reference to the subjects' MRI scans (see Fig. 7).

MANOVA revealed a significant diagnosis effect for the set of regional metabolic values. Subsequent *post hoc* ANOVAs revealed significant ($p < .01$) metabolic reductions in AD relative to control in the hippocampus, lateral temporal lobe, and frontal lobe. The visual cortex, white matter, basal ganglia, thalamus, cerebellum, and midbrain were not significantly different between groups (p's > .05). From these results it appears that PET hippocampal evaluations show promise as a new measure of the effects of AD on the brain.

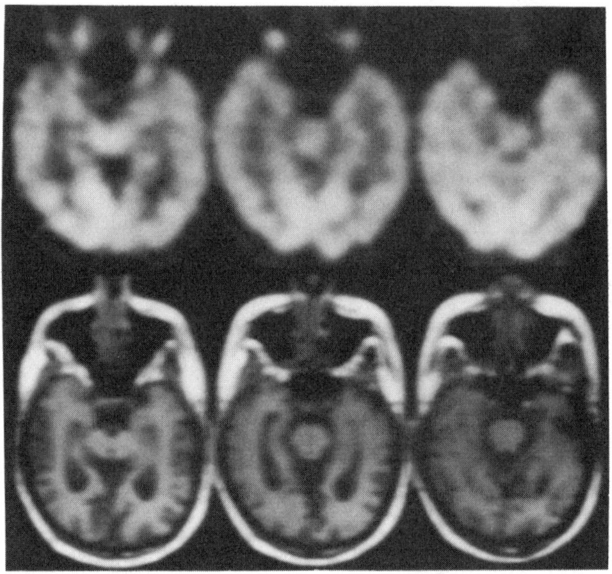

Figure 7. Hippocampal studies using high resolution PET (top) and MRI (bottom).

Postmortem Studies

On seven confirmed AD cases, we directly compared the postmortem hippocampal examination with both the antemortem CT and the postmortem MRI. The neuropathology protocol included coronal hippocampal samples taken at the level of the lateral geniculate nucleus ($N = 7$) and at multiple levels ($N = 5$). Also, two hemispheres were cut in the special negative angulation. Following standard processing with paraffin, the formalin-fixed sections were stained with Luxol fast blue and counterstained with H and E, also Congo red and Bodian stains and 4G8 anti-β-amyloid protein immunostain were used. The AD patients showed at the lateral geniculate level, extensive neurofibrillary tangles, amyloid deposition, and neuronal loss in the hippocampal formation and parahippocampal gyrus when compared against two controls. All AD cases demonstrated neuropathologic and neuroimaging evidence of dilated transverse fissures including choroidal and hippocampal fissures.

Conclusions

In vivo examination of the hippocampal region appears to be of diagnostic and predictive value in the clinical study of AD. Hippocampal changes appear early in the natural history of AD and when present in the absence of other structural or metabolic brain changes, appear to be associated with the minimal cognitive deficits so often ascribed to the normal aging process. Early hippocampal atrophy appears to be related to the AD process pathologically, although it should be noted that there are no reports of autopsies on early AD patients who died of other causes. Nevertheless, it is universally observed that hippocampal atrophy occurs in AD, and we extend these findings to show that the *in vivo* imaging changes are directly related to degenerative parenchymal changes in the hippocampus.

Our longitudinal data indicate that hippocampal atrophy is predictive of the clinical dementia associated with AD. For the first time, these findings now permit the identification of patients at risk for AD and therefore provide an early marker for potentially suitable drug study patients. Hippocampal atrophy as a potential early marker also provides an opportunity to investigate the emergence of pathophysiologic phenomena and to evaluate their clinical significance in AD. In particular, neuroendocrine disturbances are related to hippocampal degeneration in both animal models (Sapolsky et al., 1986) and in AD (de Leon et al., 1988). The extent to which these changes are related to the hippocampal or subsequent neocortical damage remains unknown. However, we have renewed optimism that with earlier diagnosis the search for relevant mechanisms of disease and effective treatment will be facilitated.

References

Ball MJ, Hachinski V, Fox A, Kirshen AJ, Fisman M, Blume W, Kral VA, Fox H (1985): A new definition of Alzheimer's disease: A hippocampal dementia. *Lancet*, 5 January 1985:14–16

Benson DF, Kuhl DE, Hawkins RA, Phelps ME, Cummings JL, Tsai SY (1983): The fluorodeoxyglucose 18F scan in Alzheimer's disease and multi-infarct dementia. *Archives of Neurology* 40:711–714

Cutler NR, Haxby JV, Duara R, Grady CL, Kay AD, Kessler RM, Sundaram M, Rapoport SI (1985): Clinical history, brain metabolism, and neuropsychological function in Alzheimer's disease. *Annals of Neurology* 18:298–309

de la Monte SM (1989): Quantitation of cerebral atrophy in preclinical and end-stage Alzheimer's disease. *Annals of Neurology* 25:450–459

de Leon MJ, Ferris SH, George AE, Christman DR, Fowler JS, Gentes C, Reisberg B, Gee B, Emmerich M, Yonekura Y, Brodie J, Kricheff II, Wolf AP (1983): Positron emission tomography studies of aging and Alzheimer's disease. *American Journal of Neuroradiology* 4:568–571

de Leon MJ, George AE, Stylopoulos LA, Smith G, Miller DC (1989): Early marker for Alzheimer's disease: The atrophic hippocampus. *Lancet*, September 16, 1989:672–673

de Leon MJ, McRae T, Tsai JR, George AE, Marcus DL, Freedman M, Wolf AP, McEwen B (1988): Abnormal hypercortisolemic response in Alzheimer's disease linked to hippocampal atrophy. *Lancet*, August 13:391–392

Duara R, Grady C, Haxby J, Sundaram M, Cutler NR, Heston L, Moore A, Schlageter N, Larson S, Rapoport SI (1986): Positron emission tomography in Alzheimer's disease. *Neurology* 36:879–887

Ferris SH, de Leon MJ, Wolf AP, Farkas T, Christman DR, Reisberg B, Fowler JS, MacGregor R, Goldman A, George AE and Rampal S (1980): Positron emission tomography in the study of aging and senile dementia. *Neurobiology of Aging* 1:127–131

Foster NL, Chase TN, Mansi L, Brooks R, Fedio P, Patronas NJ, Di Chiro G (1984): Cortical abnormalities in Alzheimer's disease. *Annals of Neurology* 16:649–654

Frackowiak RSJ, Pozzilli C, Legg NJ, DuBoulay GH, Marshall J, Lenzi GL, Jones T (1981): Regional cerebral oxygen supply and utilization in dementia–a clinical and physiological study with oxygen-15 and positron tomography. *Brain* 104:753–778

Friedland RP, Budinger TF, Koss E, Ober BA (1985): Alzheimer's disease: Anterior-posterior and lateral hemisphere alterations in cortical glucose utilization. *Neuroscience Letters* 53:235–240

George AE, de Leon MJ, Stylopoulos LA, Miller J, Kluger A, Smith G, Miller DC (1990): CT diagnostic features of Alzheimer disease: Importance of the coronal/hippocampal fissure complex. *American Journal of Neuroradiology* 11:101–107

Haxby JV, Grady CL, Duara R, Schlageter N, Berg G, Rapoport SI (1986): Neocortical metabolic abnormalities precede nonmemory cognitive defects in early Alzheimer's-type dementia. *Archives of Neurology* 43:882–885

Hyman BT, Van Hoesen GW, Damasio AR, Barnes CL (1984): Alzheimer's disease: Cell-specific pathology isolates the hippocampal formation. *Science* 225:1168–1170

Jagust WJ, Friedland RP, Budinger TF, Koss E, Ober B (1988): Longitudinal studies of regional cerebral metabolism in Alzheimer's disease. *Neurology* 38:909–912

Kemper T (1984): Neuroanatomical and neuropathological changes in normal aging and in dementia. In: *Clinical Neurology of Aging*, Albert ML, ed., New York: Oxford University Press

Mann DMA, Esiri MM (1988): The site of the earliest lesions of Alzheimer's disease. *New England Journal of Medicine* 318:12:789–790

Reisberg B, Ferris SH, de Leon MJ, Crook T (1982): The global deterioration scale for assessment of primary degenerative dementia. *American Journal of Psychiatry* 139:1136–1139

Sapolsky R, Krey LC, McEwen BS (1986): The neuroendocrinology of stress and aging: The glucocorticoid cascade hypothesis. *Endocrine Review* 7:284–301

Seab JP, Jagust WS, Wong SFS, Roos MS, Reed BR, Budinger TF (1988): Quantitative NMR measurements of hippocampal atrophy in Alzheimer's disease. *Magnetic Resonance* 8:200–228

Selkoe D, Kosik K (1983): Neurochemical changes with aging. In: *Clinical Neurology of Aging*, Albert ML, ed., New York: Oxford University Press, pp. 53–75

Is Alzheimer's Dementia a Treatable Disease?

B. Winblad, R. F. Cowburn, C. O'Neill, and C. J. Fowler

In recent years, there have been a large number of clinical trials aimed at alleviating the symptoms of dementia of the Alzheimer type (DAT), a disorder affecting more than 5% of the population over the age of 65 years. Despite a wide variety of treatment strategies, ranging from neurotransmitter (preferentially cholinergic) replacement therapy through hormonal treatments to metabolic activators (see Table 1 for examples), results of clinical trials have generally been disappointing (for reviews, see Bagne et al., 1986; Gottfries, 1989; Whalley, 1989). Such poor outcomes of treatments raise the question as to whether dementia of the Alzheimer type is a treatable condition. In the present chapter, this question has been examined with respect to the neurotransmitter replacement strategies.

Clinical Trial Design

In general, clinical trials of drugs in DAT have been difficult to interpret, for two reasons. First, the patient population used has generally been in an advanced stage of DAT, in itself a rate-limiting factor in treatment outcome. Second, rating scales and other neuropsychological tests to be used have to cover a broad spectrum of cognitive and behavioral symptoms, and have to be theoretically anchored in experimental psychology. The criteria tasks required must cover the symptoms that are assumed to be related to the neurochemical deficit that is the target for the remedial therapy in question. Interpretation of rating scales can be obfuscated by unequal baselines between differentially affected patients and normal control subjects (for review, see Backman et al., 1990).

Table 1

Strategy / Examples	**Strategy** / Examples
Neurotransmitter replacement	**Vasodilators**
1. *Cholinergic neurones*	Papaverine, Cinnarizine
Precursors: choline, lecithin	
Releasers: 4-aminopyridine	**"Metabolic enhancers"**
Agonists: arecoline, pilocarpine	Piracetam, Vincamine
AChE inhibitors:	
Tetraminoacridine, physostigmine	**Multi-action**
2. *Monoaminergic neurones*	Hydergine
Precursors: L-DOPA, L-tryptophan	
Agonists: Bromocryptine, clonidine	**Vitamins, etc.**
Reuptake inhibitors: Amitriptyline,	Vitamin B12
alaproclate, citalopram	Zinc sulphate
MAOIs: L-selegiline	Niacin-pentylenetetrazol
3. *Neuromodulators*	
Neuropeptides: ACTH(4-19),	**Other treatments**
vasopressin	Gerovital H3
Opioid receptor antagonists:	Hyperbaric oxygen
naloxone, naltrexone	Phosphatidylserine
	Captopril

Treatment strategies in DAT: some clinically tested examples (adapted and partially updated from Table 1 of Bagne et al., 1986).

As well as these two constraints, other factors such as trial design, dose regime, and duration of treatment are critical to the validity of the treatment (for review see Petrie, 1986). While it is beyond the scope of the present chapter to deal with all these factors, patient selection and dose regimes used will be considered below.

It is now clear that the clinical diagnosis of DAT, based essentially on exclusion criteria (for review, see Gustafson, 1990), has considerable shortcomings. Several studies have demonstrated the rather poor rate of histopathological verification of clinically diagnosed DAT (for review, see Winblad et al., 1986). Use of such a heterogeneous patient population renders successful outcome of treatment rather unlikely, since improvement in a subgroup of patients will be masked by nonresponders with other dementia disorders. Recently, however, considerable advances have been made towards providing diagnostic markers for DAT. Thus, for example, the use of functional imaging techniques (see Risberg, 1990), CSF markers for protein abnormalities (Iqbal and Grundke-Iqbal, 1990) and measurement of neurochemical abnormalities in cultured skin fibroblasts

(Blass et al., 1990) may provide useful methods for the accurate differential diagnosis of DAT at an early stage of the disease, thus increasing the likelihood of a successful outcome of therapy. In addition, discovery of the "Alzheimer gene" for patients with the familial form of the disease will enable genetic counselling and early treatment to be applied to this group of individuals (for review, see St. George-Hyslop et al., 1989).

A second complication associated with treatment of DAT is that the dose regimes used may not be appropriate. Most drugs are evaluated clinically in younger individuals, both in terms of pharmacokinetics and efficacy. Age-related changes in pharmacokinetics and pharmacodynamics (as well as in brain neurochemistry, as exemplified in Table 2), may result in incorrect dosages or problems with adverse drug reactions (for review, see Nolan and O'Malley, 1989) leading to a negative clinical outcome of the trial. However, the increasing number of reports of pharmacokinetic analyses of drugs in elderly subjects (see, e.g., Fleishaker et al., 1989) may help to ensure correct dose regimes in future studies.

Table 2

Neurotransmitter system	Alteration
Acetylcholine	↓ ChAT in cortex and hippocampus Cell loss in nucleus basalis
Serotonin	↓ density of 5-HT_1 and 5-HT_2 receptors in frontal cortex
Noradrenaline	Cell loss in locus coeruleus loss of NA
Dopamine	↓ D-2 binding in striatum ↑ D-1 binding in caudate and putamen ↓ nigrostriatal dopamine neurones ↓ DA and tyrosine hydroxylase
GABA	↓ GAD in thalamus and cerebral cortex

Some examples of neurotransmitter changes in normal aging in the human brain (for detailed review, see O'Neill et al., 1987).

DAT as a Multi-Transmitter Disorder

Implicit in many of the neurotransmitter replacement treatment strategies shown in Table 1 is the notion that there is a "selective lesion" in DAT analogous to the dopaminergic deficit hypothesis of Parkinson's disease. Dominant among the neurotransmitter replacement strategies for DAT shown in Table 1 have been those trials aimed at enhancing cholinergic neurotransmission in the brains of sufferers (for reviews, see Bartus et al., 1986; Collerton, 1986). Although the degeneration of ascending cholinergic pathways in DAT has convincingly been demonstrated to correlate with the progression of cognitive decline in the disorder (Perry, 1986), it has become apparent that DAT represents a disease of multiple transmitter systems rather than a "selective cholinergic lesion" (see Table 3 for summary of neurochemical and neuropathological findings). There remains, however, a poor understanding of the relationship between the degeneration of noncholinergic pathways and the progression of cognitive decline in DAT (Cowburn et al., 1989; Wester et al., 1988). In addition, the importance of cholinergic neurones in memory functions, in particular with respect to the deleterious effects of anticholinergic agents, should not be minimized (see Collerton, 1986), in view of the anticholinergic side effects of many of the psychoactive drugs (see, e.g., Williams, 1989), which may negate the therapeutic usefulness of compounds being tested for activity towards DAT. On the other hand, a compound such as tetrahydroaminoacridine has a number of cellular actions in addition to inhibition of acetyl cholinesterase, which may contribute positively to clinical outcome (see Elinder et al., 1989; Nordberg et al., 1989;).

From Table 3, it can be seen that those neurotransmitter systems affected in DAT are either intrinsic to the cerebral cortex (e.g., GABA, somatostatin) or provide projections to this area (acetylcholine, serotonin, noradrenaline) (see Hardy et al., 1986, for further discussion). Furthermore, recent histopathological studies of DAT have demonstrated that cortical pathology spreads along cortico-cortical association pathways (Pearson and Powell, 1989). Taken together, these findings have led to the visualization of DAT pathology as having a neuroanatomical rather than neurotransmitter-specific basis, thus limiting the therapeutic usefulness of transmitter-based strategies.

Signal-Transduction Abnormalities in DAT

As discussed above, multiple neurotransmitter deficits would limit the therapeutic usefulness of transmitter replacement treatment strategies di-

Table 3

Transmitter	Evidence for degeneration Neurochemical	 Histopathological
Acetylcholine	↓ ACh levels ↓ ChAT and AChE activity ↓ Choline uptake ↓ synthesis and release	Cell loss and tangle accumulation in nucleus basalis; Cholinergic markers in cortical neuritic plaques
Serotonin	↓ 5-HT and S-HIAA levels ↓ Uptake and binding to uptake sites ↓ Release	Cell loss and tangle accumulation in Raphe nucleus
Noradrenaline	↓ NA and MHPG levels ↓ DBH activity ↓ NA uptake	Cell loss and tangle accumulation in locus coeruleus; Noradrenergic markers in cortical neuritic plaques
GABA	↓ GABA levels ↓ GAD activity ↓ GABA uptake	GAD immunoreactivity in cortical neuritic plaques
Excitatory amino acids	↓ glutamate and aspartate levels ↓ uptake and binding to uptake sites	Tangle accumulation in, and loss of cortical pyramidal neurons
Neuropeptides	↓ somatostatin and CRF	Tangle accumulation in cortical somatostatin neurons; SLIR immunoreactivity in cortical neuritic plaques

Neurotransmitter degeneration in DAT (for review and references, see Cowburn et al., 1989).

rected towards a single transmitter system. A further constraint upon this approach is that its success depends upon the preservation of intracellular signalling mechanisms. Although little is known about signal transduction systems in DAT, evidence is accumulating that there are disturbances in the cellular responses to receptor stimulation (see Table 4). In particular, the phosphoinositide signalling pathway, where activation of coupled receptors (including receptor systems for neurotransmitter systems known to be affected in DAT) leads to a stimulation of the activity of phosphoinositide-specific phospholipase C (PLC), is affected. Thus, the recognition site density for the PLC-derived second messenger Ins(1,4,5)P_3, a structure homologous to the ryanodyne receptor (Furuichi et al., 1989; Mignery et al., 1989) and involved in the mobilization of calcium from intracellular stores (for review, see Irvine, 1989), is reduced by 50–70% in the DAT temporal cortex and hippocampus (Young et al., 1988). The other second messenger derived from PLC activation, diacylglycerol, acts to activate protein kinase C activity. Protein kinase C-mediated processes are involved, among others, in secretion and exocytosis, modulation of ion conductance, gene expression, cell proliferation, and hippocampal long-term potentiation (for reviews, see Anwyl, 1989; Nishizuka, 1989). Thus, the 50% reduction in frontal cortical particulate protein kinase C levels, with a concomitant reduction in its phosphorylating activity found in DAT (Cole et al., 1988), is likely to have important cellular consequences.

Table 4

Adenylyl cyclase	PLC*-derived second messengers
↑/0 basal activity	↓ Particulate protein kinase C and
↑/↓ forskolin-stimulated	PKC phosphorylation of P 86
↓ VIP (+ GTP)-stimulated	substrate protein in frontal cortex;
↓ Isoprenaline-stimulated	↓ Ins(1,4,5)P_3 binding sites in
(Hippocampus)	temporal cortex and hippocampus

Alterations in brain signal transduction in DAT (for details, see Fowler et al., 1990)

Conclusions

From the above discussion, it is clear that several explanations can be afforded to explain the lack of therapeutic efficacy of drugs so far tested in DAT. These explanations per se suggest new avenues for drug development and testing. Thus, existing drugs may prove useful when given in correct dosages to DAT patients at early stages of the disease. The visualization of DAT as a multiple transmitter deficit suggests that a more beneficial approach would be the targeting of neuronal populations at risk and the development of strategies to minimize neuronal loss (Hefti et al., 1989), rather than to supplement neurotransmitter levels following cell death and the disruption of neuronal circuitry. Of relevance to this, it is worth remembering that DAT sufferers typically first present clinically with treatable noncognitive behavioral abnormalities before developing a progressive less easily treatable cognitive dysfunction. With respect to the deficits in signal transduction found in DAT, research in this area will provide a more integrated picture of the neurochemical imbalances found in this disease, by indicating which signal transduction systems are intact and which are damaged in DAT. Such information will greatly aid the designing of clinical trials aimed at enhancing particular neurotransmitter systems affected in this disease. Thus, it is reasonable to conclude that despite the lack of success of treatment strategies for DAT hitherto, the disease should still be considered as treatable.

References

Anwyl R (1989): Protein kinase C and long-term potentiation in the hippocampus. *Trends Pharmacol Sci* 10:236–239

Bagne CA, Pomara N, Crook T, Gershon S (1986): Alzheimer's disease: strategies for treatment and research. In: *Treatment Development Strategies for Alzheimer's Disease*, Crook T, Bartus R, Ferris S, Gershon S, eds. Madison: Mark Powley Associates Inc., pp. 585–638

Bartus RT, Dean RL, Fisher SK (1986): Cholinergic treatment for age-related memory disturbances: dead or barely coming of age? In: *Treatment Development Strategies for Alzheimer's Disease*, Crook T, Bartus R, Ferris S, Gershon S, eds. Madison: Mark Powley Associates Inc., pp. 421–450

Backman L, Mantyla T, Herlitz A (1990): The optimization of episodic remembering in old age. In: *Successful Aging: Perspectives from the Behavioral Sciences*, Baltez PB, Baltez MM, eds. New York: Cambridge University Press, in press

Blass JP, Gibson GE, Black RS, Ko L-W, Sheu K-FR (1990): Cultured skin fibroblasts, "Alzheimer antigens" and cellular calcium homeostasis in Alzheimer's disease. In: *Biological Markers in Dementia of Alzheimer Type*, Fowler CJ, Carlson LA, Gottfries C-G, Winblad B, eds. London: Smith-Gordon and Co. Ltd., pp. 203–214

Cole G, Dobkins KR, Hansen LA, Terry RD, Saitoh T (1988): Decreased levels of protein kinase C in Alzheimer brain. *Brain Res* 452:165–174

Collerton D (1986): Cholinergic function and intellectual decline in Alzheimer's disease. *Neuroscience* 19:1–28

Cowburn RF, Hardy JA, Roberts PJ (1989): Neurotransmitter deficits in Alzheimer's disease. In: *Alzheimer's Disease: Towards an Understanding of the Aetiology and Pathogenesis*, Davies DC, ed. London and Paris: John Libbey, pp. 9–32

Elinder F, Mohammed AK, Winblad B, Arhem P (1989): Effects of THA on ionic currents in myelinated axons of Xenopus laevis. *Eur J Pharmacol* 164: 599–602

Fleishaker JC, Phillips JP, Smith TC, Smith RB (1989): Multiple-dose pharmacokinetics and pharmacodynamics of adinazolam in elderly subjects. *Pharmaceut Res* 6:379–386

Fowler CJ, O'Neill C, Garlind A, Cowburn RF (1990): Alzheimer's disease: is there a problem beyond recognition? *Trends Pharmacol Sci* 11:183–184

Furuichi T, Yoshikawa S, Miyawaki A, Wada K, Maeda N, Mikoshiba K (1989): Primary structure and functional expression of the inositol 1,4,5-triphosphate binding protein P400. *Nature* 342:32–38

Gottfries CG (1989): Pharmacological treatment strategies in dementia disorders. *Pharmacopsychiat* 22:129–134

Gustafson L (1990): The clinical diagnosis of dementia of Alzheimer type: prospects and limitations. In: *Biological Markers in Dementia of Alzheimer Type*, Fowler CJ, Carlson LA, Gottfries C-G, Winblad B, eds. London: Smith-Gordon and Co. Ltd., pp. 1–14

Hardy JA, Mann DMA, Wester P, Winblad B (1986): An integrative hypothesis concerning the pathogenesis and progression of Alzheimer's disease. *Neurobiol Aging* 7:489–502

Hefti F, Hartikka J, Knusel B (1989): Function of neurotrophic factors in the adult and aging brain and their possible use in the treatment of neurodegenerative disorders. *Neurobiol Aging* 10:515–533

Iqbal K, Grundke-Iqbal I (1990): Cytoskeletal protein abnormalities as indicators of Alzheimer disease. In: *Biological Markers in Dementia of Alzheimer Type*, Fowler CJ, Carlson LA, Gottfries C-G, Winblad B, eds. London: Smith-Gordon and Co. Ltd., pp. 81–87

Irvine RF (1989): How do inositol 1,4,5-trisphosphate and inositol 1,3,4,5-tetrakisphosphate regulate intraceullular Ca^{2+}? *Biochem Soc Trans* 17:6–9

Mignery GA, Sdhof TC, Takei K, De Camili P (1989): Putative receptor for inositol 1,4,5-trisphosphate similar to ryanodyne receptor. *Nature* 342:192–195

Nishizuka Y (1989): The family of protein kinase C for signal transduction. *JAMA* 262:1826–1833

Nolan L, O'Malley K (1989): Adverse drug reactions in the elderly. *Br J Hosp Med* 41:446–457

Nordberg A, Nilsson-Hkanson L, Adem A, Lai Z, Winblad B (1989): Multiple actions of THA on cholinergic neurotransmission in Alzheimer brains. In: *Alzheimer's Disease and Related Disorders*, Iqbal K, Wisniewski H, Winblad B, eds. New York: Alan R Liss Inc., pp. 1169–1178

O'Neill C, Marcusson J, Nordberg A, Winblad B (1987): The influence of age on neurotransmitters in the human brain. In: *Modification of Cell to Cell Signals During Normal and Pathological Ageing*. NATO ASI Series, Vol H9, Govoni S, Battaini F, eds. Berlin: Springer-Verlag, pp. 183–198

Pearson RCA, Powell TPS (1989): The Neuroanatomy of Alzheimer's disease. *Rev Neurosci* 2:101–122

Perry EK (1986): The cholinergic hypothesis—ten years on. *Br Med Bull* 42: 63–69

Petrie WM (1986): Pharmacologic treatment—from the viewpoint of the clinical investigator. In: *Treatment Development Strategies for Alzheimer's Disease*, Crook T, Bartus R, Ferris S, Gershon S, eds. Madison: Mark Powley Associates Inc., pp. 639–654

Risberg J (1990): Regional cerebral blood flow abnormalities in dementia of Alzheimer type. In: *Biological Markers in Dementia of Alzheimer Type*, Fowler CJ, Carlson LA, Gottfries C-G, Winblad B, eds. London: Smith-Gordon and Co. Ltd., pp. 61–68

St. George-Hyslop PH, Myers RH, Haines JL, Farrer LA, Tanzi RE, Abe K, James ME, Connealy PM, Polinsky RJ, Gusella JF (1989): Familial Alzheimer's disease: progress and problems. *Neurobiol Aging* 10:417–425

Wester P, Eriksson S, Forsell A, Puu G, Adolfsson R (1988): Monoamine metabolite concentrations and cholinesterase activities in cerebrospinal fluid of progressive dementia patients: relation to clinical parameters. *Acta Neurol Scand* 77:12–21

Whalley LJ (1989): Drug treatments of dementia. *Br J Psychiat* 155: 595- 611

Williams GO (1989): Management of depression in the elderly. *Prim Care* 16: 451–474

Winblad B, Wallace W, Hardy J, Fowler C, Bucht G, Alafuzoff I, Adolfsson R (1986): Neurochemical, genetic and clinical aspects of Alzheimer's disease. In: *Dimensions in Aging*, Bergener M, Ermini M, Sthelin MB, eds. London: Academic Press, pp 183–205

Young LT, Kish SJ, Li PP, Warsh JJ (1988): Decreased brain [3H]inositol 1,4,5-trisphosphate binding in Alzheimer's disease. *Neurosci Letts* 94:198–202

Dementia of the Alzheimer Type Recapitulates Ontogeny Inversely on Specific Ordinal and Temporal Parameters

Barry Reisberg, Angela Pattschull-Furlan, Emile Franssen, Steven G. Sclan, Alan Kluger, Louis Dingcong, and Steven H. Ferris

The phenomenology of Alzheimer's disease (AD) has been described in increasing detail over the past decade. These increasingly detailed descriptions of the evolution of AD have manifest diagnostic, prognostic, and management importance. Additionally, striking homologies between normal human development and the degenerative course of AD are now apparent. These homologies may be more than fortuitous. Accordingly, it is reasonable at this time to speculate that the course of AD may result directly or indirectly from the disruption of CNS developmental factors and/or processes.

Background

The earliest societal recognition of age-related cognitive disturbance, and the limitations of cognition in some aged individuals, appears to have come from legal, philosophical, and literary sources. Solon, circa 500 BC, listed impaired judgment resulting from old age as a condition in which a will might be declared invalid (Plutarch, 1967 translation). More than a century later Plato also noted that "the commission of certain crimes (sacrilege, treachery, treason) is excusable in a state of madness or when affected by disease or under the influence of extreme old age or in a fit of childish wantonness" (Plato, 1921 translation; Torack, 1983). Medical recognition of old age as a cause of dementia is credited to Aretaeus of Cappadoccia in the second century (Adams, 1861; Lipowski, 1980), who noted an age-related condition characterized by "a torpor of the senses, and a stupefaction of the gnostic and intellectual faculties"

(Adams, 1861). Galen (ca. 130–200 AD), who is believed to have been a contemporary of Aretaeus, also adopted these conceptions, defining morosis (dementia) as a condition in which "the knowledge of letters and other arts are totally obliterated, indeed they can't even remember their own names.... Even now it seems that on account of extreme debility in old age, some are afflicted with similar symptoms" (Galen, 1821–1833 translation; Torack, 1983). An early accurate, albeit brief, clinical description of what we now recognize as Alzheimer's disease comes from Benjamin Rush, who noted that "It would be sufficiently humbling to human nature if our bodies exhibited in old age the marks only of a second childhood, but human weakness descends still lower. I met with an instance of a woman between 80 and 90 who exhibited the marks of a second infancy, by such a total decay of her mental faculties as to lose all consciousness in discharging her alvine and urinary excretions" (Rush, 1793).

Hence, relevant medical homologies between the degenerative course of AD and normal developmental functional capacity can be traced back more than two centuries. Knowledge of the clinical syndrome of AD until approximately 1980 was summarized in the American Psychiatric Association's (1980) Diagnostic and Statistical Manual, Third Edition (DSM III) definition of primary degenerative dementia. This noted in part that "The onset is insidious and the course is one of uniform, gradual progression. In the early stages memory impairment may be the only apparent cognitive deficit. There may also be subtle personality changes, such as the development of apathy, lack of spontaneity, and a quiet withdrawal from social interaction. Individuals usually remain neat and well groomed and, aside from an occasional irritable outburst, are cooperative and behave in the socially appropriate way. With progression to the middle stage of the disease, various cognitive disturbances become quite apparent and behavior and personality are more obviously affected. By the late stage, the individual may be completely mute and inattentive. At this point he or she is totally incapable of caring for himself or herself. This stage leads inevitably to death. With senile onset, the average duration of symptoms, from onset to death is about five years" (American Psychiatric Association, 1980).

This 1980 definition was considered to be quite sketchy and nonspecific by the authors. Hence AD (or primary degenerative dementia, as it was then termed) was considered to be a "diagnosis of exclusion." All other major illnesses were diagnosed by recognizing the characteristic illness process, and then differentially diagnosed by excluding other illnesses with which the entity might be confused; AD, alone amongst major medical, psychiatric, and neurologic entities, was said to be exclusively "a diagnosis of exclusion." Only after ruling out more than 50 other possible causes of dementia were physicians confident in diagnosing AD.

Global and Hierarchic Descriptions of Clinical Changes in AD

In 1982, three independent, more detailed clinical descriptive schemata of aging and progressive primary degenerative dementia (AD) were published (Hughes et al., 1982; Pfeffer et al., 1982; Reisberg et al., 1982). The scale of Hughes et al. described five stages from normality to most severe AD, those of Pfeffer et al. and Reisberg et al. distinguished seven functional and global stages from normality to most severe AD. In utilizing a hierarchic approach to staging, each of these measures was consistent with efforts dating back more than 20 years. For example, Leeds (1960) described six very broad conditions, which he referred to as "stages," of an entity that he termed "senile recession." Similarly, de Ajuriaguerra and his co-workers in Switzerland emphasized a hierarchic approach to dementia assessment (de Ajuriaguerra et al., 1964; de Ajuriaguerra and Tissot, 1968). Cole and co-workers, in Canada, built upon the ideas of de Ajuriaguerra and his associates in developing a hierarchic dementia scale (Cole et al., 1983; Cole and Dastoor, 1987). In its current form the Hierarchic Dementia Scale contains 20 categories assessing cognitive abilities, praxis, recent and remote memory, orientation, reading and writing abilities, calculation skills, ability to follow instructions, and prefrontal neurologic reflexes. An initial objective of the Hierarchic Dementia Scale was the development of measures that would be concordant relative to the progression of particular forms of dementia disorder, and, consequently, diagnostic of dementia of the Alzheimer's type, in particular. The authors do not at this time profess to have achieved their goal of concordance. Rather, it is their belief that different dementias will proceed at varying rates on the twenty measures, and, consequently, the Hierarchic Dementia Scale may ultimately prove to be of differential diagnostic utility when employed longitudinally.

Concordant Ordinal (Hierarchic) Description of Changes in AD

Beginning in 1983, Reisberg et al. divided their 7-stage Global Deterioration Scale (GDS) into separate 7-stage ordinal "constituents" or measures.

These component scales were empirically designed to: (1) mirror the progression of deficits in normal aging and progressive dementia of the Alzheimer type; and (2) to be optimally concordant with the corresponding GDS stages (Reisberg et al., 1982). Initially, concordant constituents or "axes" describing progressive changes in concentration, recent memory, past memory, orientation, and functioning-and-self care were described in the form of a new, more explicitly detailed measure, the Brief Cognitive Rating Scale, or BCRS (Reisberg et al., 1983a).

Subsequently, concordant ordinal stages were described for language, motor capacity, and mood (Reisberg et al., 1983b), and later for praxis and calculation (Reisberg et al., 1985). One of these ten "axes" proved particularly useful in describing the progression of deficit in AD: "functioning and self care" (BCRS Axis V).

Ultimately, by expanding the description of the late stages of AD, a total of 16 ordinal "functioning and self care" stages in aging and AD were identified (Reisberg 1988; Reisberg et al.,1984; see Table 1). These 16 Functional Assessment Stages (FAST stages) of aging and AD appeared to be sufficiently detailed as to describe a characteristic syndrome.

Although deficit in patients with AD characteristically followed this progression (Borenstein and Reisberg, 1987; Reisberg, 1986), other disorders associated with the presence of dementia do not necessarily follow this characteristic pattern (see Table 2). For example, gait disturbance and loss of ambulatory capacity commonly precede overt evidence of cognitive disturbance (dementia) in normal pressure hydrocephalus. In Alzheimer's disease, loss of ambulatory capacity occurs much later in the course of the disease, long after the occurrence and progression of cognitive disturbance. Similarly, in normal pressure hydrocephalus, urinary incontinence is commonly an early sign, which may precede overt cognitive disturbances; in Alzheimer's disease urinary incontinence occurs much later in the course of the illness, long after other functional losses resulting from progressive cognitive disturbance have become manifest. As illustrated in Table 2, dementias associated with Creutzfeldt-Jacob disease, multi-infarct dementia, depressive affective disorder, hyponatremia, and diffuse brain metastases all may commonly follow functional sequences of loss very different from that of AD.

Most recently, we have also described progressive changes in meal preparation and self-feeding skills in AD. These changes have also been enumerated so as to be optimally concordant with the corresponding GDS and FAST stages. These "feeding concomitants of Alzheimer's disease" are described in detail in Table 3. Independent data from our laboratory support the weighting of these measures as illustrated in Table 3. Recent

Table 1

Sequence of Functional Loss in Normal Aging and Alzheimer's Disease (Functional Assessment Staging)[a] *

FAST Stage	Characteristics	Clinical Diagnosis	Estimated Duration in AD[b]
1	No decrement	Normal adult	
2	Subjective deficit in word finding or recalling location of objects	Normal aged adult	
3	Deficits noted in demanding employment settings	Compatible with incipient AD	7 years
4	Requires assistance in complex tasks (e.g., handling finances, planning dinner party)	Mild AD	2 years
5	Requires assistance in choosing proper attire	Moderate AD	18 months
6 (a)	Requires assistance dressing	Moderately severe AD	5 months
(b)	Requires assistance bathing properly		5 months
(c)	Requires assistance with mechanics of toileting (such as flushing, wiping)		5 months
(d)	Urinary incontinence		4 months
(e)	Fecal incontinence		10 months
7 (a)	Speech ability limited to about a half-dozen words	Severe AD	12 months
(b)	Intelligible vocabulary limited to a single word		18 months
(c)	Ambulatory ability lost		12 months
(d)	Ability to sit up lost		12 months
(e)	Ability to smile lost		18 months
(f)	Ability to hold head up lost		12 months or longer

[a] Adapted from Reisberg, B. (1986) *Geriatrics* 41:30–46. [b] In subjects without other complicating illnesses who survive and progress to the subsequent deterioration stage.

Table 2. Examples of Abbreviated Sequences of Functional Loss in Non-Alzheimer Disorders Associated with Progressive or Gradual Onset of Dementia and FAST AD Distinctions

Common Abbreviated Sequences of Functional Loss in Non-Alzheimer Disorders			FAST AD Distinctions	
Disorder — Pathology or Presumptive Etiology	Abbreviated Sequence of Functional Loss in Non-AD Disorder*	Equivalent FAST Stage	Abbreviated Sequence of Functional Loss in AD as Described in the FAST	FAST Stages in AD Corresponding to Abbreviated Sequence
Normal Pressure Hydrocephalus	1. Gait disturbance	7c	1. Loss of ability to perform complex tasks	4
— Dilated cerebral ventricles	2. Urinary incontinence	6d	2. Urinary incontinence	6d
	3. Loss of ability to perform complex tasks	4	3. Ambulatory (gait) disturbance	7c
Creutzfeldt-Jacob Disease	1. Gait disturbance	7c	1. Loss of ability to perform complex tasks	4
— "Slow" virus	2. Loss of ability to perform complex tasks	4	2. Gait (ambulatory) disturbance	7c
Multi-Infarct Dementia	1. Loss of speech	7a–7b	1. Loss of ability to perform complex tasks	4
— Multi-cerebral infarctions	2. Loss of urinary continence	6d	2. Loss of ability to pick out clothing properly	5
	3. Loss of ability to perform complex tasks	4	3. Loss of ability to put on clothing unassisted	6a
	4. Loss of ability to put on clothing	6a	4. Loss of ability to bathe without assistance	6b
	5. Loss of ability to bathe without assistance	6b	5. Loss of urinary continence	6d
	6. Loss of ability to pick out clothing	5	6. Loss of fecal continence	6e
	7. Loss of ambulatory capacity	7c	7. Loss of speech	7a–7b
	8. Fecal incontinence	6e	8. Loss of ambulatory capacity	7c

Dementia Syndrome of Depression — Affective disorder with neurotransmitter imbalance	1. Loss of ability to perform complex tasks 2. Refusal to put on clothing (associated with negativity) 3. Refusal to bathe (associated with negativity) 4. Loss of ability to pick out clothing properly	4 6a 6b 5	1. Loss of ability to perform complex tasks 2. Inability to pick out clothing properly 3. Inability to put on clothing without assistance 4. Inability to bathe without assistance	4 5 6a 6b
Dementia Associated with Hyponatremia — Electrolyte disturbance	1. Loss of ability to perform complex tasks 2. Loss of ability to pick out clothing 3. Loss of ability to dress, bathe, and toilet independently 4. Loss of ambulation capacity 5. Loss of urinary and fecal continence 6. Loss of speech	6 5 6a–6b –6c 7c 6d–6e 7a–7b	1. Loss of ability to perform complex tasks 2. Loss of ability to pick out clothing 3. Loss of ability to dress, bathe, and toilet independently 4. Loss of urinary and fecal continence 5. Loss of speech 6. Loss of ambulatory capacity	4 5 6a–6b –6c 6d–6e 7a–7b 7c
Dementia Associated with Diffuse CNS Metastasis — Neoplastic diffuse cerebral trauma	1. Loss of ability to perform complex tasks 2. Loss of ability to dress, bathe, and toilet independently 3. Loss of ambulation capacity 4. Loss of urinary and fecal continence 5. Loss of speech	4 6a–6b –6c 7c 6d–6e 7a–7b	1. Loss of ability to perform complex tasks 2. Loss of ability to dress, bathe, and toilet independently 3. Loss of urinary and fecal continence 4. Loss of speech 5. Loss of ambulatory capacity	4 6a–6b –6c 6d–6e 7a–7b 7c

* The sequences of functional loss shown are typical for normal pressure hydrocephalus and Creutzfeldt-Jacob disease; the sequence for multi-infarct dementia is one of various common presentations; the sequences in the dementia syndrome of depression, dementia associated with hyponatremia and dementia associated with diffuse CNS metastasis are previously observed examples of the presentation of these dementias.

Table 3

Feeding Concomitants of Alzheimer's Disease*

Stage	
1–2	No objective or subjective decrement in the ability to adequately prepare meals, order food and beverages in a restaurant setting, and/or in table etiquette.
4	Decreased facility in preparing and/or serving relatively complex meals, and/or decreased facility in ordering food and beverages in restaurant setting.
5	Decreased ability in preparing simple foods or beverages (e.g., coffee or tea). May occasionally make mistakes in eating food (e.g., improper use of seasoning or condiments).
6 (a)	Occasional difficulty with proper manipulation or choice of eating utensils.
(b)	Meat and similar foods must be cut up for the patient.
(c)	No longer trusted to properly use a knife. May also eat foods that would have formerly been refused.
(d)	No longer trusted to use a knife and decreased ability to use a fork, but can still properly use a spoon. May also display occasional misrecognition of dietary substances (pica).
(e)	Capable of going to the refrigerator or cupboard but has difficulty discerning and choosing food, may have difficulty chewing hard food.
7 (a)	Capable of picking up spoon or fork. Will occasionally drop food or misutilize silverware (e.g., may attempt to drink soup or other liquids with a fork). Capable of reaching for a cup when desirous of fluid.
(b)	Must be assisted in actual feeding. Generally patients are not permitted to handle a knife or fork. May not be able to properly lift a cup.
(c)	Can reach for and pick up food with hands. Cannot properly pick up a fork or a spoon, but can grasp a spoon or another utensil. Must be spoon-fed, but can chew.
(e)	Cannot distinguish foods from nondietary substances. Will reach out for objects, including food.

observations also indicate that the progression of these disturbances in meal preparation and self-feeding skills can be characteristic of the dementia of AD, just as the progression of functional disturbance described using the FAST is characteristic of AD.

As noted previously, in 1985 we described progressive praxis changes in the course of CNS aging and AD (Reisberg et al., 1985). These praxis changes were initially developed from a previously described hierarchy of praxis change on the Hierarchic Dementia Scale (Cole et al., 1983). This hierarchy was modified, expanded, and enumerated so as to be optimally concordant with the GDS and FAST stages. The praxis "axis" is illustrated in Table 4.

Table 4

Praxis Concomitants of Normal Aging and Alzheimer's Disease [a] *

Stage	Praxis Ability [b]
1–2	Can draw a cube (prompt "like an ice cube")
3	Difficulty drawing a cube with proper perspective
4	Can draw a rectangle (prompt "like a square"; square credited if drawn)
5	Can draw a circle inside a circle
6 (a)	Can draw a circle
(b)	Can draw a straight line
(c)	Can write own name (with more or less acceptable signature)
(d)	Can write some parts of own name
7	Will not write anything, but may grasp a pen or pencil in a usable fashion

[a] Modified from Reisberg B, Ferris SH, de Leon MJ (1985): Senile dementia of the Alzheimer type: Diagnostic and differential diagnostic features with special reference to functional assessment staging. In: *Senile Dementia of the Alzheimer Type* (Vol. 2). Traber J, Gispen, WH, eds. Berlin: Springer-Verlag, pp. 18–37.
[b] N.B. Praxis ability is assessed by the subjects' ability to draw upon request, not copy.

Degenerative Course of AD: Developmental Homologies

Even some of the earliest descriptions of the progressive pathology of what we now recognize as AD have emphasized homologies between AD and human development. These homologies, in their simplest form, can be traced back to antiquity. For example, in his play, "Clouds," Aristophanes (423 BC) noted that "old age is but a second childhood." Present day Greeks have a lay word for AD, "xemoranthy," which literally translates into "return to infancy." It is conceivable that this lay usage may be traceable to antiquity. As noted above, in an early accurate description of AD, Benjamin Rush (1793) noted that this condition is accompanied not only by "childhood changes" but also by a "return to infancy," including urinary and fecal incontinence. Linden and Courtney (1953) and Linden (Linden, 1957) note the phenomenon of "senile recession" as a part of the "life cycle." Linden (1957) defined this entity as follows: "In recession, the aged person appears to proceed through a series of retrogressive steps through level after level of the infantile development stages all the way back to psychological infantility. In a sense, these are the stages of childhood development up to early adolescence observed in reverse. The last stage of recession is the last stage of dementia before death, and corresponds in many ways to foetal [*sic*] and neonatal organization." Interestingly, Linden believed that this psychological deterioration was due to social rejection and could be treated by creating the correct social attitudes toward the aging process. Leeds (1960), described six stages of "senile recession," citing Linden (1957) and Linden and Courtney's (1953) earlier work. Recession stage 1 was briefly described as one of reminiscence; recession stage 2 was forgetfulness; recession stage 3 "withdrawal from the world." Despite Leeds's emphasis on the developmental recession model, the developmental homologies in these first three recession stages are not clearly inferred from his paper. Recession stages 4, 5, and 6 are however described by Leeds with defined developmental correlates.

Recession stage 4, "puberty back to about 18 months," is further described as "here the tie with current reality is slowly weakened, and a resemblance to some aspects of schizophrenia appears." In recession stage 5, "18 months back to 2 months," Leeds provides a description of a stage of senile recession that does appear to be accurate and which clearly corresponds to the early seventh stage of the Global Deterioration Scale and, more specifically, to stages 7a to 7d on the FAST. Leeds notes that:

> At this stage, the person comes closest to a second childhood. Often, when he feels threatened or fearful, he cries "Mama." Ego structure is so deteriorated that frequently groaning sounds take the place of speech. Moans and animal cries are uttered, usually in an unending series, broken only by sleep. The person soils himself often, even though toilet facilities or commodes are immediately available. There is disorientation for day and night, familiar people are often unrecognized, and reference points begin to slip away one by one.

In a brief accompanying case history, Leeds also notes some feeding concomitants of this stage. In the case example, an 87-year-old woman, the former founder of the institution fifty-five years previously, now "sits in a chair by her bed most of the day.... She moans 'Ohhhhh, ohhhhh,' much of the day." Leeds notes that "Since it is difficult to eat with utensils, she manages better with her hands." Both in terms of this patient's behavior and the specific feeding aspects described, this case of Leeds appears to be recognizable as stage 7c. On the FAST (Table 1), the woman appears to be nonverbal and nonambulatory but able to sit up, behavior consistent with stage 7c. In terms of the feeding concomitants of AD described in Table 3, she appears to be unable to handle utensils and to prefer to eat food with her hands, behavior corresponding to stage 7c in Table 3.

Leeds's final stage 6 of senile recession is described as corresponding to "2 months to fetal stages." He notes that "at this stage, incontinence is the rule, and the patient goes back into diapers like a baby. He must be fed, too, or starvation and dehydration will carry him off quickly." He continues, "The knees and elbows draw up, the spine curves forward, the head turns inward, and the shell of the person that is left resembles nothing so much as the caricature of an overgrown fetus."

Leeds cited Gesell and Ilg (1943) for normative data regarding infant and childhood development. Subsequently, French-speaking Swiss investigators were influenced by the work of Piaget (e.g., Piaget, 1952, 1960, 1970, 1973) and drew analogies between Piaget's childhood developmental stages and the degeneration observed in senile dementia (Constantinidis et al., 1978; de Ajuriaguerra et al., 1964; de Ajuriaguerra and Tissot, 1968, 1975). This work, in turn, influenced Cole and associates in their development of the Hierarchic Dementia Scale (Cole and Dastoor 1980, 1987; Cole et al., 1983).

In 1985 and 1986 we noted that the FAST stages of aging and AD (Reisberg et al., 1984) represented a precise reversal of the order of acquisition of these same stages in the course of normal human development (Reisberg et al., 1985, 1986; Reisberg, 1986). This developmental correspondence with regard to the FAST stages is shown in Table 5. In describing these striking homologies, it was also noted that in other ways, clearly, the degeneration of AD was notably different from normal human development. For example, AD patients never go back to crawling or crying as infants do.

Nevertheless, it does appear that these homologies between the degenerative course of AD and normal human development can, in certain ways, be further extended. Thus we report in Table 6 the relationship between the previously described progressive losses in praxis capacity in AD and the approximate time course and order of acquisition of these praxis capacities in normal development. As was previously noted for the FAST assessments, an inverse ordinal correspondence is observed between the loss of praxis capacity in AD and the acquisition of praxis capacity in normal development. Similarly, the time course of acquisition of the feeding concomitants of AD shown in Table 3 can be determined from the developmental literature. These relationships, described in Table 7, again demonstrate a precise inverse ordinal correspondence between the loss of feeding capacity in AD and the acquisition of feeding capacity in normal human development.

Although the precise time course of the progressive loss of capacity that occurs in AD has not yet been definitively described, empirical observations have permitted us to estimate the mean temporal course of loss of function, measured using the FAST, which occurs in uncomplicated AD (Reisberg 1986, see Table 1). This temporal course of loss of capacity in AD can be compared with the time course of acquisition of the same functions in normal human development as shown in Table 5. The same temporal estimates would apply for the loss of capacity in uncomplicated AD for the corresponding praxis and feeding stages in Tables 4 and 3, respectively.

Consequently, we can compare the estimated time course of temporal loss of all of these capacities in AD with the time course of acquisition of these same capacities in normal human development as shown in Tables 5, 6, and 7. The resulting comparisons are displayed in Fig. 1, and illustrate the striking ordinal and temporal similarities between the degenerative course of AD and the course of normal human development.

Table 5

Correspondence between Normal Human Development and FAST Deficits in Normal Aging and Alzheimer's Disease (AD)

FAST stage of normal aging and AD[a]	Estimated duration of FAST stage in normal aging and in AD[b]	Approximate age at which function is acquired in normal development
1	50 years	adult
2	15 years	aged adult
3	7 years	young adult
4	2 years	7 years to adolescence
5	18 months	5–7 years
6 (a)	5 months	around 5 years (Eisenberg, 1975)
(b)	5 months	around 4 years (Eisenberg, 1975)
(c)	5 months	around 48 months (Vaughn, 1969)
(d)	4 months	around 36–54 months (Pierce, 1975)
(e)	10 months	around 24–36 months (Eisenberg, 1975; Vaughn, 1969; Pierce, 1975)
7 (a)	12 months	around 12 months (Eisenberg, 1975; Vaughn 1969)
(b)	18 months	around 12 months (Eisenberg, 1975; Vaughn 1969)
(c)	12 months	around 12 months (Eisenberg, 1975; Vaughn 1969)
(d)	12 months	around 6–9 months (Eisenberg, 1975, Vaughn 1969)
(e)	18 months	around 8–16 weeks (Eisenberg, 1975; Vaughn 1969)
(f)		around 4–12 weeks (Eisenberg, 1975; Vaughn 1969)

[a] See Table 1 for brief descriptions and Reisberg (1988), for detailed description of these stages.

[b] In subjects who survive and progress to the subsequent deterioration stage without additional pathologic medical complications.

Table 6
Loss of Praxis Ability in Alzheimer's Disease and Corresponding Developmental Ages of Acquisition of Drawing Ability in Infancy and Childhood

Praxis stage of Normal Aging and AD	Approximate age at which ability is acquired in normal development	References / Citations
1–2 Can draw a cube (prompt "like an ice cube")	12 years	Cox (1986): 12-year-olds have been taught how to draw a cube in perspective.
3 Difficulty drawing a cube with proper perspective	8 to 10 years	Chen (1985): 10-year-olds had difficulty drawing a cube in perspective; were successful in drawing a cylinder. At 8 years, 50% of children can draw a cube. Mitchelmore (1978) 9-year-old children can by no means complete an incomplete drawing of a cube.
4 Can draw a rectangle (prompt "like a square"; square is credited if drawn)	4–5 years	Gassier (1984): The 4- to 5-year-old is able to use rectangles and squares in own drawings. Broderick et al. (1987): The 4- to 5-year-old can draw a square. Piaget and Inhelder (1956): The 4- to 5-year-old can distinguish a square from a triangle.
5 Can draw a circle inside a circle	4 years (?)	Gesell (1940): The 4-year-old can draw an elongated or misshapen circle.
6 (a) Can draw a circle	3–4 years	Cratty (1986): When asked, the 3- to 4-year-old can draw a circle. Gassier (1984): The 3- to 4-year-old can draw a circle. Broderick et al. (1987): The 3- to 4-year-old can draw a circle. Gesell (1940): The 3-year-old can copy a circle.
(b) Can draw a straight line	30 months	Gesell (1940): The 30-month-old imitates horizontal lines.
(c) Can write own name more or less accurately	*24 months	*Gesell (1940): The 2-year-old can confine scribbles to the page.
(d) Can write some parts of own name	*13 to 15 months	*Gesell (1940): The 13-month-old child can be expected to make marks on the paper. The 15-month-old scribbles spontaneously. *Cratty (1986): The 15-month-old child attempts to scribble with a whole-hand grasp of the writing instrument.
7 Will not write anything, but may grasp a writing instrument in a usable form.	0 to 12 months	Gesell (1940): At 1 year, a child, if asked, brings the crayon to the paper and hits more than marks it.

* The residual ability to write one's personal name or parts of one's personal name appears to correspond developmentally with the ability to produce a scribble with a writing instrument. This is probably related to the overlearned aspects of signing one's personal name. The correspondences shown are based upon this empirically developed assumption.

A Utilitarian Implication of the Homologous Relationship between Normal Human Development and AD: Development of Improved Psychological Assessment of AD

A major problem in AD assessment has been that conventional psychometric measures uniformly result in "bottom" or zero scores by the 6th GDS stage. This "floor" effect has been noted by at least three groups of investigators (Mohs et al., 1986; Reisberg et al., 1985; Wilson and Kaszniak, 1986). On the basis of our observations, described above, of the homologous relationships between normal human development and AD, we hypothesized that psychological tests that have been previously developed for usage in infants and small children might be useful also in the assessment of the patient with moderately severe and severe AD (i.e., GDS stages 6 and 7 corresponding to developmental ages of five years or less). [This methodologic approach to the assessment of dementia has been awarded a United States patent, No. 5,082,446, on January 21, 1992.]

In an initial approach to this problem, we adapted scales from the Ordinal Scales of Psychological Development (OSPD; Uzgiris and Hunt, 1975) for use in the GDS stage 6 and GDS stage 7 AD patient. The OSPD test was designed to assess an infant's level of sensorimotor development and parallels Piaget's hierarchic theory of early cognitive development. We then studied the relationship between the adapted OSPD scales and the progressive functional stages of AD patients in GDS stages 6 and 7.

The results of these investigations have recently been reported (Sclan et al., 1990). We observed that all psychometric tests studied indeed evidenced uniform bottom scores in 18 GDS stage 7 patients. In contrast, the mean total score for the 18 GDS stage 7 patients on the five Uzgiris and Hunt OSPD measures (object permanence, means–ends, causality, spatial relations, and schemes) was 9.7. Only FAST stage 7f subjects sometimes demonstrated bottom (i.e., zero) scores on all of the five OSPD scales studied.

The immediate practical result of these studies is that the exploitation of the homologous relationship between the deteriorating course of AD and the inverse homologous course of normal development has facilitated the search for and subsequent adaptation of psychological measures for the assessment of cognitive capacity even in some of the most severe GDS stage 7 AD patients. The ability to elicit residual cognitive capacity can now be applied in efforts to optimize the cognition and alertness of patients in the final stages of AD. For example, over-medication,

Table 7
Sequence of Loss of Feeding Capacity in Alzheimer's Disease and Corresponding Sequence of Acquisition of Feeding Capacity in Normal Human Development

Feeding Capacity Stage of Normal Aging and AD		Approximate age at which ability is acquired in normal development	References / Citations
1–2	No objective or subjective deficit in the ability to adequately prepare meals, order food and beverages in a restaurant, and/or in table etiquette.	Young adult	Estimate
4	Deficit in preparing and/or serving complex meals; deficit in ordering food in a restaurant setting.	7 yrs–adolescence	Estimate
5	Deficit in ability to prepare simple foods and beverages such as coffee and tea. May occasionally make mistakes in eating food.	6 years	Hurlock (1978): At 6 years of age, most children have mastered all the tasks used in self-feeding.
6 (a)	Occasional difficulty with proper manipulation or choice of utensils.	5 years	Lowrey (1978): By age 5, children can manipulate or choose eating utensils although they still need occasional help.
(b)	Meat and similar foods must be cut for the patient.	2.5–5 years	Grand (1987): At 36 months, children can manage a knife with some help on hard food. Illingworth (1987): By age 2.5 to 3 years, most children can manage a knife and fork if helped on hard articles. Hurlock (1978): If given the opportunity for practice, children are able at 4 years of age to cut tender meat with a knife. Simko (1989): A knife can be added for the 4- to 6-year-old.
(c)	No longer trusted to properly use a knife.	23–24 months	Hurlock (1978): By the end of th second year, children can use forks as well as spoons.
(d)	No longer trusted to use a knife and deficits in using a fork but can properly use a spoon.	18–24 months	Illingworth (1987): Most children who are given a chance early can manage a spoon with a minimum of spilling by 18 months. Grand (1987): At 18 months, a child used a spoon well. At 24 months, children self-feed with a spoon without turning it in their mouths. Pipes (1981): By 2 years of age, spilling seldom occurs.

Table 7 Continued
Sequence of Loss of Feeding Capacity in Alzheimer's Disease and Corresponding Sequence of Acquisition of Feeding Capacity in Normal Human Development

Feeding Capacity Stage of Normal Aging and AD	Approximate age at which ability is acquired in normal development	References / Citations
6 (e) Capable of going to the cupboard, but has difficulty discerning and choosing food.		
7 (a) Capable of picking up a spoon or fork; however, will occasionally drop food. Capable of reaching for a cup.	15–18 months	Pipes (1981): Spilling and messiness are marked during the first half of the second year. By 16–18 months, only moderate spilling occurs. By 15 months, children can manage the cup, but not expertly. Nelms (1982): By 15 months, most infants can manipulate a cup quite well, although there may be difficulty replacing the cup on the table after drinking.
(b) Must be assisted in actual feeding. Generally, not permitted to handle knife or fork. May not be able to properly lift a cup.	11–12 months	Grand (1987): At 12 months, an infant holds a cup and is likely to lift it, causing spilling. Hurlock (1978): By the time they are 11 or 12 months old, babies can hold their cups temporarily and try to feed themselves with their spoons.
(c) Can reach for and pick up food with hands. Cannot properly pick up a fork or a spoon, but can grasp a spoon or other utensil. Must be spoon fed, but can chew.	6–8 months	Alford (1982): At 8–10 months, infants can pick up a chunk of food with their hands. A 7-month-old grasps a spoon. Illingworth (1983): At the age of 6 months, infants' eyes are so well coordinated with their hands, they can feed themselves with a biscuit. Pipes (1981): The beginning of chewing movements occurs between 24 and 28 weeks. Nelms (1982): By age 5 or 6 months, the infant is interested in chewing and solid food is well tolerated. Grand (1987): The infant chews solid food at 12 months.
(e) Cannot distinguish food from nondietary substances. Will reach out for objects, including food.	5 months	Illingworth (1987): From age 5 months, almost everything that the infant picks up is taken into the mouth. Pipes (1981): At age 20 weeks, infants reach for and grasp an object on sight. In almost every instance, the object goes into the mouth.

over-sedation, the absence of a stimulatory environment, can now all be examined in terms of their effects on cognition in the final stages of AD, and consequently avoided.

The ability to assess cognition even in the final seventh stage of AD also provides an opportunity to examine the relationships between the functional degenerative stage in AD, corresponding cognitive capacity, and the reciprocal relationships between functioning and cognition in normal human development. An examination of these relationships using the five OSPD scales can be seen in Table 8. The results clearly indicate that in the early seventh GDS stage, cognitive capacity measured using the OSPD scales appears to be at a lower developmental level than the corresponding functional capacity.

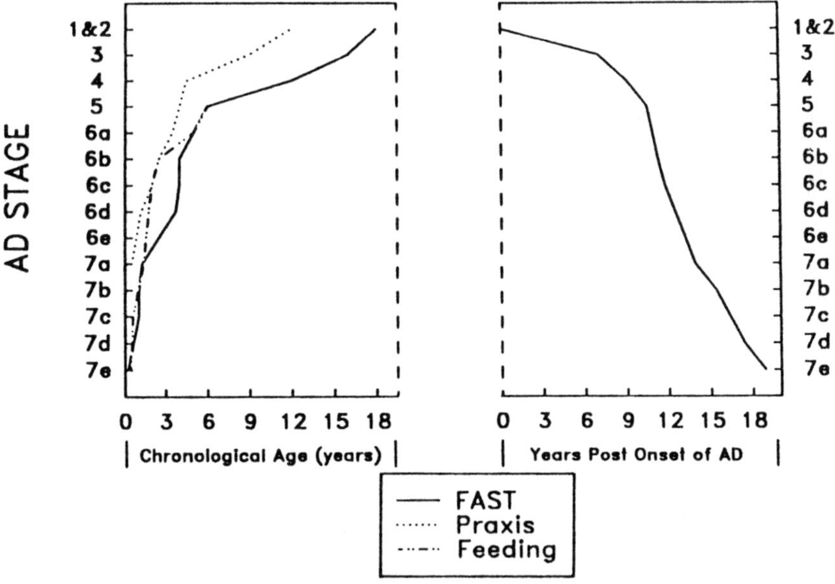

Figure 1. Relationship between the temporal course of acquisition of functional, praxic, and feeding capacities in normal human development and the typical temporal course of degeneration in Alzheimer's disease (AD).

Possible reasons for this disparity might include social as well as physiologic explanations. Specifically, virtually all of the stage 7 patients studied on the OSPD measures were in nursing homes and it is possible that these AD patients were not being stimulated up to their optimal

Table 8
Relationship between FAST Stages, OSPD Scores, and Developmental Reciprocals in GDS Stage 7 AD Patients

FAST Stage	N	Approximate Age Function Acquired in Normal Development	OP (EDA)	M-E (EDA)	C (EDA)	S-R (EDA)	Sc (EDA)
7(a)	5	12 months	3.4 (5 mo)	3.6(4.5 mo)	2.0(3 mo)	5.2(6 mo)	2.6(4 mo)
(b)	3	12 months	2.0 (3 mo)	2.0 (3 mo)	0.7(<2 mo)	3.3(4 mo)	1.3(3 mo)
(c)	2	12 months	2.0(3 mo)	3.5(4.5 mo)	1.5(2.5 mo)	3.5(4.5 mo)	1.5(3 mo)
(d)	4	6–9 months	1.5(2.5 mo)	2.0(3 mo)	0.8(<2 mo)	4.3(5 mo)	1.3(3 mo)
(e)	1	2–4 months	1.0(2 mo)	1.0(2 mo)	0.0(<2 mo)	4.0(5 mo)	0.0(<3 mo)
(f)	3	1–3 months	0.0(<2 mo)	0.7(<2 mo)	0.7(<2 mo)	0.0(<2 mo)	0.0(<3 mo)

GDS = Global Deterioration Scale stage (Reisberg et al., 1982); FAST = Functional Assessment Stages of AD (Reisberg, 1986, 1988); OSPD = Ordinal Scales of Psychological Development (Uzgiris and Hunt, 1975); OP = Object Permanence (OSPD Scale); M-E = Means-Ends (OSPD Scale); C = Causality (OSPD Scale); S-R = Spatial Relations (OSPD Scale); Sc = Schemes (OSPD Scale); EDA = Estimated Developmental Age; mo = months (Dunst, 1980).

cognitive capacity or that these AD patients were being over-medicated. Naturally, it is also possible that cognition is relatively more affected than functioning in the early seventh GDS stage. Future studies may usefully resolve these clinically relevant interrelationships between dementia and development and between functioning and cognition. The recent advent of an independent neurologic series of parameters for addressing these relationships between the degeneration of dementia and normal human development may also prove useful in separating out psychological and physiological aspects of the degenerative course of AD (Franssen et al., 1991).

Possible Etiologic Implications of the Degenerative Course of AD

The seemingly remarkable reciprocal homologies between the degenerative course of AD and normal human development may, of course, be fortuitous. However, it seems more likely that this homology may be the result of similar primary or secondary processes. Recent discoveries regarding neurotrophic factors and their possible relevance for the pathology of AD permit the formulation of specific, empirically verifiable hypotheses, of common relevance for normal CNS development and the characteristic degenerative dementia of AD. Nerve growth factor (NGF), in particular, has been hypothesized to be of possible relevance for both AD and normal neuronal development (Phelps et al., 1989). A decrease in NGF receptor-positive cells in AD has been reported, in addition to a reduced intracellular staining intensity for NGF in normal aging (Hefti and Mash, 1989). Close relationships between NGF staining and the CNS cholinergic system have also been noted (Hefti and Mash, 1989; Phelps et al., 1989). Other possible relationships between NGF and the pathology of AD can also be logically hypothesized. For example, NGF has been demonstrated to effect various microtubule-associated proteins including their expression (Brugg and Matus, 1988) and phosphorylation (Aletta et al., 1988). Possible roles of NGF in microtubule assembly, stability, cross-linking and nuclear signaling have been hypothesized (Aletta et al., 1988). NGF and dexamethasone have been reported to have opposing actions in the regulation of an RNA encoding intermediate filament protein (Leonard et al., 1988), a finding of interest in light of de Leon et al.'s (1988) observation of an abnormal cortisol response to glucose in AD. Additionally, recent work has strongly implicated NGF in neuronal repair mechanisms (Rosenberg et al., 1988). The evidence for a possible role of NGF in normal human neuronal development is even more convincing. In addition to NGF, other growth factors of relevance for central nervous system development are also being identified (Lin et al., 1989). Consequently, it is possible that central nervous system growth factors directly or indirectly play an important role in both normal neuronal development and in the, in many ways analogous, degenerative process of AD. These etiologic commonalities, in turn, may account for the distinctly similar ordinal and temporal behavioral patterns observed in both AD and normal human development.

If AD in fact does ultimately prove to be a developmentally (i.e., ontogenetically) relevant progressive pathologic process, it is likely to be one of a class of such ontogenic degenerative processes existing in other species as well as humans. Study of nonhuman aged species is likely to ultimately reveal other examples of these phenomena.

Conclusion

There has been a "progressive discrimination," which has resulted in the definition of 16 stages and substages of AD. This discrimination, in turn, has resulted in "an opposite and complementary trend.... as differences multiply, new relationships are noted, which connect up the multiplicity" (King, 1963). In the present instance, striking developmental reciprocal relationships and homologies are noted, and the changes in CNS aging and AD appear to recapitulate these ontogenic homologous phenomena. It is tempting to speculate that this "ascending spiral" is leading to a new understanding of the etiology and treatment of AD. Speculations along these lines result in the hypothesis that common physiologic mechanisms may be operative, and one such mechanism in particular, neurotrophic factor-related CNS changes, appears particularly applicable. The validity of these speculations can be subjected to empirical evaluation at the present time.

Acknowledgments. This work was supported in part by grants AG03051, AG09127, and AG08051 from the National Institute on Aging, and by grant MH43486 from the National Institute of Mental Health.

This chapter is a modified and expanded version of a chapter by the same authors entitled "Cognition Related Functional, Praxis and Feeding Changes in CNS Aging and Alzheimer's Disease and Their Developmental Analogies," published in: *Molecular Mechanisms of Aging*, Beyreuther K, Schettler G, eds., Berlin: Springer-Verlag, 1990; pp. 18–40. The original chapter has been copyrighted by Springer-Verlag, and has been adapted with permission.

References

Adams F, ed. (1861): *The Extant Works of Aretaeus, the Cappadoccian*. London: Syndenham Society

Aletta JM, Lewis SA, Cowan NJ, Greene LA (1988): Nerve growth factor regulates both the phosphorylation and steady-state levels of microtubule-associated protein 1.2 (MAP1.2). *J Cell Biol* 106:1573–1581

Alford BB, Bogle ML (1982): *Nutrition during the Life Cycle*. Englewood Cliffs, NJ: Prentice-Hall

American Psychiatric Association (1980): *Diagnostic and Statistical Manual of Mental Disorders*, 3rd Edition. Washington, DC: American Psychiatric Association

Aristophanes (423 BC): *Clouds*, translated by Thomas Mitchell [1783–1845]. Cited in: *Familiar Quotations*, Bartlett J, 13th ed. (1955), Little, Brown and Co., p. 23

Borenstein J, Reisberg B (1987): Functional deficits in Alzheimer's disease. 1987 New Research Program and Abstracts. American Psychiatric Association, 140th Annual Meeting, 56 (Abstract)

Broderick P, Laszlo JI (1987): The drawing of squares and diamonds: A perceptual-motor task analysis. *J Exper Child Psychol* 43:44–61

Brugg B, Matus A (1988): PC12 cells express juvenile microtubule-associated proteins during nerve growth factor-induced neurite outgrowth. *J Cell Biol* 107:643–650

Chen MJ (1985): Young children's representational drawings of solid objects: A comparison of drawings and copying. In: *Visual Order*, Freedman NH, Cox MV, eds. New York: Cambridge University Press

Cole MG, Dastoor D (1980): Development of a dementia rating scale: Preliminary communication. *J Clin Exp Gerontol* 2:46–63

Cole MG, Dastoor DP (1987): A new hierarchic approach to the measurement of dementia. *Psychosomatics* 28:298–304

Cole MG, Dastoor DP, Koszycki D (1983): The hierarchic dementia scale. *J Clin Exp Gerontol* 5:219–234

Constantinidis J, Richard J, de Ajuriaguerra J (1978): Dementias with senile plaques and neurofibrillary changes. In: *Studies in Geriatric Psychiatry*, Isaacs AD, Post F, eds. New York: John Wiley, pp. 119–152

Cox MV (1986): Cubes are difficult things to draw. *British Journal of Developmental Psychology* 4:341–345

Cratty BJ (1986): *Perceptual and Motor Development in Infants and Young Children*. Macmillan, New York

de Ajuriaguerra J, Rey M, Bellet-Muller M (1964): A propos de quelques problemes posees par le deficit operatoire de viellards atteints de demence degenerative en debut d'evolution. *Cortex* 1:232–256

de Ajuriaguerra J, Tissot R (1968): Some aspects of psychoneurologic disintegration in senile dementia. In: *Senile Dementia*, Mueller CH, Ciompi L, eds. Switzerland: Huber, pp. 67–79

de Ajuriaguerra J, Tissot R (1975): Some aspects of language in various forms of senile dementia: Comparisons with language in childhood. In: *Foundations of Language Development*, Vol 1. Lennenberg EH, Lennenberg E, eds. New York: Academic Press

de Leon MJ, McRae T, George AE, Marcus D, Freedman M, Wolf AP, McEwen B (1988) Abnormal cortisol response in Alzheimer's disease with CT and PET temporal lobe changes. *Lancet* 2(8607): 391–392

Dunst C (1980): *A Clinical and Educational Manual for Use with the Uzgiris and Hunt Scales of Infant Psychological Development.* Austin: Pro-Ed Pubs., Inc.

Eisenberg L (1975): Normal child development. In: *Comprehensive Textbook of Psychiatry II*, Vol 2. Freedman AM, Kaplan HR, Sadock BJ, eds. Baltimore: Williams and Wilkins, pp. 2036–2054

Franssen EH, Reisberg B, Kluger A, Sinaiko E, Boja C (1991): Cognition independent neurologic symptomology in normal aging and probable Alzheimer's disease. *Archives of Neurology*, 48:148–154

Galen (1821–1833 translation): De symptomatum differentis liber. In: *Opera Omnia*, Kuhn CG, ed. Leipzig: Knobloch, pp. 200–201

Gassier J (1984): *A Guide to the Psychomotor Development of the Child.* New York: Churchill Livingstone

Gesell AL (1940): *The First Five Years of Life; A Guide to the Study of the Preschool Child, from the Yale Clinic of Child Development.* New York: Harper Bros

Gesell AL, Ilg FL (1943): *Infant and Child in the Culture of Today.* New York: Harper Bros, pp. 141–245

Grand RJ, ed. (1987): *Pediatric Nutrition: Theory and Practice.* Boston: Butterworth

Hefti F, Mash DC (1989): Localization of nerve growth factor receptors in the normal human brain and in Alzheimer's disease. *Neurobiology of Aging* 10:75–87

Hughes CP, Berg L, Danziger WL, Cohen LA, Martin RL (1982): A new clinical scale for the staging of dementia. *Br J Psychiatry* 140:566–572

Hurlock EB (1978): *Child Development*, 6th ed. New York: McGraw-Hill

Illingworth RS (1983): *The Development of the Infant and Young Child: Normal and Abnormal.* Edingburgh: Churchill Livingstone

Illingworth RS (1987): *The Normal Child. Some Problems of the Early Years and Their Treatment.* Edinburgh: Churchill Livingstone

King LS (1963): *The Growth of Medical Thought.* Chicago: University of Chicago Press

Leeds M (1960): Senile recession: A clinical entity? *J of the Am Ger Soc* 8:122–131

Leonard DGB, Gorham JD, Cole P, Greene LA, Ziff EB (1988): A nerve growth factor regulated messenger RNA encodes a new intermediate filament protein. *J Cell Biol* 106:181–193

Lin L-FH, Mismer D, Lile JD, Armes LG, Butler ET, Vannice JL, Collins F (1989): Purification, cloning, and expression of ciliary neurotrophic factor (CNTF). *Science* 246:1023–1025

Linden M (1957): Regression and recession in the psychoses of the aging. In: *The Kirkpatrick Memorial Programs on Gerontology for 1955 and 1956*, pp. 35–37

Linden M, Courtney D (1953): The human life cycle and its interruptions. A psychologic hypothesis. Studies in gerontologic human relations I. *Am J Psychiatry* 109:906–915

Lipowski ZJ (1980): Organic mental disorders: Introduction and review of syndromes. In: *Comprehensive Textbook of Psychiatry*, Vol. 2, Kaplan HI, Freedman AM, Sadock BJ, eds. Baltimore: Williams and Wilkins, pp. 1359–1392

Lowrey GH (1978): *Growth and Development of Children*, 7th ed. Chicago: Year Book Medical Publishers

Mitchelmore MG (1978): Developmental stages in children's representation of regular solid figures. *J Genet Psychol* 133:229–239

Mohs R, Kim G, Johns C, Dunn D, Davis K (1986): Assessing changes in Alzheimer's disease: Memory and language. In: *Handbook for Clinical Memory Assessment*, Poon LW, ed. Washington, DC: American Psychological Association, pp. 149–155

Nelms BC (1982): *Growth and Development: A Primary Health Care Approach*. Englewood Cliffs, NJ: Prentice-Hall

Pfeffer RI, Kurosaki TT, Harrah CH, Chance JM, Filos S (1982): Measurement of functional activities in older adults in the community. *J Gerontol* 37:323–329

Phelps CH, Gage FH, Growdon JH, Hefti F, Harbaugh R, Johnston MV, Khachaturian ZS, Mobley WC, Price DL, Raskind M, Simpkins J, Thal LJ, Woodcock J (1989): Potential use of nerve growth factor to treat Alzheimer's disease. *Science* 243:11

Piaget J (1952): *The Origins of Intelligence in Children*. London: Routledge and Kegan Paul

Piaget J (1960): *The Psychology of Intelligence*. Totowa (NJ): Littlefield, Adams and Co

Piaget J (1970): *Genetic Epistemology*. New York: Columbia University Press

Piaget J (1973): *The Child and Reality: Problems of Genetic Psychology*. New York: Grossman Pub

Piaget J Inhelder B (1956): *The Child's Conception of Space*. London: Routledge and Kegan Paul

Pierce CM (1975) Enuresis and encopresis. In: *Comprehensive Textbook of Psychiatry II*, Vol. 2, Freedman AM, Kaplan HR, Sadock BJ, eds. Baltimore: Williams and Wilkins, pp. 2116–2125

Pipes PL (1981): *Nutrition in Infancy and Childhood*, 2nd ed. St. Louis: CV Mosby Co

Plato (1921): *The Laws* (ed. trans.), Book IX. Manchester University Press

Plutarch (1967): *Lives* North T, trans. New York: AMS Press

Reisberg B (1986): Dementia: A systematic approach to identifying reversible causes. *Geriatrics* 41:30–46

Reisberg B (1988): Functional assessment staging (FAST). *Psychopharmacology Bulletin* 24:653–659

Reisberg B, Ferris SH, Anand R, de Leon MJ, Schneck MK, Buttinger C, Borenstein J (1984): Functional staging of dementia of the Alzheimer's type *Ann NY Acad Sci* 435:481–483

Reisberg B, Ferris SH, de Leon MJ, Crook T (1982): The global deterioration scale for assessment of primary degenerative dementia. *Am J of Psychiatry* 139:1136–1139

Reisberg B, Ferris SH, de Leon MJ (1985): Senile dementia of the Alzheimer type: Diagnostic and differential diagnostic features with special reference to functional assessment staging. In: *Senile Dementia of the Alzheimer Type*, Vol. 2, Traber J, Gispen WH, eds. Berlin: Springer-Verlag, pp. 18–37

Reisberg B, Ferris SH, Franssen E (1986): Functional degenerative stages in dementia of the Alzheimer's type appear to reverse normal human development. In: *Biological Psychiatry*, 1985, Vol 7, Shagass C et al., eds. New York: Elsevier Science Publishing Co, pp. 1319–1321

Reisberg B, London E, Ferris SH, Borenstein J, Scheier L, de Leon MJ (1983b): The Brief Cognitive Rating Scale: Language, motoric, and mood concomitants in primary degenerative dementia. *Psychopharmacology Bulletin* 19:702–708

Reisberg B, Schneck MK, Ferris SH, Schwartz GE, de Leon MJ (1983a): The brief cognitive rating scale (BCRS): Findings in primary degenerative dementia (PDD). *Psychopharmacology Bulletin* 19:47–50

Rosenberg MB, Friedman T, Robertson RC, Tuszynski M, Wolff JA, Breakefield XO, Gage FH (1988): Grafting genetically modified cells to the damaged brain: restorative effects of NGF expression. *Science* 242:1575–1578

Rush B (1793): An account of the state of mind and body in old age. In: *Medical Inquiries and Observations*, Vol. 2, Philadelphia: Dobson, p. 311

Sclan SG, Foster JR, Reisberg B, Franssen E, Welkowitz J (1990): Application of Piagetian measures of cognition in severe Alzheimer's disease. *The Psychiatric Journal of the University of Ottawa* 15:223–228

Simko MD, Cowell C, Hetza MS (1989): *A Quick Reference for the Health Care Practitioner*. Rockville, Maryland: Aspen Publishers, Inc.

Torack RM (1983): The early history of senile dementia. In: *Alzheimer's Disease*, Reisberg B, ed. New York: Free Press, Macmillan, pp. 23–28

Uzgiris I, Hunt JMcV (1975): *Assessment in Infancy: Ordinal Scales of Psychological Development*. Urbana: University of Illinois

Vaughn VC (1969): Growth and development. In: *Textbook of Pediatrics*, 9th ed. Nelson WE, Vaughn VC, McKay RJ, eds. Philadelphia: Saunders, pp. 15–57

Wilson R, Kaszniak A (1986): Longitudinal changes: Progressive idiopathic dementia. In: *Handbook for Clinical Memory Assessment*, Poon LW, ed. Washington, DC: American Psychological Association, pp. 285–293

Psychopharmacological Aspects of Antidementia Agents

Stephen Curran and Ian Hindmarch

Psychopharmacologists are interested in measuring psychoactive drug-induced changes in cognitive and/or psychological behavior (Hindmarch and Stonier, 1987; 1989; Hindmarch et al., 1988). The number of tests included in a test battery should be kept to a minimum to avoid identifying significant results by chance, and individual tests should be appropriate for the particular psychological function under investigation, e.g., psychomotor performance, information processing capacity (IPC)/CNS integration, and memory (Hindmarch and Wattis, 1988). It is frequently unclear what a particular test is supposed to be measuring (Adams, 1974; Lahtinen et al., 1978). Pre-experimental screening of subjects and the control of confounding variables will also increase the sensitivity of psychometric measures and facilitate the interpretation of results.

Assessment of drug effects needs special consideration when dealing with elderly and especially demented patients. In addition, once a test has been chosen it should be established that the test is valid and reliable, pharmacosensitive, easy for both the investigator and the subject to use, relatively quick to administer, free from learning effects, and an objective measure of performance. Tests should also be well within the intellectual range and sensory/memory capacity of the subject (Hindmarch and Wattis, 1988). In the elderly, many psychological functions are impaired compared to young volunteers, e.g., reaction time (Botwinik and Storandt, 1974), learning ability (Botwinik, 1973), critical flicker fusion threshold (Curran et al., 1990), and a variety of other psychological functions (Busse, 1978). Care must therefore be exercised when examining elderly subjects with tests that have been mainly developed in young normal volunteers. This is particularly important in patients with Alzheimer's disease where cognitive abilities are severely impaired; choice of test becomes even more crucial.

Although many tests are able to satisfy some of the criteria outlined above, Critical Flicker Fusion Threshold (CFFT) appears unique in being able to satisfy all these criteria.

Critical Flicker Fusion Threshold (CFFT)

Measurement of CFFT continues to be one of the most popular techniques in psychopharmacology (Ott and Kranda, 1982). There is abundant evidence that it is a valid and reliable technique (Curran, 1990) as well as being a pharmacosensitive measure (Hindmarch, 1980, 1982; Simonson and Brozek, 1952; Smith and Misiak, 1976; Turner, 1968).

The neuroanatomical basis of CFFT is still the subject of considerable theoretical debate. However, there is considerable evidence that it reflects, at least in part, central and/or cortical function (Curran et al., 1990). The CFFT is undoubtedly a complex phenomenon. It appears to reflect CNS function rather than being related to a specific anatomical structure or structures and may be the "the final outcome" or response of many interconnected regions within the CNS. However, its precise anatomical and physiological nature remains unknown. It is regarded as a measure of IPC/Central Integrative Function (Alford and Hindmarch, 1990; Hindmarch and Wattis, 1988).

Subthreshold intermittent light is perceived as flicker. If the frequency is gradually increased the flicker becomes gradually less distinct until flicker is no longer perceived. This point of fusion is known as the fusion threshold (ascending method). Alternatively, if the flickering is suprathreshold and is gradually decreased, the point at which flicker is perceived is known as the flicker threshold (descending method). These are variations of the continuous Method of Limits (Woodworth and Schlosberg, 1958). The mean of the fusion and flicker thresholds is the Critical Flicker Fusion Threshold (CFFT). A number of different techniques can be used to measure CFFT (Curran, 1990). The method of limits is particularly appropriate in psychopharmacological research because it allows for rapid data collection. The CFFT can be measured using four red-light-emitting diodes in binocular foveal fixation at 1 m under microprocessor control. A detailed account of methodological procedures is given by Curran et al. (1990).

Other psychometric measures described include the Sternberg Memory Scanning Task (SMST) and Choice Reaction Time (CRT). CRT is taken as a measure of psychomotor performance (Hindmarch and Wattis, 1988) and is measured using the Leeds Psychomotor Tester (Frewer and

Hindmarch, 1988). Using this apparatus it is possible to measure both the Motor Reaction Time (MRT) and the Recognition Reaction Time (RRT), this latter being a measure of the processing time within the CNS. A central touch pad is surrounded by six identical pads arranged in a semicircle, each having an adjacent red light. One of the red lights appears at random and is under microprocessor control. The subject is required to extinguish the appropriate light by touching the adjacent response pad as quickly as possible.

The SMST is a measure of high speed scanning ability and retrieval from short-term memory (Sternberg, 1969, 1975). This test is under microcomputer control, thus facilitating the presentation of "sets" and recording of reaction time and accuracy. Sternberg (1969) proposed that the reaction process in such a task occurs in a series of at least three stages. The "set" is stored in short-term memory and the subject must first perceive and encode the test stimulus (stimulus encoding). The stimulus is then compared to the members of the memory set (memory scanning) and a response is finally made on the basis of these comparisons (response selection organization). "Sets" containing sequences of words, letters, or digits of different lengths are presented to the subject to be memorized. The subject is then required to indicate whether a subsequent list presented to the subject contains elements that were in the previous list. Subject response is via a yes/no button press. Accuracy and response latency are recorded. The initial set size can be varied in size depending on the test situation.

It is therefore possible using these simple psychometric measures to evaluate short-term memory (SMST), psychomotor performance (CRT), and IPC/Central Integrative Function (CFFT).

Antidementia Agents

The concept of nootropics has been reviewed by Herrmann and Scharer (1987). These can be defined as drugs that exert a positive effect on brain functions in the sense that they lead to an activation of adaptation capacity (i.e., the capacity of centrally controlled functions to regulate themselves). Nootropics are therefore substances that can stabilize disorders of centrally regulated functions, restore reduced activity and combat excessive fatigability (Herrmann and Scharer, 1987). A more useful working definition has been proposed by Giurgea (1982). Nootropics are regarded as compounds that operate by direct functional activation of the higher integrative brain mechanisms that enhance cortical vigilance and the restoration of deficient higher nervous activity. In contradistinction to other psychotropic drugs, nootropics do not induce direct reticular, limbic, or other subcortical events. Nootropics are related to other psychotropic drugs in the following way: psycholeptics (hypnotics, neuroleptics, and anxiolytics), psychoanaleptics (CNS stimulants, antidepressants, and nootropics) and psychodysleptics (hallucinogens such as opiates, LSD, and cannabinoids) (Giurgea, 1982). Skondia (1979) has suggested some basic criteria for defining nootropic agents, which include: no direct vasoactivity, no effect on the basic EEG rhythm, must be able to pass the blood brain barrier, and have demonstrable metabolic activity.

There is still no general agreement on the definition of nootropic agents and their mechanisms of action need further clarification. Although little is known about the neurochemical mechanisms of nootropics, a fuller understanding will only emerge if the approach is multifaceted, e.g., neurochemistry, neurophysiology, and psychopharmacology. The neurochemical and pharmacological aspects of nootropics have been reviewed (Nicolaus, 1982).

Psychopharmacological Aspects of Nootropics

A large variety of nootropics have been evaluated in both volunteer and patient studies. However, it is often difficult to make direct comparisons between studies because of differences in assessment procedures and methodologies. This review therefore concentrates predominantly on nootropics that have been investigated using identical methodologies and test procedures. The measures chosen represent subjective assessments of drug effects, IPC/Central Integrative Function, psychomotor performance, and short-term memory.

Hydergine

This is a dihydrogenated ergot alkaloid. It is an α-antagonist and inhibits cAMP phosphodiesterase (Hindmarch et al., 1979). Clinical studies have demonstrated improvements in behavioral symptoms (Roubicek et al., 1972) and the EEG (Venn, 1976). It has also been demonstrated to improve mental alertness, confusion and lack of motivation (Bazo, 1973).

Hindmarch et al. (1979) examined Hydergine in normal young volunteers (3 mg, q.i.d. for two weeks). Critical flicker fusion threshold was significantly increased after one week of treatment and remained significantly elevated two weeks after discontinuation. This change was mirrored by subjective measures of arousal. In a follow-up study (Parrott and Hindmarch, 1982), 12 elderly normal volunteers received 3 mg per day for 12 weeks. After two weeks significant improvements were seen in CRT, measures of subjective alertness, and CFFT. Two weeks following discontinuation these measures remained significantly improved.

Tucker (1982) extended this work and examined Hydergine in patients with dementia. Eight patients were included in the study and received 0.3 mg b.d. (i.m.) for two weeks. Clinical improvement was noted in five patients; in the same patients the CFFT was also found to be increased. However, this study was a small open study and no discussion of the diagnosis of dementia was given. Knox et al. (1982) examined 31 patients with dementia. Seventeen received Hydergine 3 mg t.d.s. and 14 were given Praxilene 200 mg t.d.s. for three weeks. Hydergine produced improvements in orientation, constructional ability, continence, washing ability, motivation, dressing skills, communication, understanding, and depression compared with Praxilene. However apart from changes in "communication," these just failed to reach significance, probably because of the small numbers used. The clinician's overall impression was unable to distinguish between the two drugs emphasizing the need for objective and sensitive psychometric measures.

Ginkgo biloba extract

This is a complex product prepared from dried leaves of *Ginkgo biloba*. It is believed to be a combination of flavanoid glycosides, pro-anthocyanides, and terpene substances. Subhan and Hindmarch (1984) examined this compound in young volunteers (120, 240 and 600 mg) in a double-blind, crossover design, with each subject receiving all four treatment conditions (including placebo) following a one week wash-out period between treatments. This "compound" was found to have no effect on CFFT, CRT, or subjective measures of alertness but significantly decreased the overall reaction time on the SMST one hour following treatment. This was only significantly decreased for the 600 mg dose. Only memory function was affected, with no effect on IPC/Central Integrative Function or psychomotor performance.

HWA-285

Hindmarch and Subhan (1985) examined HWA-285, 150, 300, and 450 mg (1-5-oxohexyl-3-methyl-7-propyl-xanthine) in young healthy volunteers. The mechanism of action is thought to be inhibition of ADP platelet aggregation. The drug has also been found to increase cerebral oxygen consumption and glucose turnover (Hindmarch and Subhan, 1985). This compound also produced no effect on subjective measures of arousal, CFFT, or CRT, but the overall reaction time on the SMST was significantly decreased for all active treatments, i.e., 150, 300, and 450 mg HWA-285. The magnitude of the change was not found to be dose related. These data suggest a potential use for HWA-285 in the treatment of impaired memory function.

Vinpocetine

This is a vincamine derivative (ethyl apovincaminate) and has been used for the treatment of disorders arising from cerebrovascular and cerebrodegenerative diseases. In animals, it produces increased cerebral blood flow (Benscath et al., 1976) and Hadjiev and Yancheva (1976) have reported improvements in memory function. It has also been demonstrated to improve neurotransmitter turnover (Rosdy et al., 1976) and cerebral blood flow (Karpati and Szporny, 1976). Subhan and Hindmarch (1984) examined 10, 20, and 40 mg t.d.s. in normal young volunteers. As with *Ginkgo biloba* extract and HWA-285, the overall reaction time of the SMST was significantly decreased one hour following treatment but only with the 40 mg dose. There was no effect on CFFT, CRT, or subjective measures of arousal. These data are in agreement with other reports that vinpocetine improves memory function, e.g., Szobor and Klein (1976).

More recently this initial study has been replicated (Bhatti and Hindmarch, 1987) using 40 mg t.d.s. of vinpocetine. There was no effect on CFFT and subjective measures of arousal. Choice reaction time was not examined. Vinpocetine 40 mg t.d.s. was found not only to improve overall reaction time on the SMST but was also able to reverse memory impairment induced by flunitrazepam 1 mg, which was administered prior to "sleep onset." Coleston and Hindmarch (1988) examined CFFT, CRT, and the SMST in 12 normal volunteers following vinpocetine 10, 20, and 40 mg. Subjective measures were not examined. The only improvement was in the SMST following 40 mg of vinpocetine confirming previously reported findings.

Pyritinol

This is a pyridoxine derivative devoid of B_6 activity (Knežević et al., 1989) and has been examined in patients with Alzheimer's disease (e.g., Hamouz, 1977). It is thought to exert its effect in part by improving cerebral glucose metabolism but does not appear to affect cerebral blood flow (Hoyer et al., 1977). It has also been suggested that it increases endogenous acetylcholine concentrations in the brains of old rats (Martin and Vyas, 1987). A large multi-centre study reported by Tazaki et al. (1980) examined the compound in 458 patients with cerebral arterial sclerosis (now an obsolete concept) in 55 different centers. The only significant improvements seen were in clinical global impression. Objective measures such as the EEG were not affected. This study illustrates the need to be aware of relevant concepts, e.g., cerebral arteriosclerosis is not now regarded as a cause of dementia and the inclusion of 55 centers into the study leaves the study open to criticism regarding inter-rater reliability.

A recent study in 12 healthy male volunteers examined pyritinol 600 and 1200 mg and placebo t.d.s. for three days according to a randomized double-blind, crossover design with each subject acting as his own control (Hindmarch and Coleston, 1990). Pyritinol was found to significantly reduce the recognition component of the CRT ("thinking time") suggesting that this was not a motor learning effect. In addition, the CFFT was found to be significantly improved but there was no effect on the SMST. This appears to have a similar profile to Hydergine and appears to be operating by improving IPC/Central Integrative Function and central aspects of psychomotor performance rather than on memory function.

Nicotine

Nicotine gum has been demonstrated to be a reliable medium for the delivery of systemic nicotine (Sherwood et al., 1989) and has been the subject of considerable research. It exerts its sympathomimetic effect by the same mechanisms as acetylcholine, i.e., facilitation of transmission across sympathetic ganglia, adrenal medullary stimulation and release of norepinephrine from chromaffin tissue in the cardiac atria and arterioles. Muscarinic effects due to the stimulation of parasympathetic ganglia occur especially on the gastrointestinal tract and eye (Meyers et al., 1980). Hindmarch et al. (1990) investigated normal young volunteers wherein smokers used 0, 2, and 4 mg of nicotine polacrilex gum, whereas nonsmokers used 0 and 2 mg of nicotine polacrilex gum using methodologies similar to those outlined above, i.e., double-blind, crossover designs, etc. Additional nicotine improved speed and accuracy among the smokers (motor reaction time and tracking task) but central effect as measured by CFFT and RRT was not affected. Memory function was also unaffected. There were no significant change in nonsmokers. One possible reason for this is that non- smokers are not used to the adverse effects of systemic nicotine, which may impair their performance. The lack of improvement in the CFFT is of interest. One possible reason for this may be the small number of volunteers used (six smokers and five nonsmokers). Other authors have reported increases in the CFFT following administration of nicotine (Barlow and Baer, 1967; Fabricant and Rose, 1951; Warwick and Eysenck, 1963).

A recent study reported by Sahakian et al. (1989) found that subcutaneous nicotine (0, 0.4, 0.6, and 0.8 mg) administered to patients with dementia of the Alzheimer type, produced significantly quicker reaction times and improved the CFFT. No beneficial effects on memory function were demonstrated. Nicotine may therefore have a role in the management of IPC and psychomotor function deficits in patients with Alzheimer's disease rather than in the treatment of memory dysfunction.

Dementia Research: Reasons for Delay

Alzheimer's disease is now a major medical, social, and economic problem. Despite enormous effort the etiology still remains unknown (Curran and Wattis, 1989; Davies, 1989). Lack of progress can be partly explained by confusion and disagreement about definitions and diagnostic criteria. The difficulty of making an accurate clinical diagnosis (Eisdorfer et al., 1980; Marsden and Harrison, 1972; Morris et al., 1987; Nott and Fleminger, 1975; Ron et al., 1979) has aggravated the situation. When based on isolated symptoms and/or signs (e.g., memory impairment alone) the diagnosis becomes unreliable (Ballinger et al., 1982; Hare, 1978; Rabins et al., 1984). In addition, Alzheimer's disease is increasingly being conceptualized as a heterogeneous group of related conditions rather than a single disease entity (Chui et al., 1985; Mayeux et al., 1985). Longitudinal studies are needed to clarify the concept of clinical subgroups (Wilcock, 1988) as different patterns of cognitive decline may have important implications for diagnosis and treatment (Chui, 1987) and prognosis (Neary et al., 1986).

Prior to commencing a clinical trial to evaluate a potentially new therapeutic compound for Alzheimer's disease, careful attention should be directed at all aspects of the study as this will have enormous implications for the interpretation of efficacy. There should be some evidence that the drug is likely to be efficacious; its pharmacological action should be well defined. It should also be able to reach its site of action, that is, be well absorbed, be able to cross the blood–brain barrier and reach an adequate concentration in the CNS. The drug should also be relatively cheap, well tolerated, and require only once- or twice-daily administration.

Studies from different centers, especially large multicenter studies, and variations in study treatment, route of administration, diagnostic criteria, study duration and design, test batteries, the number of different assessments included, and statistical procedures, all complicate the interpretation of results. A review of clinical trial methodology is given in Spriet and Simon (1985).

Lack of knowledge about Alzheimer's disease is reflected in the diversity of drug strategies investigated in the search for a treatment for this syndrome. In general, there has been only very limited success and a satisfactory treatment is still awaited. The drug literature is extensive, confusing and generally unhelpful.

Numerous treatment strategies have been tried including manipulations of neurotransmitters, especially the cholinergic system, cerebral vasodilators, neuropeptides, narcotic antagonists, and more recently specific antidementia agents (nootropic compounds) (Giacobini and Becker, 1988; Kendall, 1987). Central to the development of new therapeutic compounds is a clear understanding of the neurochemistry of Alzheimer's disease (Cowburn et al., 1989; Fraser, 1987; Rossor, 1987). Unfortunately, more progress is urgently needed.

If pharmacotherapeutic research on Alzheimer's disease is to make any meaningful progress it is essential that the neuropathology, neurochemistry, and etiology of the syndrome are clearly defined in order to have a rational basis for treatment. In addition, clinical features and diagnosis require considerably more refinement. Agreement on standardized research protocols to facilitate comparisons between research groups would be the next logical step. Finally, it is the role of psychopharmacologists to develop pharmacosensitive measures that are reliable and valid, in addition to being quick and easy to administer. If progress is to be achieved, a multidisciplinary approach is essential.

Conclusion

In the management and treatment of patients with Alzheimer's disease it may be more productive to manage specific areas of dysfunction, e.g., memory, psychomotor performance, or information processing capacity. It is therefore important for psychopharmacologists to develop psychometric tests that are able to profile a potential antidementia agent. It is unlikely that a single drug will improve all aspects of this complex syndrome. This may be one reason why single drug strategies have on the whole failed whereas combination treatments have been generally more successful.

The use of objective and psychologically relevant psychometric measures like the ones described will improve the chances of identifying possible therapeutic agents and provide a more rational basis for treatment.

Finally, there may be two broad groups of antidementia agents: those with an effect predominantly on memory function and those acting primarily on central integrative processes/ information processing capacity.

Acknowledgment. The authors would like to thank the Violet M. Richards charity for their generous equipment grant, which has made the preparation of this manuscript possible.

References

Adams RG (1974): Pre-sleep ingestion of two hypnotic drugs and subsequent performance. *Psychopharmacologica* 40:185–190

Alford C, Hindmarch I (1990): Measuring the effects of psychoactive drugs, with particular reference to anti-depressants. In: *Microcomputers, Psychology and Medicine*, West R, Christie M, Weinman J, eds. John Wiley and Sons, pp. 85–95

Ballinger BR, Reid AH, Heather BB (1982): Cluster analysis of symptoms in elderly demented patients. *British Journal of Psychiatry* 140:257–262

Barlow DH, Baer DJ (1967): Effect of cigarette smoking on the critical flicker frequency of heavy and light smokers. *Perception and Motor Skills* 24:151–155

Bazo AJ (1973): An ergot alkaloid preparation (Hydergine) versus papaverine in treating common complaints of the aged: a double-blind study. *Journal of the American Geriatric Society* 11:63

Benscath P, Debreczeni L, Takacs L (1976): Effect of ethyl apovincaminate on cerebral circulation of dogs under normal conditions and in arterial hypoxia. *Arzneimittelforschung* (Drug Research) 26:1920–1923

Bhatti JZ, Hindmarch I (1987): Vinpocetine effects on cognitive impairments produced by flunitrazepam. *International Clinical Psychopharmacology* 2(4): 325–331

Botwinik J (1973): *Cognitive Processes in Maturity and Old Age*. New York: Springer

Botwinik J, Storandt M (1974): Cardiovascular status, depressive affect and other factors in reaction time. *Journal of Gerontology* 29:343–348

Busse EW (1978): Duke longitudinal study: senescence and senility. In: *Alzheimer's Disease, Senile Dementia and Related Disorders*, Katz R, Terry RD, Bick KL, eds. New York: Raven Press, pp. 59–69

Chui HC, Teng EL, Henderson VW, Moy AC (1985): Clinical subtypes of dementia of the Alzheimer type. *Neurology* 35:1544–1550

Chui HC (1987): The significance of clinically defined subgroups of Alzheimer's disease. *Journal of Neural Transmission* 24:57–68

Coleston DM, Hindmarch I (1988): Possible memory enhancing properties of Vinpocetine. *Drug Development Research* 14:191–193

Cowburn RF, Hardy JA, Roberts PJ (1989): Neurotransmitter deficits in Alzheimer's disease. In: *Alzheimer's Disease: Towards an Understanding of the Pathogenesis*, Davies DC, ed. London and Paris: John Libbey, pp. 9–32

Curran S, Wattis JP (1989): Round-up: searching for the cause of Alzheimer's disease. *Geriatric Medicine* 19(3):13–14

Curran S, Wattis JP, Shillingford C, Hindmarch,I (1990): Critical flicker fusion in normal elderly subjects; a cross-sectional community study. *Current Research and Reviews in Psychology* (in press).

Curran S (1990): Critical flicker fusion techniques in psychopharmacology. In: *Human Psychopharmacology: Measures and Methods*, Volume 3, Hindmarch I, Stonier PD, eds. Chichester: John Wiley and Sons

Davies DC (1989): *Alzheimer's Disease: Towards an Understanding of the Aetiology and Pathogenesis*. London and Paris: John Libbey

Eisdorfer C, Cohen D, Veith R (1980): *Current Concepts: The Psychopathology of Aging*, Washington: Upjohn, pp. 8–12

Fabricant ND, Rose IW (1951): Effects of smoking cigarettes on flicker fusion threshold of normal subjects. *Eye, Ear, Nose and Throat Monthly* 30:541–543

Fraser M (1987): *Dementia: Its Nature and Management*, Chichester: John Wiley and Sons, pp. 38–59

Frewer LJ, Hindmarch I (1988): The effects of time of day, age and anxiety on a choice reaction task. In: *Psychopharmacology and Reaction Time*, Hindmarch I, Aufdembrinke B, Ott H, eds. Chichester: John Wiley and Sons, pp. 103–114

Giacobini E, Becker R (1988): *Current Research in Alzheimer Therapy*. New York: Taylor and Francis

Giurgea CE (1982): The nootropic concept and its prospective implications. *Drug Development Research* 2:441–446

Hadjiev D, Yancheva S (1976): Rheoencephalographic and psychological studies with ethyl apovincaminate in cerebral vascular insufficiency. *Arzneimittelforschung* (Drug Research) 26:1947–1950

Hamouz W (1977): The use of pyritinol in patients with moderate to severe organic psychosyndrome. *Pharmatherapeutica* 1:398–404

Hare M (1978): Clinical check-list for diagnosis of dementia. *British Medical Journal*, ii:266–267.

Herrmann WM, Scharer E (1987): Effects and efficiency of nootropics discussed with the example of Pyritinol. In: *Higher Nervous Functions*, Herrmann WM, ed. Vieweg, pp 81–112.

Hindmarch I, Parrott AC, Lanza M (1979): The effects of an ergot alkaloid derivative (Hydergine) on aspects of psychomotor performance, arousal, and cognitive processing ability. *Journal of Clinical Pharmacology* 19(11 and 12):726–732

Hindmarch I (1980): Psychomotor function and psychoactive drugs. *British Journal of Clinical Pharmacology* 10(3):189–209

Hindmarch I (1982): Critical flicker fusion frequency (CFFF): The effects of psychotropic compounds. *Pharmacopsychiatrica* 15(Suppl. 1):44–48

Hindmarch I, Subhan Z (1985): A preliminary investigation of "Albert 285" (HWA 285) on psychomotor performance, mood, and memory. *Drug Development Research* 5(4):379–386

Hindmarch I, Stonier PD (1987): *Human Psychopharmacology: Measures and Methods*, Vol. 1. Chichester: John Wiley and Sons

Hindmarch I, Aufdembrinke B, Ott H (1988): *Psychopharmacology and Reaction Time*. Chichester: John Wiley and Sons

Hindmarch I, Wattis JP (1988): Measuring effects of psychotropic drugs. In: *Psychological Assessment of the Elderly*, Wattis JP, Hindmarch I, eds., Edinburgh: Churchill Livingstone, pp. 180–197

Hindmarch I, Stonier PD (1989): *Human Psychopharmacology: Measures and Methods*, Volume II. Chichester: John Wiley and Sons

Hindmarch I, Kerr JS, Sherwood N (1990): Effects of nicotine gum on psychomotor performance in smokers and non-smokers. *Psychopharmacology* 100:535–541

Hindmarch I, Coleston DM (1990): The psychopharmacological effects of pyritinol in normal healthy subjects. *Neuropsychobiology* (in press)

Hoyer S, Oesterreich K, Stoll KD (1977): Effects of Pyritinol-HCl on blood flow and oxidative metabolism of the brain in patients with Dementia. *Arzneimittelforschung* (Drug Research) 27:671–674

Karpati E, Szporny L (1976): General and cerebral haemodynamic activity of ethyl apovincaminate. *Arzneimittelforschung* (Drug Research) 26:1908–1912

Kendall M (1987): Drugs for dementia. In: *Dementia*, Pitt B, ed. Edinburgh: Churchill Livingstone, pp. 265–280

Knežević S, Mubrin Z, Risberg J, Vucinic G, Spilich G, Gubarev N, Wannenmacher W (1989): Pyritinol treatment of SDAT patients: Evaluation by psychiatric and neurological examination, psychometric testing and rCBF measurements. *International Clinical Psychopharmacology* 4:25–38

Knox J, Hindmarch I, Wallace MG (1982): Visual analogue scales in assessment of severe mental impairment. *British Journal of Clinical Practice* 516:12–15

Lahtinen V, Lahtinen A, Pekkola P (1978): The effect of nitrazepam on grip strength, reaction time and subjective evaluation of sleep. *Acta Pharmaceutica Toxicologia* 42:130–134

Marsden CD, Harrison MJG (1972): Outcome of investigations of patients with pre-senile dementia. *British Medical Journal* 1:249–252

Martin KJ, Vyas S (1987): Increase in acetylcholine concentration in the brain of old rats following treatment with pyrithioxin (Encephabol). *British Journal of Pharmacology*, 90:561–565

Meyers FH, Jawetz E, Goldfien A (1980): *Review of Medical Pharmacology*, California, Lange, p. 69

Mayeux R, Stern Y, Spanton S (1985): Heterogeneity in dementia of Alzheimer's type—Evidence of subgroups. *Neurology* 35:453–461

Morris JC, Berg L, Fulling K, Torack RM, McKeel DW (1987): Validation of clinical diagnostic criteria in senile dementia of the Alzheimer type. *Annals of Neurology* 22:122

Neary D, Snowden JS, Bowden DM (1986): Neuropsychological syndromes in presenile dementia due to cerebral atrophy. *Journal of Neurology, Neurosurgery and Psychiatry* 49:163–174

Nicolaus BJR (1982): Chemistry and pharmacology of nootropics. *Drug Development Research* 2:463–474

Nott PN, Fleminger JJ (1975): Pre-senile dementia: The difficulties of early diagnosis. *Acta Psychiatrica Scandinavica* 51:210–217

Ott H, Kranda K (1982): *Flicker Techniques in Psychopharmacology*, Weinheim and Basel: Beltz Verlag

Parrott AC, Hindmarch I (1982): The effects of hydergine upon psychomotor performance, indices of alertness and behavioural ratings in studies involving normal and geriatric subjects. *British Journal of Clinical Practice* 16:18–20

Rabins PV, Starr LB, Price TR (1984): A two year longitudinal study of mood disorders following stroke: Prevalence and duration at six months follow-up. *British Journal of Psychiatry* 144:488–492

Ron MA, Toone BK, Garralda ME, Lishman WA (1979): Diagnostic accuracy in pre-senile dementia. *British Journal of Psychiatry* 134:161–168

Rosdy B, Balazs M, Szporny L (1976): Biochemical effects of ethyl apovincaminate. *Arzneimittelforschung* (Drug Research) 26:1923–1926

Rossor M (1987): Alzheimer's disease: neurobiochemistry. In: *Dementia*, Pitt B, ed. Edinburgh: Churchill Livingstone, pp 140–153

Roubicek J, Geiger C, Abt K (1972): An ergot alkaloid preparation (Hydergine) in geriatric therapy. *Journal of the American Geriatric Society* 20:222

Sahakian B, Jones G, Levy R, Gray J, Warburton D (1989): The effects of nicotine on attention, information processing, and short-term memory in patients with dementia of the Alzheimer's type. *British Journal of Psychiatry* 154:797–800

Simonson E, Brozek J (1952): Flicker fusion frequency: Background and applications. *Physiological Reviews* 32:349–378

Sherwood N, Kerr JS, Hindmarch I (1989): The use of nicotine gum in psychopharmacological research. *Medical Science Research* 17:891–892

Skondia, V (1979): Criteria for clinical development and classification of nootropic drugs. *Clinical Therapeutics* 2:316

Smith JM, Misiak H (1976): Critical flicker fusion (CFF) and psychotropic drugs in normal human subjects—a review. *Psychopharmacology* 47:175–182

Spriet A, Simon P (1985): *Methodology of Clinical Drug Trials*. Basel: Karger

Sternberg S (1969): Memory Scanning: Mental processes revealed by reaction time experiments. *American Scientist* 57:421–457

Sternberg S (1975): Memory Scanning: New findings and current controversies. *Quarterly Journal of Experimental Psychology* 27:1–32

Subhan Z, Hindmarch I (1984): The psychopharmacological effects of *Ginkgo biloba* extract in normal healthy volunteers. *International Journal of Clinical Pharmacology* IV(2):89–93

Szobor A, Klein M (1976): Ethyl apovincaminate therapy in neurovascular disease. *Arzneimittelforschung* (Drug Research) 26:1984–1989

Tazaki Y, Omae T, Kuromaru S et al. (1980): Clinical effect of Encephabol (Pyritinol) in the treatment of cerebrovascular disorders. *Journal of International Medical Research* 8:118–126

Tucker JS (1982): As assessment of the effects of intramuscular hydergine in elderly patients with early dementia using critical flicker fusion threshold. *British Journal of Clinical Practice*, Symposium Supplement, 16:16–17

Turner P (1968): Critical flicker frequency and centrally acting drugs. *British Journal of Ophthalmology* 52:245–250

Venn D (1976): Electroencephalogram and ergot alkaloids. *Postgraduate Medical Journal* 52(Suppl 1):55

Warwick KM, Eysenck HJ (1963): The effects of smoking on the CFF threshold. *Life Sciences* 4:219–225

Wilcock GK (1988): Recent research into dementia. *Age and Ageing* 17:73–86

Woodworth RS and Schlosberg H (1958): *Experimental Psychology*. London: Methuen

Pharmacological Treatment of Alzheimer's Dementia: A Review

Stevo Knežević, Nenad Bogdanović, George Spilich, and Darko Chudy

For most individuals suffering from dementia, the prognosis is bleak. A small percentage (about 10%) of patients who present with such a clinical picture have potentially reversible dementia, and about one-third of that group will be cured. For those individuals with a probable diagnosis of Alzheimer's disease (AD), the disease progresses with variable speed despite our best efforts (Clarfield, 1988). In spite of the irreversible nature of the disease, some treatment or assistance can still be offered. Pharmacological treatment may have a palliative effect upon the patient while social services, counseling and similar activities aid the patient and their family to cope with AD. In that respect, dementia is an example of a disease whose treatment requires a holistic medical approach.

Our ability to discover more effective treatments for Alzheimer's dementia would be greatly facilitated by deepening our understanding of the disease. Therefore, the goal of this chapter is to briefly review what we know about such efforts and to mark some paths that clinical researchers might explore in the future treatment of dementia.

In the last decade, the medical profession along with the neuroscience community has focused its attention upon dementia, and much of the results of this research are summarized in this text. One catalyst for this sudden and intensive interest was the discovery that decreased cholinergic transmission occurred in the brain of patients with AD. The apparent sharing by AD and Parkinson's disease of a cholinergic deficit stimulated concerted investigation into the role of the cholinergic system in cognitive processes. Some cholinergic treatment is usually considered an important component in the development of possible treatments for AD, although we now know that AD is a multisystem disease and that other neurotransmitter changes also result as a consequence of the degenerative process.

Cholinergic Agents

Precursors. In light of the apparent role of the cholinergic system in defining AD, it is not surprising that clinical researchers have long considered cholinergic interventions to be a path likely to lead to success (Decombe et al., 1990). One avenue taken has been to deliver the building blocks of ACh to AD individuals. Choline and lecithin, as precursors in the synthesis of acetylcholine, have been given in various combinations and doses in several clinical trials but improvement was noted only in a few patients (Roberts and Lazareno, 1989).

Blockers. Acetylcholinesterase (AChE) blocking agents have also been tried but the improvements reported have been inconsistent and small in magnitude. Physostigmine was the most frequently used agent of this group (14 negative studies and 23 with some positive effect) (Ashford et al., 1989). After the report by Summers (Summers et al., 1986) tacrine or THA (tetrahydroaminoacridine) quickly became a very popular experimental drug. Although the methods and conclusions of the original article were later heavily criticized, the work of Summers et al did succeed in raising the general level of interest in THA and many clinical trials were stimulated by that publication (Gauthier et al., 1990). Although much of this work (especially larger trials) is still in progress, preliminary reports suggest that the results of these studies will be similar to the initial results (Eagger et al., 1991). Some patients on THA will improve, especially when that improvement is judged by psychological tests, but there will not be any significant effect on the course of the disease (Roberts and Lazareno, 1989). Eagger et al. summarize the results nicely by saying that "the clinical relevance of this improvement is a matter for individual clinical judgment."

It is worth mentioning some of the more "exotic" anticholinesterase agents that have been used experimentally or considered to be potentially effective. Galanthamine is the active ingredient of the *Galanthus nivalis* (snowdrop), which has been known for centuries to have a mild effect similar to physostigmine but with fewer side effects. Even nerve gases and long-acting organophosphates were considered as potential candidates for AD therapy agents. Lastly, huperzines, a new class of alkaloids, have recently been isolated in China from the herb *Huperzia serrata. In vitro* they seem to be very selective and potent AChE blockers, and it is also claimed that in clinical trials they produce significant improvement in memory in patients with MID (Ashford et al., 1989; Decombe et al., 1990).

Sympathetic Agents. Acetylcholine acts on both nicotinic and muscarinic receptors and so agents with similar action might have an efficacious effect. Infusion of nicotine has been shown to have a small positive effect measured with tests of cognition (Newhouse et al., 1988) and so nicotine has been touted as potentially therapeutic for the cognitive decline characteristic of AD. However, recent work investigating the effects of nicotine upon cognition have questioned the oft-reported tale that nicotine improves cognitive performance (Spilich et al., 1992) Other studies were done with direct-acting muscarinic agonists. For example, pylocarpine, oxotremorine, RS-86 (Mouradian et al., 1988) were not beneficial, but arecoline infusion resulted in some improvement in the picture recognition task (Christie et al., 1981). A rather aggressive therapy with intraventricular infusions of bethanecol chloride was tried several times and some success has been reported (Penn et al., 1988; Roberts and Lazareno, 1989). It is worth mentioning under the heading of sympathetic agents that THA probably also acts directly on muscarinic and nicotinic receptors (Ashford et al., 1989).

An interesting approach to ACh therapy is that of neural transplantation. While not yet an established procedure, neural transplants may also offer an important avenue for future research (Gash, 1987; Decombe et al., 1990).

It is easy to wonder why, if cholinergic mechanisms are so central to AD, ACh-related therapy has provided so little return and so many problems. It appears that some of the drugs have a very narrow therapeutic window, some have significant side effects, etc. There are also other problems in understanding and treating AD besides the pharmacological difficulties. We believe that Alzheimer's disease will be found to be a syndrome with several clinical subgroups and that each subgroup will differ subtly but importantly in clinical presentation, pathophysiological mechanisms, and possibly etiology. It may be necessary to individualize various treatments for different subgroups of AD and searching for diagnostic criteria through biological-pharmacological-response or cognitive markers might be very useful. Within the context of differentiating potential subgroups based upon underlying differences, the question of identifying responders is relevant. It is obvious that small effects found in clinical trials are not equally distributed among individuals in target populations. Overall group marginal responses to a drug therapy can represent either lack of response, or can be the statistical mean of some patients who do not react to therapy at all and some who do show a clearly positive effect. Identifying these responders might be a very good avenue for future research, especially if different treatments will have different responders. P. Riekkinen of Finland claims that EEG could be helpful in in identifying the patients who would respond to cholinergic agents (Ashford et al., 1989), but it is surprising that very little work on responders has been published so far. We are not aware at this time of any clinical controlled study that has attempted to define responders.

Other Replacement Therapies

Although everybody seems to agree that the cholinergic deficit is a constant and prominent aspect of AD, if AD is truly a multisystem disease, then other replacement therapies should be considered and tried either alone or in combination with other drugs.

Cortical noradrenergic innervation was found to be deficient in AD (Mohr et al., 1989). This deficiency includes both the depletion of noradrenalin along with its synthesizing enzyme, and also neuronal loss in the locus coeruleus, which provides noradrenergic innervation. Given that experimental data has shown that the noradrenergic system is important in the learning process and also that noradrenergic receptors are preserved in the Alzheimer cortex, the rationale for including NE therapy as a component of AD therapy seems clear. However, clonidine hydrochloride, a centrally-active noradrenergic receptor agonist, has failed to show any clinical or cognitive improvement in several clinical studies. Nevertheless this new stream of research deserves further trials with some other NE agonists and especially in combination with cholinergic drugs (Schlegel et al., 1989).

Attempts to decrease symptoms of AD by increasing the neurotransmitter serotonin using imipramine hydrochloride and zimeldine hydrochloride have been unsuccessful (Reifler et al., 1989).

Dopamine enjoys a critical role in behavioral arousal and, in light of the observation that Parkinson's disease is sometimes accompanied by dementia, a number of anti-Parkinsonists were tried in the treatment of demented patients. Some positive findings were reported but the results are not yet convincing (Ambrozi and Danielczyk, 1988; Orgogozo and Spiegel, 1987). The most successful has been the use of selegiline, a monoamine oxidase β-enzyme inhibitor, which is employed in Parkinson patients to increase brain serotonin and noradrenalin. Improvements noted in some individual patients have been relatively minor (Tariot et al.,1988).

Clinical trials have also been performed using some of the neuropeptides like vasopresin, ACTH, and somatostatin, but they have also failed to show any significant cognitive improvement in patients with AD (Decombe et al., 1990).

In summary, it hardly seems surprising that a disease as tenacious and pernicious as AD would resist falling to the first pharmacological intervention we attempt. We must persist in our efforts because the cost of AD in terms of lost economic and human resources is so high.

Neuroprotective Agents

In the search for etiologic factors of AD, endogenous and exogenous toxins have been thought to be responsible for neuron loss. Although endogenous neurotoxicity is not limited to the excitatory amino acids (EAAs), this neurotransmitter class has attracted the most attention and has most clearly been implicated in the pathogenesis of stroke and also Huntington's chorea. A great deal of progress has been achieved in understanding the components that comprise the macroreceptor complex involved in glutamate transmission and stimulation in the cortex. With this theory in mind, several novel pharmaceuticals have been developed for neurodegeneration mediated by EAAs, such as: direct competitive protective agents, glycine antagonists, adenosine agonists, free radical scavengers, and noncompetitive inhibitors (e.g., MK-801). These neuroprotective agents may be found to be useful as possible treatments for AD, but they have not yet reached the state of clinical trials, which would also be difficult to design and perform (Bartus, 1990).

It is known that nerve cell membranes are disrupted in AD patients and consequently membrane fluidity is increased. The relationship between cell membrane fluidity and dementia symptoms is not clear. Results with phosphatydilserin, a phospholipid that alters neuronal membrane fluidity, have been promising (Amaduccie et al., 1988). Aluminum chelating agents are also of interest to those who believe that Al toxicity produces AD, but reports of their successful use have not been found.

Neurotrophic Factors

Nerve growth factor (NGF) was known to be effective as a stimulator of growth and differentiation of certain peripheral neurons. It was later discovered that NGF can also act as a neurotrophic factor for cholinergic neurons in the basal forebrain magnocellular complex. Even so, NGF has been shown in laboratory models to improve cholinergic functioning both by preventing the death of cholinergic neurons and also by augmenting the function of the remaining cholinergic neurons in aging animals. These and many other experiments present evidence that NGF might be useful in treatment early in the course of AD when a larger population of cholinergic neurons remains intact to act as a substrate for NGF action (Phelps et al., 1989; Tuszynski et al., 1990).

Another interesting feature of NGF is the possibility that it might promote "aberrant" neurite changes in the brain (Mobley et al., 1988) in the same fashion as epidermal growth factor (EGF), another neurotrophic factor that upregulates the expression of amyloid precursor protein (APP), which could than act as a neurite extensive factor. Aberrant proteolytic cleavage of APP during this process could result in the production of β-amyloid protein and so cause additional neural damage (Robakis et al., 1991). Careful examination of the effects of chronic NGF infusion in the aged nonhuman primate may lead to the development of drugs that would inhibit this proteolytic activity.

It may be that NGF has an additive or even synergistic effect with thyroid hormones on cholinergic neurons. As the link between cognitive and thyroid function is well known, it is possible that thyroid hormones may be useful in attenuating the atrophy of cholinergic neurons.

Although both NGF and thyroid hormones appear to show potential for AD therapy, they also present many complicated clinical and practical questions because therapies with these substances in their present form would hardly be possible. But it is to be expected that in the near future, pharmacologists, chemists and technologists will offer us interesting and novel ways of applying these drugs or their analogs to the problems of AD patients (Hefti et al., 1989).

Other avenues for future exploration include genetic research on chromosome 21, understanding the biochemical and molecular nature of senile plaques and tangles, and exploring amyloid deposition with abnormalities in the distribution and phosphorylation of cytoskeletal proteins. Any of these avenues may help us obtain a clearer understanding of the spatio-temporal sequence of neuropathologic events that comprise AD. Advances in molecular genetics may establish a new therapeutic strategy that can slow the rate of progression of the disease and perhaps try to ameliorate some structural and molecular defects in AD brains (Davies, 1991).

Antagonists of Benzodiazepine Receptors

There is some evidence to suggest that basal forebrain cholinergic neurons that innervate the cortex can be influenced by GABAergic drugs and also benzodiazepine acting on the same receptor complex. This observation is also supported by clinical reports that diazepam induces in volunteers a state that is in some ways similar to patients with senile dementia. β-carbolines were found to have antagonist or partial inverse agonist properties at the GABA benzodiazepine receptor complex and so by disinhibiting the remaining cholinergic neurons of the basal forebrain they may offer a unique way of treating senile dementia. Some studies in volunteers have been performed with promising results (Sarter et al., 1988; Sarter and Stephens, 1989).

Immunomodulant Therapy

One stream of thought in developing a therapy for AD is the approach that defines at least four subgroups of AD patients according to their different immunological properties. In two of these groups, immunomodulant therapy has been tried and apparently pyrolidone and some leucocyte extracts do show some positive effects in patients with dementia. It is interesting that the advocates of this approach hypothesize that any success of immunomodulant therapy demonstrates that the neurons involved in memory and cognitive functions are not dead but only inactive or suppressed and so can be reactivated by an appropriate therapy (Singh and Fudenberg, 1988). In the future it would be of interest to determine if the process of amyloidogenesis results from an interleukin-mediated acute phase reaction in the brain (Vandenabeele and Fiers, 1991).

Cerebral Vasodilators, Metabolic Enhancers, and Nootropics

All of the above-mentioned therapies are based on some evidence of the etiology and pathophysiology of dementias. Some of these therapies were not found to be useful; others offer promise but due to their experimental nature very few are available in pharmacies for our patients to use every day. The pharmaceutical niche represented by vasodilators, enhancers, and nootropics is mostly filled by older drugs that have been in use for some time as a treatment for such maladies as organic brain syndromes, atherosclerotic senilities, and so forth. Do these drugs still offer some therapeutic possibilities for the AD patient?

Cerebral vasodilators have been used since the times when dementia among the elderly was considered to be a result of vascular insufficiency. The most popular drug in this group is Hydergine, which is a combination of three hydrogenated alkaloids. Hydergine has shown a remarkable clinical persistence despite the passage of time and the discovery of new drugs; its continued use is due, among other things, to its registration by the FDA in the USA. In many clinical trials using Hydergine, some global improvement was noted, especially in social functions, but no clear improvement on memory and cognitive functions was typically shown.

Trental or pentoxiphiline is a drug similar to Hydergine; its action is based upon a more current conceptualization of the hemorheology and fluid properties of blood. These properties seem to be disturbed in old age and dementia, especially that of the multi-infarct type. A very large multicenter study to investigate the effect of pentoxiphiline in patients with MID is about to be completed at the end of 1991.

Nootropics are considered to be a type of nonspecific mental or cognitive enhancer that generally increases neuron metabolic activity. The mechanism by which these drugs may affect memory is unknown, but most stimulate phospholipid turnover and protein synthesis and enhance cholinergic transmission. Piracetam is the prototype of this class of drugs. Moderately positive results as well as some negative results have been reported with piracetam in combination with acetylcholine precursors such as lecithin (Hollander et al., 1986). Preliminary studies indicate oxiracetam, a related drug, increases vigilance and can partially overcome the cognitive deficits (Itil et al., 1986; Waters, 1988).

Cerebral metabolic enhancers are an interesting class of AD drugs. They seem to share some of the properties of the vasodilators and of nootropics, and their general action seems to promote an energetic cerebral metabolism. In addition to the observed metabolic enhancement, at least one, pyritinol, has been shown to increase ACh in the brain. We have tested the effects of pyritinol in 26 patients with SDAT and we have found a modest but statistically reliable improvement in their cognitive functions (Knežević et al., 1989). This improvement was reflected in both patient's self-assessment scores and in the "blind" assessment of a neurologist.

It is interesting to note that these old drugs have stood the test of time very well and that they compare well with the new classes of drugs that are the result of a lengthy scientific enquiry. The big advantage of drugs like pyritinol, piracetam, etc. is that they are readily available and also that their extended use indicates them to be relatively safe even for use in populations in which pharmacological treatment is not without difficulties.

References

Amaduccie L, SMID Group (1988): Phosphatydilserine in the treatment of Alzheimer's disease: results of multicenter study. *Psychopharmacol Bull* 24: 130–134

Ambrozi L, Danielczyk W (1988): Treatment of impaired cerebral function in psychogeriatric patients with memantine-results of a phase II: double-blind study. *Pharmacopsychiatry* 21:144–146

Ashford JW, Sherman KA, Kumar V (1989): Advances in Alzheimer therapy: Cholinesterase inhibitors. *Neurobiology of Aging* 10:99–105

Bartus RT (1990): Drugs to treat age-related neurodegenerative problems. The final frontier of medical science? *Jour American Geriatrics Society* 38:680–695

Clarfield AM (1988): The reversible dementias: Do they reverse? *Annals of Internal Medicine* 109:479–486

Christie JE, Shering A, Ferguson J, Glen AIM (1981): Physostigmine and arecoline: Effects of intravenous infusions in Alzheimer presenile dementia. *Br J Psychiatry* 138:46–50

Davies P (1991): Therapy for Alzheimer's disease: choosing a target. *Clinical Neuropharmacology* (Suppl 1) 14:S24–S33

Decombe R, Bentue-Ferrer D, Reyman JM, Allain H (1990): Nouvelles perspectives thérapeutiques dans la maladie d'Alzheimer. *Médecine et Hygiène* 48:1825–1832

Eagger SA, Levy R, Sahakian BJ (1991): Tacrine in Alzheimer's disease. *Lancet* 337:989–992

Gash DM (1987): Neuronal transplantation: potential therapy for Alzheimer's disease. *J Neural Transm* (suppl) 24:301–308

Gauthier S, Bouchard R, Lamontagne A, Bailey P et al. (1990): Tetrahydroaminoacridine-lecithin combination treatment in patients with intermediate-stage Alzheimer's disease. *N Eng J Med* 322:1272–1276

Hefti F, Hartikka J, Knusel B (1989): Function of neurotrophic factor in the adult and aging brain and their possible use in the treatment of neurodegenerative diseases. *Neurobiology of Aging* 10:515–533

Hollander E, Mohs RC, Davis KL (1986): Cholinergic approaches to the treatment of Alzheimer's disease. *Br Med Bull* 42:97–100

Itil TM, Menon GN, Songar A, Itis KZ (1986): CNS pharmacology and clinical therapeutic effects of oxiracetam. *Clin Neuropharmacology* (Suppl 3), 9:S70–S72

Knežević S, Mubrin Z, Risberg J, Vucinic G, Spilich G, Gubarev N, Wannenmacher W (1989): Pyritinol treatment of SDAT patients: Evaluation by psychiatric and neurological examination, psychometric testing and rCBF measurements. *Int Clinical Psychopharmacology* 4:25–38

Mobley WC, Nerve RL, Prusiner SB, McKinley MP (1988): Nerve growth factor induces gene expression for prion- and Alzheimer's β-amyloid proteins. *Proc Natl Acad Sci USA* 85:9811–9815

Mohr E, Schlegel J, Fabbrini G et al. (1989): Clonidin treatment of Alzheimer's disease. *Arch Neurol* 46:376–378

Mouradian MM, Mohr E, Williams JA, Chase TN (1988): No response to high dose muscarinic agonist therapy in Alzheimer's disease. *Neurology* 38:606–608

Newhouse PA, Sunderland T, Tariot PN, Blumhardt CL, Weingartner H, Mellow A,Murphy DL (1988): Intrevenous nicotine in Alzheimer's disease: a pilot study. *Psychopharmacology* 95:171–175

Orgogozo JM, Spiegel R (1987): Critical review of clinical trials in dementia II. *Postgraduate Medical Journal* 63:337–343

Penn RD, Martin EM, Wilson RS, Fox JH, Savoy SM (1988): Intraventricular bethanechol infusion for Alzheimer's. *Neurology* 38:219–222

Phelps CH, Gage FH, Growdon JH et al. (1989): Potential use of Nerve Growth Factor to treat Alzheimer's disease. *Neurobiology of Aging* 10:205–207

Reifler BV, Teri L, Raskind M et al. (1989): Double-blind trial of imipramine in Alzheimer's disease patients with and without depression. *Am J Psychiatry* 146:45–49

Roberts F, Lazareno S (1989): Cholinergic treatments for Alzheimer's disease. *Biochemical Society Transactions* 17:76–79

Robakis NK, Anderson JP, Refolo LM, Wallace W (1991): Expression of the Alzheimer amyloid precursor in brain tissue and effects of NGF and EGF on its metabolism. *Clinical Neuropharmacology* (Suppl 1) 14:S15–S23

Sarter M, Schneider HH, Stephens DN (1988): Treatment strategies for senile dementia: antagonist β carbolines. *TINS* 11:13–17

Sarter M, Stephens DN (1989): Disinhibitory properties of β carboline antagonists of benzodiazepine receptors: a possible therapeutic approach for senile dementia? *Biochemical Society Transactions* 17:81–83

Schlegel J, Mohr E, Williams J, Mann U, Gearing M, Chase TN (1989): Guanfacine treatment of Alzheimer's disease. *Clinical Neuropharmacology* 12: 124–128

Singh VK, Fudenberg HH (1988): Implications of immunomodulant therapy in Alzheimer's disease. *Prog Drug Research* 32:21–42

Spilich GJ, June L, Renner J (1992): Cigarette smoking and cognitive performance. *British Journal of Addiction*, in press

Summers WK, Majovski LV, Marsh GM, Tachiki K, Kling A (1986): Oral tetrahydroaminoacridine in long term treatment of senile dementia, Alzheimer type. *N Eng J Med* 315:1241–1245

Tariot PN, Sunderland T, Cohen RM, Newhouse PA, Mueller EA, Murphy EL (1988): Tranylcipromine compared with L-deprenyl in Alzheimer's disease. *J Clin Psychopharmacology* 8:23–27

Tuszynski MH, Hoi Sang U, Amaral DG, Gage FH (1990): Nerve growth factor infusion in the primate brain reduces lesion-induced cholinergic neuronal degeneration. *J Neuroscience* 10:3604–3614

Vandenabeele P, Fiers W (1991): Is amyloidogenesis during Alzheimer's disease due to an IL-1-/IL-6-mediated "acute phase response" in the brain. *Immunology Today* 12:217–219

Waters C (1988): Cognitive Enhancing Agents: Current status in the treatment of Alzheimer's disease. *Can J Neurol Sci* 15:249–256

Index